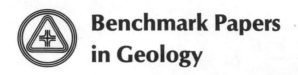

Benchmark Papers in Geology

Series Editor: Rhodes W. Fairbridge
Columbia University

A selection from the published volumes in this series

Volume
60 RIFT VALLEYS: Afro-Arabian / *A. M. Quennell*
61 MODERN CONCEPTS OF OCEANOGRAPHY / *G. E. R. Deacon and Margaret B. Deacon*
62 OROGENY / *John G. Dennis*
63 EROSION AND SEDIMENT YIELD / *J. B. Laronne and M. P. Mosley*
64 GEOSYNCLINES: Concept and Place Within Plate Tectonics / *F. L. Schwab*
65 DOLOMITIZATION / *Donald H. Zenger and S. J. Mazzullo*
66 OPHIOLITIC AND RELATED MELANGES / *G. J. H. McCall*
67 ECONOMIC EVALUATION OF MINERAL PROPERTY / *Sam L. VanLandingham*
68 SUNSPOT CYCLES / *D. Justin Schove*
69 MINING GEOLOGY / *Willard C. Lacy*
70 MINERAL EXPLORATION / *Willard C. Lacy*
72 PHYSICAL HYDROGEOLOGY / *R. Allan Freeze and William Back*
73 CHEMICAL HYDROGEOLOGY / *William Back and R. Allan Freeze*
74 MODERN CARBONATE ENVIRONMENTS / *Ajit Bhattacharyya and G. M. Friedman*
75 FABRIC OF DUCTILE STRAIN / *Mel Stauffer*
76 TERRESTRIAL TRACE-FOSSILS / *William A. S. Sarjeant*

Related titles

THE ENCYCLOPEDIA OF ATMOSPHERIC SCIENCES AND ASTROGEOLOGY / *Rhodes W. Fairbridge*
QUATERNARY STRATIGRAPHY OF NORTH AMERICA / *W. C. Mahaney*

A complete listing of volumes published in this series begins on p. 395.

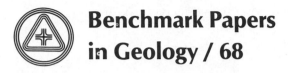

Benchmark Papers
in Geology / 68

A BENCHMARK® Books Series

SUNSPOT CYCLES

Edited by

D. JUSTIN SCHOVE

St. David's College
Beckenham, England

Hutchinson Ross Publishing Company

Stroudsburg, Pennsylvania

Copyright ©1983 by **Hutchinson Ross Publishing Company**
Benchmark Papers in Geology, Volume 68
Library of Congress Catalog Card Number: 82-15657
ISBN: 0-87933-424-X

85   84   83   1   2   3   4   5
Manufactured in the United States of America.

LIBRARY OF CONGRESS CATALOGING IN PUBLICATION DATA
Main entry under title:
Sunspot cycles.
   (Benchmark papers in geology; 68)
   Bibliography: p.
   Includes index.
   1. Sun-spots. 2. Solar cycle.   I. Schove, D. Justin
(Derek Justin)   II. Series.
QB525.S86   1983   523.7'4   82-15657
ISBN 0-87933-424-X

Distributed worldwide by Van Nostrand Reinhold Company Inc.,
135 W. 50th Street, New York, NY 10020.

CONTENTS

Series Editor's Foreword   ix
Preface   xi
Contents by Author   xiii

Introduction   1

PART I: EARLY SUNSPOTS

Editor's Comments on Papers 1 Through 6   30

1   **SCHOVE, D. J.:** The Earliest Dated Sunspot   36
British Astron. Assoc. Jour. **61:**22, 22-23, 23-24, 25 (1950)

2   **SCHOVE, D. J.:** Eclipses, Comets and the Spectrum of Time in Africa   39
British Astron. Assoc. Jour. **78:**96, 98 (1968)

3   **SCHOVE, D. J.:** Sunspots, Aurorae and Blood Rain: The Spectrum of Time   40
Isis **42:**133-134, 135-136 (1951)

4   **GOLDSTEIN, B. R.:** Some Medieval Reports of Venus and Mercury Transits   43
Centaurus **14:**50, 51-54, 55-59 (1969)

5   **CLARK, D. H., and F. R. STEPHENSON:** An Interpretation of the Pre-Telescopic Sunspot Records from the Orient   51
Royal Astron. Soc. Quart. Jour. **19:**387, 388-402, 405, 406, 408, 409-410 (1978)

6   **GALILEO, G.:** Letters on Sunspots   70
Discoveries and Opinions of Galileo, S. Drake, ed. and trans., Doubleday, New York, 1957, pp. 106-107, 116-117, 119

PART II: THE ELEVEN-YEAR CYCLE

Editor's Comments on Papers 7 Through 11   74

7   **SCHWABE, H.:** Solar Observations During 1843   81
Early Solar Physics, A. J. Meadows, ed. and trans., Pergamon, London, 1970, pp. 95-97

8   **WOLF, R.:** Universal Sunspot Numbers: Sunspot Observations in the Second Part of the Year 1850   83
Translated from *Naturf. Gesell. Bern. Mitt.* **1:**89-95 (1851)

Contents

9   **WOLF, R.:**   Sunspot Epochs Since A.D. 1610: The Periodic Return of Sunspot Minima   **88**
Translated from *Acad. Sci. Comptes Rendus* **35**:704–705 (1852)

10   **MAUNDER, E. W.:**   Note on the Distribution of Sun-spots in Heliographic Latitude, 1874 to 1902   **90**
Royal Astron. Soc. Monthly Notices **64**:760, Figure 8 (1904)

11   **SCHOVE, D. J.:**   Sunspot Cycles   **92**
The Encyclopedia of Atmospheric Sciences and Astrogeology,
R. W. Fairbridge, ed., Encyclopedia of Earth Sciences series, vol. II, Dowden,
Hutchinson & Ross, Stroudsburg,
Pa., 1967, pp. 963, 964, 965–966, 967–968

PART III: EARLY AURORAE

Editor's Comments on Papers 12 Through 19   **96**

12   **De MAIRAN, J. J. d'O.:**   Auroral Fluctuations   **101**
Translated from *Traité Physique et Historique de L'Aurore Boréale,* 2nd ed.,
Académie Royale des Sciences, Paris, 1754, p. 264

13   **SCHOVE, D. J.:**   Sunspots and Aurorae, 500–250 B.C.   **102**
British Astron. Assoc. Jour. **58**:178–180, 185, 190 (1948)

14   **SCHOVE, D. J.:**   The Spectrum of Time   **106**
Abridged from *Croydon Astron. Assoc. Jour.* **9**:1–6 (1967)

15   **SCHOVE, D. J., and P.-Y. HO:**   Chinese Records of Sunspots and Aurorae in the Fourth Century A.D.   **109**
Am. Orient. Soc. Jour. **87**:105–107, 110–111 (1967)

16   **SCHOVE, D. J.:**   Visions in North-West Europe (A.D. 400–600) and Dated Auroral Displays   **113**
British Archaeol. Assoc. Jour., ser. 3, **13**:34–38, 43–46, 48–49 (1950)

17   **SCHOVE, D. J., and P.-Y. HO:**   Chinese Aurorae: I, A.D. 1048–1070   **123**
British Astron. Assoc. Jour. **69**:295–299, 302–304 (1959)

18   **STOTHERS, R.:**   Ancient Aurorae   **130**
Isis **70**:90–95 (1979)

19   **SCHOVE, D. J.:**   Auroral Numbers Since 500 B.C.   **136**
British Astron. Assoc. Jour. **72**:30–35 (1962)

PART IV: SYNTHESIS

Editor's Comments on Papers 20 Through 23   **144**

20   **SCHOVE, D. J.:**   The Sunspot Cycle, 649 B.C. to A.D. 2000   **149**
Jour. Geophys. Research **60**:127–146 (1955)

21   **MAUNDER, E. W.:**   The Prolonged Sunspot Minimum, 1645–1715   **169**
British Astron. Assoc. Jour. **32**:140–145 (1922)

22   **EDDY, J. A.:**   The Maunder Minimum   **175**
Science **192**:1189–1202 (1976)

23   **ANONYMOUS:**   Junk in Space: What Goes Up   **189**
The Economist, April 28, 1979, pp. 113–114

PART V: LONGER CYCLES

Editor's Comments on Papers 24 Through 28   192

· **24**   HALE, G. E.:   The Law of Sun-Spot Polarity   199
           Natl. Acad. Sci. (USA) Proc. **10:**53–55 (1924)

25   GLEISSBERG, W.:   The Eighty-Year Solar Cycle in Auroral Frequency
           Numbers   202
           British Astron. Assoc. Jour. **75:**227–228, 230–231 (1965)

26   KIRAL, A.:   Autocorrelation and Solar Cycles   206
           Translated from *Istanbul Univ. Obs. Publ. No. 70,* pp. 12–21 (1961)

27   COLE, T. W.:   Periodicities in Solar Activity   211
           Solar Physics **30:**103, 104, 105–106, 106–107, 108, 110 (1973)

28   DAMON, P. E.:   Solar Induced Variations on Energetic Particles at
           One AU   216
           The Solar Output and Its Variation, O. R. White, ed., Colorado Associated
           University Press, Boulder, 1977, pp. 434–441, 446–448

PART VI: SUNSPOTS IN HISTORY AND THEIR EFFECT ON CLIMATE

Editor's Comments on Papers 29 Through 36   226

· **29**   HERSCHEL, W.:   Observations tending to investigate the Nature of the
           Sun, in order to find the Causes or Symptoms of its variable
           Emission of Light and Heat; with Remarks on the Use that
           may possibly be drawn from Solar Observations   239
           Royal Soc. London Philos. Trans. **91:**306, 310, 315 (1801)

30   JEVONS, W. S.:   The Periodicity of Commercial Crises, and its Physical
           Explanation   241
           British Assoc. Adv. Sci. Rep., pp. 666–667, 1878

· **31**   SCHOVE, D. J.:   Solar Cycles and the Spectrum of Time Since 200 B.C. 242
           New York Acad. Sci. Ann. **95:**107, 109, 110, 115–118,
           119–120, 122–123 (1961)

32   SCHOVE, D. J.:   Tree Rings and Climatic Chronology   250
           New York Acad. Sci. Ann. **95:**605–622 (1961)

· **33**   SCHOVE, D. J.:   The Biennial Oscillation, Tree Rings and Sunspots   268
           Weather **24:**390–395 (1969)

34   SCHOVE, D. J.:   Biennial Oscillations and Solar Cycles, AD 1490–1970   274
           Weather **26:**201–209 (1971)

35   SCHOVE, D. J.:   African Droughts and the Spectrum of Time   283
           Internat. Afr. Inst. (London) Environ. Spec. Rep. 6, D. Dalby, R. J. Harrison-
           Church, and F. Bezzaz, eds., 1977, pp. 38–40, 45–48, 50–52

· **36**   SISCOE, G. L.:   Solar-Terrestrial Influences on Weather and Climate 293
           Nature **276:**348–350, 351 (1978)

PART VII: VARVE AND GEOLOGICAL CYCLES

Editor's Comments on Papers 37 Through 40   298

Contents

37 **ANDERSON, R. Y.:** Solar-Terrestrial Climatic Patterns in Varved
Sediments   **302**
New York Acad. Sci. Ann. **95:**424–426, 427, 429–435, 437–439 (1961)

38 **SCHOVE, D. J.:** Solar Cycles and Equatorial Climates   **316**
Geol. Rundschau **54:**448–449, 451–453, 456–462, 462–463, 464–466,
468, 470, 474–477 (1964)

39 **SCHOVE, D. J.:** Tree-Ring and Varve Scales Combined, c. 13500 B.C.
to A.D. 1977   **333**
Palaeogeography, Palaeoclimatology, Palaeoecology **25:**209–233 (1978)

40 **SCHOVE, D. J.:** Varve-Chronologies and Their Teleconnections,
14000–750 B.C.   **358**
Moraines and Varves, C. Schlüchter, ed., Balkema, Rotterdam,
Netherlands, 1979, pp. 319–325

Conclusion   **365**
APPENDIX A: Sunspot Minima 653 B.C. to A.D. 1501   **368**
APPENDIX B: Sunspot Maxima c. 649 B.C. to A.D. 1506   **370**
APPENDIX C: Sunspots and Radiocarbon Since A.D. 5   **372**
APPENDIX D: Cycle Index   **378**
Author Citation Index   **381**
Subject Index   **385**
About the Editor   **393**
Benchmark Papers in Geology Series List   **395**

SERIES EDITOR'S FOREWORD

The philosophy behind the Benchmark Papers in Geology is one of collection, sifting, and rediffusion. Scientific literature today is so vast, so dispersed, and, in the case of old papers, so inaccessible for readers not in the immediate neighborhood of major libraries that much valuable information has been ignored by default. It has become just so difficult, or so time consuming, to search out the key papers in any basic area of research that one can hardly blame a busy person for skimping on some of his or her "homework."

This series of volumes has been devised, therefore, as a practical solution to this critical problem. The geologist, perhaps even more than any other scientist, often suffers from twin difficulties—isolation from central library resources and immensely diffused sources of material. New colleges and industrial libraries simply cannot afford to purchase complete runs of all the world's earth science literature. Specialists simply cannot locate reprints or copies of all their principal reference materials. So it is that we are now making a concerted effort to gather into single volumes the critical materials needed to reconstruct the background of any and every major topic of our discipline.

We are interpreting "geology" in its broadest sense: the fundamental science of the planet Earth, its materials, its history, and its dynamics. Because of training in "earthy" materials, we also take in astrogeology, the corresponding aspect of the planetary sciences. Besides the classical core disciplines such as mineralogy, petrology, structure, geomorphology, paleontology, and stratigraphy, we embrace the newer fields of geophysics and geochemistry, applied also to oceanography, geochronology, and paleoecology. We recognize the work of the mining geologists, the petroleum geologists, the hydrologists, and the engineering and environmental geologists. Each specialist needs a working library. We are endeavoring to make the task of compiling such a library a little easier.

Each volume in the series contains an introduction prepared by a specialist (the volume editor)—a "state of the art" opening or a summary of the object and content of the volume. The articles, usually some twenty to fifty reproduced either in their entirety or in significant extracts, are selected in an attempt to cover the field, from the key papers of the last century to fairly recent work. Where the original works are in foreign languages, we have endeavored to locate or commission translations. Geologists, because of their global subject, are often acutely aware of the oneness of our world. The selections cannot therefore be restricted to any one country, and whenever possible an attempt is made to scan the world literature.

To each article, or group of kindred articles, some sort of "highlight commentary" is usually supplied by the volume editor. This commentary should serve to bring that article into historical perspective and to emphasize its particular role in the growth of the field. References, or citations, wherever possible, will be reproduced in their entirety—for by this means the observant reader can assess the background material available to that particular author, or if desired, he or she too can double check the earlier sources.

A "benchmark," in surveyor's terminology, is an established point on the ground that is recorded on our maps. It is usually anything that is a vantage point, from a modest hill to a mountain peak. From the historical viewpoint, these benchmarks are the bricks of our scientific edifice.

RHODES W. FAIRBRIDGE

PREFACE

The mysterious cycles in the number of sunspots have fascinated geologists, astronomers, climatologists, economists, historians, and dendrochronologists. Meteorologists were sometimes so fascinated that they made forecasts that went wrong. In the stratosphere, however, sun-weather relationships are clear enough and sunspot cycles have to be taken into account in telecommunications and interplanetary flight. Moreover, sunspot variations are now proving important in Quaternary chronology and palaeoclimatology.

The cause of the eleven-year period is still not understood, but our knowledge of past eleven-year cycles assures us that sunspots, which reached a maximum in 1979, will be on the wane by 1984 as the *Minimum* approaches, and will again be on the increase by 1990 as the last and ninth *Maximum* of the century draws near. The cycle was not discovered until the nineteenth century, but the present proved the key to the past, and earlier cycles have since been detected and dated through two millennia. The past in turn is now used as the key to both the present and the future, and our confidence in the forecasts given above rests on interdisciplinary research. Some scientists are said to read only the latest research papers and some historians only the earliest primary chronicles, but both sources have been essential to our knowledge of sunspot cycles.

Sunspots can be seen on the solar disc if we watch a hazy sunset—perhaps from New York over the Hudson or from one of the London bridges looking upriver along the Thames. To stare at the sun is dangerous for the eyes and a better view of the spots is obtained by watching on the floor the solar image projected through a pinhole in a large card. In this way the reader will be able to check the accuracy of our predictions; as we shall illustrate from the Skylab failure of 1979, these predictions are still not precise enough for us to keep artificial satellites in their orbits.

Droughts, floods, famines, epidemics, wars, and revolutions have all been attributed to sunspots, and although we must ignore most of the extravagant claims, we can now measure the statistical significance of the implied correlations and show that, although very slight for forecasting purposes, there are real relationships between sunspot cycles and disasters.

We have made progress in tracing sunspots and climate back into the historic past, and even, with the help of tree-ring and radiocarbon studies, through the last 9000 years. With the aid of geological and paleomagnetic investigations we can go still further back. Isotopic and magnetic effects of solar cycles are under investigation, but sunspot cycles—of 11, 22, and 200 years—already provide important clues to the chronology of the Quaternary ice-cores. Periodicities are now being found in the sediments of much earlier geological epochs, and there appear to be other weather cycles—both short and long—caused by the sun.

My papers on these subjects have been scattered in many different journals and the editor has requested me to bring them together in this book. The information in my papers, however, comes especially from collaborators in various parts of the world who have participated in the *Spectrum of Time* project. The Benchmark papers selected for inclusion emphasize that continued collaboration on an international basis is needed to make our forecasts of future solar activity more accurate than they have been in the past.

My thanks are due to the series editor, Professor Fairbridge, for his interest and encouragement and to the many specialists who so kindly sent reprints for possible inclusion in this volume. My thanks are also due to Vera, my long-suffering wife, my family, and especially to Mr. A. T. Gerard and my daughter, Mrs. Ann Wagstaff, for helpful criticisms, to Mr. R. R. Harris for invaluable editing, and to Mrs. Dandridge for dealing so efficiently with the typing, the inevitable correspondence, and the production of the final copy. Those who have helped in the *Spectrum of Time* project, such as Miss C. M. Botley and Professor Peng-Yoke Ho, have made possible the translated records of aurorae (still largely unpublished) that have been used for dating sunspot cycles in the past 2000 years. Libraries are now more important than laboratories for scientists and I owe much to both local and university librarians and to those of the British Meteorological Office and the Royal Astronomical Society.

D. JUSTIN SCHOVE

CONTENTS BY AUTHOR

Anderson, R. Y., 302
Clark, D. H., 51
Cole, T. W., 211
Damon, P. E., 216
De Mairan, J. J. d'O., 101
Eddy, J. A., 175
Galileo, G., 70
Gleissberg, W., 202
Goldstein, B. R., 43
Hale, G. E., 199
Herschel, W., 239
Ho, P.-Y., 109, 123

Jevons, W. S., 241
Kiral, A., 206
Maunder, E. W., 90, 169
Schove, D. J., 36, 39, 40, 92, 102, 106, 109, 113, 123, 136, 149, 242, 250, 268, 274, 283, 316, 333, 358
Schwabe, H., 81
Siscoe, G. L., 293
Stephenson, F. R., 51
Stothers, R., 130
Wolf, R., 83, 88

SUNSPOT CYCLES

INTRODUCTION

Sunspots are the visible manifestations of convection cells near the surface of the sun's photosphere. They are characteristically found in pairs having opposite magnetic polarity and connected by tubes of high-energy particles. In the course of an 11-year cycle these tubes rise and erupt, and *openings* or *sunspots* appear first in the midlatitudes (30°–40°) of each solar hemisphere and at the maximum of the cycle in the zone 10°–20°. After some 5 to 7 years the sunspot pairs are found only near the solar equator. This apparent equatorward drift is kown as *Spoerer's Law* and is illustrated well in the so-called *Butterfly Diagram* (Paper 10) of Maunder. These spots disappear and the sun is then quiet for 2 to 3 years at the sunspot minimum until the next cycle begins.

There is no satisfactory explanation for all of these developments, although there is certainly an association with the nonuniform rotation of the sun, the equatorial region having a faster angular velocity than the midlatitude zones. The first spot to appear is the *p-spot* (preceding) and following closely behind is its companion, the *f-spot* (following). The magnetism of the *p*-spots in a given cycle matches that of the corresponding polar region, which reverses in alternate cycles. This double period of 22 years, the *Hale Cycle* (Paper 24), is used in climatic forecasting (cf. Figs. 13 through 16).

Dated records of sunspots are considered in Part I. Naked-eye observations have been made in China since before A.D. 300 and telescopic observations in Europe have been made since A.D. 1610. Nevertheless, after about 1800, the 11-year sunspot cycle (see Part II) began to behave irregularly and was not discovered until sunspot-days had been counted over many years; the sunspot cycle was not generally accepted until a method of counting *sunspot numbers* had been developed.

The discovery of the *Sunspot Cycle* by Schwabe in 1843 (see Paper 7) led Wolf in 1847 to 1851 (see Paper 8) to the concept of universal *sunspot numbers (R)*, which in 1868 he extended backward annually to 1700. At Zürich, Wolf calculated magnitudes (R_M) of

1

each cycle from 1749 and the central dates or *epochs* of *maxima* and *minima* back to about 1610. The relationship between clusters of large sunspots (as found in Papers 5 and 15) and the official maximum year is illustrated in Figure 1. The double peak reveals the intervening lull or Gnevyshev gap (see Fig. 1) at year $X + 1$, which Wittmann (1978) shows is very near the *center of gravity* of the cycle.

At Zürich Observatory Wolf's immediate successor, Wolfer, made slight revisions and the recent director, Waldmeier (1976), proved that the shape of the sunspot cycle depended almost entirely on the height of the maximum (R_M) and suggested the formula: log $R_M = 2.73 - 0.18\ T$ (see also Kane, 1978), where T is the time of rise from minimum to maximum (see Figs. 2 and 6). Curves for the individual cycles are given in Fig. 3 (see also Paper 22).

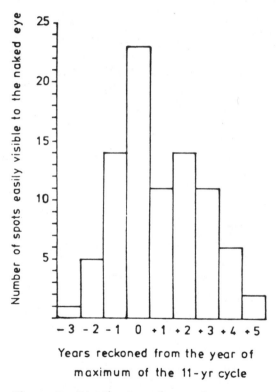

Figure 1 Distribution of conspicuous sunspots (over 1500 units, 1874–1971) in relation to the year of the Zurich Maximum. The percentages from year $X - 3$ to $X + 5$ are 1, 6, 16, 26, 13, 16, 13, 7, and 2. (*Reprinted from Gleissberg, W., 1973,* Jour. Interdisciplinary Research *3:317; copyright © 1973 by Swets & Zeitlinger, Lisse, Netherlands.*)

Figure 2 (a) Normal sunspot curves for $R_M = 60$ to 150, in relation to the year (N) of the preceding minimum. (b) Normal sunspot curves for $R_M = 60$ to 150, in relation to the year (X) of the maximum. (*Reprinted from Waldmeier, M., 1968, Astronomische Mitt.* **286**:1; *copyright © 1968 by Dr. M. Waldmeier, Swiss Federal Observatory, Zurich, Switzerland.*)

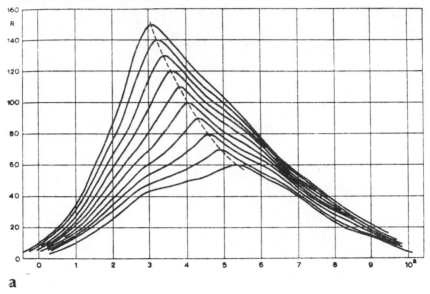

a

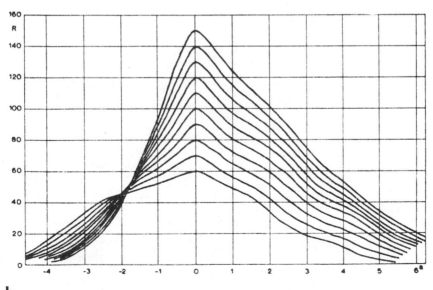

b

Figure 3 Sunspot cycles, A.D. 1755–1975. Monthly values are shown by the thin line; the thick lines are "smoothed" values (cf. Paper 25). (*Figures 3a, 3b, and 3c are reprinted from Waldmeier, M., 1961, The Sunspot Activity in the Years 1610–1960, Schulthess, Zurich, Switzerland, pp. 111–113; copyright © 1961 by Dr. M. Waldmeier, Swiss Federal Observatory, Zurich, Switzerland.) (Figure 3d is reprinted from Waldmeier, M., 1976, Astronomische Mitt. 346:13; copyright ©1976 by Dr. M. Waldmeier, Swiss Federal Observatory, Zurich Switzerland.)*

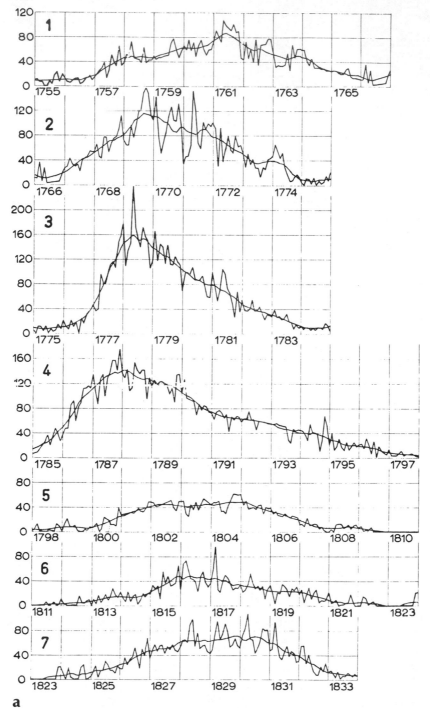

a

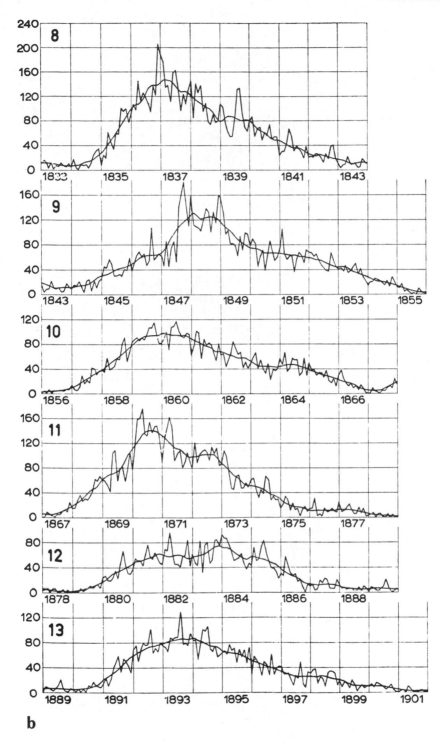

b

5

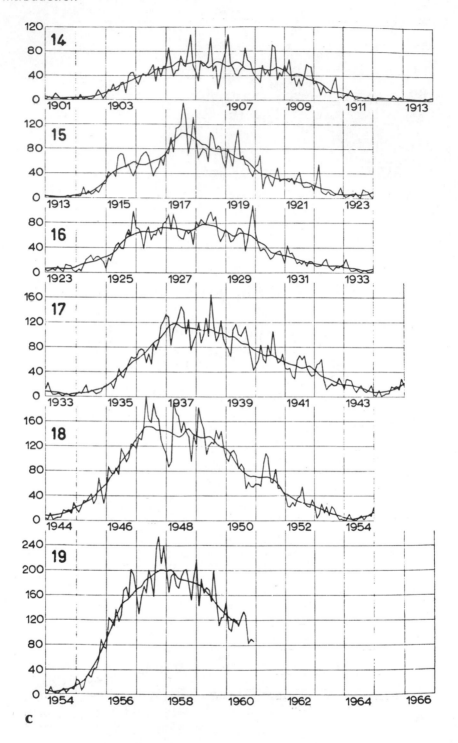

C

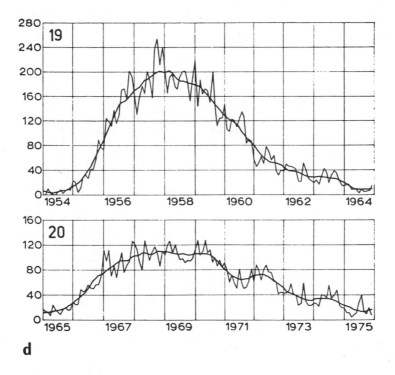

d

Early aurorae or *Northern Lights,* the theme of Part III, provide clues to the dating of early sunspot cycles. The printed catalogs (used in Figs. 4 and 5) and additional "Spectrum of Time" (Schove, in progress) material (see Papers 13, 14, 22, and 31) have made it possible to extend Wolf's tables (see Paper 8) back into earlier centuries (see Table 1 in this Introduction and Tables 1 and 2 in Paper 20).

The determination of the dates and magnitudes of early sunspot-auroral cycles is possible if we assume that they were constrained by the same rules of behavior (Schove, 1979, 1980a, 1980b) before 1710 as they have been since then. We thus suppose that the interval between adjacent spot maxima depends on the *ratio* of their magnitudes (Fig. 6a, see p. 12) and the interval between adjacent spot minima depends on the *sum* of the same two magnitudes (Fig. 6b, see p. 12). Some relations with aurorae are shown in Figure 7 (see p. 13), and Table 2 (see p. 14) gives new results for the well-established cycles back to A.D. 1500. Fluctuations of auroral activity are shown in Figure 8 (see p. 15).

Sunspot observations were scanty from 1645 to 1699 and our rules have exceptions even after 1710. Many astronomers believe that the 11-year cycle was in abeyance during that period (known as the *Maunder Minimum*), and the syntheses by Maunder and by Eddy (Papers 21 and 22) are thus included in Part IV.

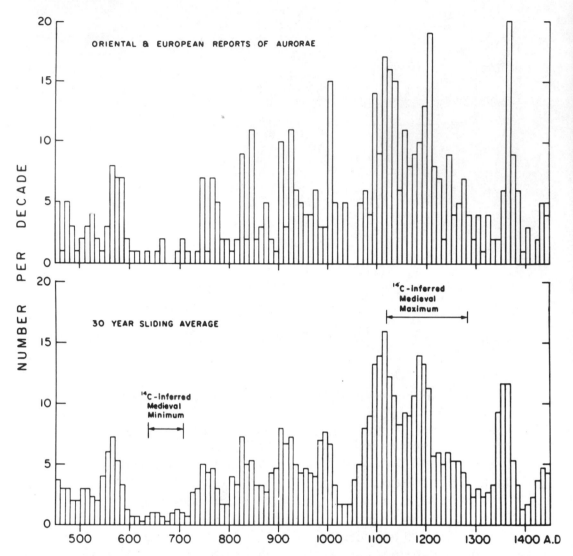

Figure 4   The number of days on which aurorae were recorded per decade in the Middle Ages as compiled by Link (1962), Newton (1972), Dall'Olmo (1979), and Keimatsu (1970–1976). The bottom panel consists of three-decade sliding averages of the data given in the top panel. (*Reprinted from Siscoe, G. L., 1980,* Rev. Geophysics and Space Physics **18:***654; copyright ©1980 by the American Geophysical Union.*)

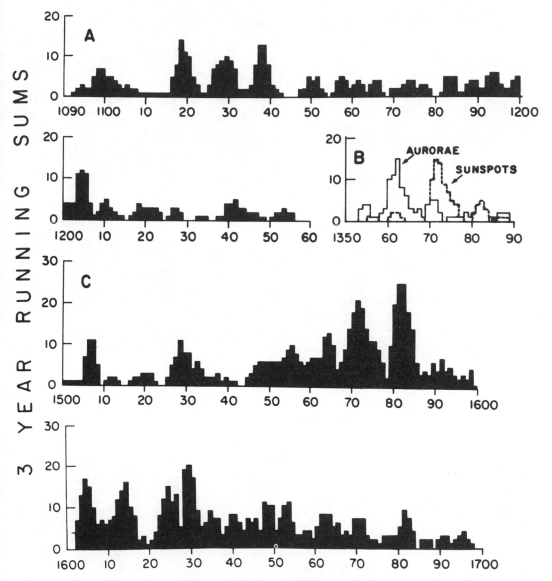

Figure 5 The number of days on which aurorae were recorded per three years, determined yearly, for the aurorally rich periods evident in this figure (Panels A and B). The interval from A.D. 1500 to A.D. 1700 is given in Panel C, which is compiled from the entries in the catalogs of Link (1962, 1964, and cf. 1978, Fig. 1). (*Reprinted from Siscoe, G. L., 1980, Rev. Geophysics and Space Physics **18**:655; copyright © 1980 by the American Geophysical Union.*)

Table 1. Annual Sunspot Numbers A.D. 1500–1980

	0	1	2	3	4	5	6	7	8	9
1500+	5	3	11	26	47	61	80	65	51	33
1510+	23	11	4	2	4	20	50	68	63	51
1520+	33	18	9	5	2	14	42	88	131	103
1530+	85	63	44	23	5	22	60	108	92	74
1540+	45	25	15	3	13	40	88	138	115	100
1550+	80	60	46	23	13	19	43	78	119	105
1560+	98	81	58	45	30	18	10	8	15	40
1570+	87	130	107	87	66	48	30	17	5	20
1580+	70	110	89	72	50	25	9	5	7	18
1590+	30	45	52	59	49	35	19	10	3	5
1600+	9	25	48	70	88	70	50	21	5	5
1610+	10	20	50	75	98	80	63	40	25	15
1620+	5	9	20	38	68	105	95	80	55	40
1630+	25	13	6	0	2	8	20	50	75	95
1640+	80	63	43	30	18	0	4	11	17	25
1650+	29	27	25	16	6	2	4	7	12	18
1660+	24	24	20	15	9	5	0	4	9	14
1670+	22	31	32	35	32	30	33	18	7	3
1680+	9	7	24	27	29	28	26	15	12	5
1690+	6	7	9	12	15	13	8	2	0	0
1700+	4	15	32	46	58	65	53	38	22	14
1710+	5	1	0	5	18	35	57	77	84	60
1720+	42	32	23	12	21	44	85	140	120	90
1730+	62	43	22	10	20	43	75	93	126	111
1740+	86	53	28	20	8	13	25	45	65	81
1750+	(83.4)	47.4	47.8	30.7	12.2	9.6	10.2	32.4	47.6	54.0
1760+	62.9	85.9	61.2	45.1	36.4	20.9	11.4	37.8	69.8	106.1
1770+	100.8	81.6	66.5	34.8	30.6	7.0	19.8	92.5	154.4	125.9
1780+	84.8	68.1	38.5	22.8	10.2	24.1	82.9	132.0	130.9	118.1
1790+	89.9	66.6	60.0	46.9	41.0	21.3	16.0	6.4	4.1	6.8
1800+	14.5	34.0	45.0	43.1	47.5	42.2	28.1	10.1	8.1	2.5
1810+	0.0	1.4	5.0	12.2	13.9	35.4	45.8	41.1	30.1	23.9
1820+	15.6	6.6	4.0	1.8	8.5	16.6	36.3	49.6	64.2	67.0
1830+	70.9	47.8	27.5	8.5	13.2	56.9	121.5	138.3	103.2	85.7
1840+	64.6	36.7	24.2	10.7	15.0	40.1	61.5	98.5	124.7	96.3
1850+	66.6	64.5	54.1	39.0	20.6	6.7	4.3	22.7	54.8	93.8
1860+	95.8	77.2	59.1	44.0	47.0	30.5	16.3	7.3	37.6	74.0
1870+	139.0	111.2	101.6	66.2	44.7	17.0	11.3	12.4	3.4	6.0
1880+	32.3	54.3	59.7	63.7	63.5	52.2	25.4	13.1	6.8	6.3
1890+	7.1	35.6	73.0	85.1	78.0	64.0	41.8	26.2	26.7	12.1

Table 1. *(continued)*

	0	1	2	3	4	5	6	7	8	9
1900+	9.5	2.7	5.0	24.4	42.0	63.5	53.8	62.0	48.5	43.9
1910+	18.6	5.7	3.6	1.4	9.6	47.4	57.1	103.9	80.6	63.6
1920+	37.6	26.1	14.2	5.8	16.7	44.3	63.9	69.0	77.8	64.9
1930+	35.7	21.2	11.1	5.7	8.7	36.1	79.7	114.4	109.6	88.8
1940+	67.8	47.5	30.6	16.3	9.6	33.2	92.6	151.6	136.3	134.7
1950+	83.9	69.4	31.5	13.9	4.4	38.0	141.7	190.2	184.8	159.0
1960+	112.3	53.9	37.5	27.9	10.2	15.1	47.0	93.8	105.9	105.5
1970+	104.5	66.6	68.9	38.0	34.5	15.5	12.6	27.5	92.5	155.4
1980+	154.6									
1981+	(International sunspot numbers will continue the Zurich series.)									

Notes: 1749–1980, official Zurich values
1700–1748, revised slightly from Wolf, 1868, and Schove, 1979
1610–1699, inferred from intermittent sunspot records
1500–1609, tentative values inferred from auroral evidence (adequate only in a relative sense: before 1640 the error must often exceed 20 when the number is over 80)

Satellite planning necessitates prediction of future sunspot activity but this is not usually successful until a cycle has actually commenced. No forecast before 1972 (Table 3, see p. 15) was successful in predicting both date and magnitude of the next sunspot maximum and the Skylab disaster (Paper 23) was partly a result of this failure. (The predictions for Cycle 21 in Table 3 were made prior to 1972. The true date was 1979.9 and the magnitude (R_M) was 164.5.)

The development of solar physics is not covered in the papers selected; it is discussed in many standard textbooks (see Kiepenheuer, 1953; Abetti, 1957; Newton, 1958; Bray and Loughhead, 1964; Cousins, 1972) and excerpts from several papers have been included (Gleissberg, 1971; Cole, 1973; Schove, 1979). (See volume 72 of *Solar Physics* for a comprehensive index to earlier papers.)

We therefore turn in Part IV to the longer sunspot cycles. There is a *medium cycle* of about 80 years in the phase (or length) and a *long cycle* of 200 years in both the phase and intensity. The medium, 80-year cycle develops in intensity provided that the data are smoothed as in Figure 8 (see p. 15) but the spectral analysis (Fig. 9, see p. 16) reveals complications, which we shall discuss.

Tree-ring studies were developed in the United States initially through the interest of an astronomer, Douglass, in sunspot cycles. The strange dearth of sunspots in the seventeenth century—the Maunder Minimum—was, he believed, the explanation of his failure to find an 11-year cycle in ring-widths in that century (cf. Schove, 1948). We now find it difficult to find unequivocable evidence in any

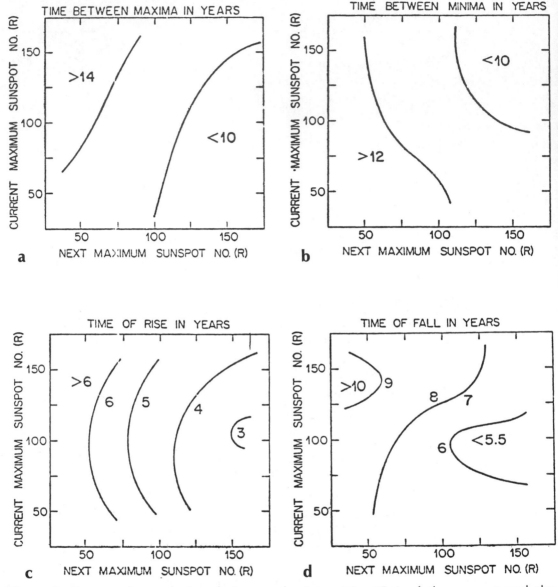

Figure 6 Sunspot intervals in relation to the intensities (R_M) of the current and the subsequent sunspot maxima. (a) Algebraically: $X_2 - X_1 = 15 - 4 (R_2/R_1) \pm 1$ (b) Algebraically: $N_2 - N_1 = 9 + \left(\dfrac{300 - R_1 - R_2}{50}\right) \pm 1$ (c) The time of rise (T) relates to the next maximum. (d) The time of fall (U). These charts are based on Zurich data since A.D. 1710: In the late seventeenth century cycles were very weak, and the values of T and U tentatively adopted in Table 1 are less then the diagrams suggest. (*Reprinted from Schove, D. J., 1980, Exploration of the Polar Upper Atmosphere, C. S. Deehr and J. A. Holtet, eds., D. Reidel, Dordrecht, Holland; copyright ©1980 by D. J. Schove.*)

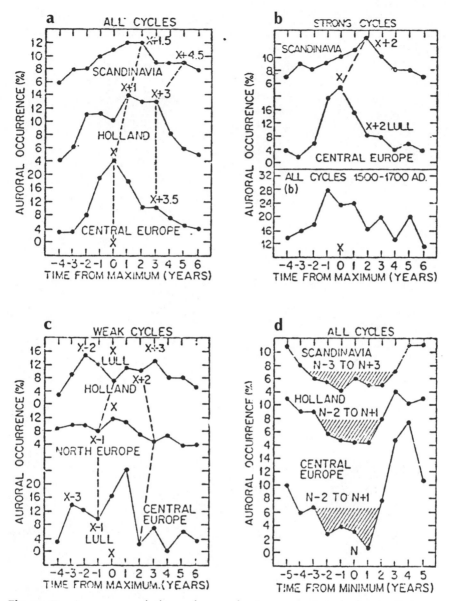

Figure 7 Aurorae and the solar cycle (curves printed also in Schove 1980a, where tentative frequency patterns for *moderately strong* and *moderate* cycles are also given; Figs. 7a, 7c, and 7d based mainly on data in Schove 1979). (a) Auroral frequency percentages for each year of the eleven-year period X − 4 to X + 6 for cycles since 1710. (b) Auroral frequency as in (a) for *strong* solar cycles ($R_M > 140$). Active or red aurorae in Link's (1962, 1964) catalogs of European aurorae are shown for comparison. (c) Auroral frequency as in (a) for *weak* solar cycles ($R_M < 87$). (d) Auroral frequencies relative to sunspot minima for cycles since 1710. (*Reprinted from Schove, D. J., 1980,* Exploration of the Polar Upper Atmosphere, *C. S. Deehr and J. A. Holtet, eds., D. Reidel, Dordrecht, Holland; copyright ©1980 by D. J. Schove.*)

13

Table 2. Sunspot Minima *(N)*, Maxima *(X)*, and Magnitude *(R$_M$)* Since A.D. 1500

Cycle Number	N	X	R$_M$
−22	1501.5 ± 1	1506.5 ± 1	80
−21	13.7 ± 1	17.9 ± ½	70
−20	24.7 ± 1	28.2 ± ½	135
−19	34.3 ± 1	37.6 ± ½	110
−18	43.7 ± 1	47.4 ± 1	140
−17	54.5 ± 1	58.3 ± 1	120
−16	67.5 ± ½	71.3 ± 1	135
−15	78.2 ± ½	81.5 ± 1	110
−14	87.5 ± 1	93.8 ± 2	60
−13	98.8 ± 1	1604.4 ± ½	90
−12	1609.2 ± ½	1614.3 ± ½	100
−11	20.2 ± ½	25.8 ± ½	115
−10	33.7 ± ½	39.3 ± ½	100
−9	45.5 ± ½	50.8 ± 1	30
−8	55.9 ± 1	61.0 ± ½	25
−7	66.7 ± 1	73.5 ± 1	35
−6	79.5 ± 2	85.0 ± 1	30
−5	89.5 ± 1	94.5 ± 1	20
−4	99.0 ± 1	1705.5 ± ½	68
−3	1712.5 ± ½	1718.2	90
−2	23.5	27.5	140
−1	34.0	38.7	130
0	45.0	50.3	92.6
1	55.2	61.5	86.5
2	66.5	69.7	115.8
3	75.5	78.4	158.5
4	84.7	88.1	141.2
5	98.3	1805.2	49.2
6	1810.6	1816.4	48.7
7	23.3	29.9	71.7
8	33.9	37.2	146.9
9	43.5	48.1	131.6
10	56.0	60.1	97.9
11	67.2	70.6	140.5
12	78.9	83.9	74.6
13	89.6	94.1	87.9
14	1901.7	07.0	64.2
15	13.6	17.6	105.4
16	23.6	28.4	78.1
17	33.8	37.4	119.2
18	44.2	47.5	151.8
19	54.2	57.9	201.3
20	64.7	68.9	110.6
21	76.5	79.9	164.5

Notes: A.D. 1500–1715, estimated (slightly revised from Schove, 1979)
A.D. 1715+, Zurich values after Waldmeier (1961, 1976)
For earlier minima see Appendix A.

Table 3. Summary of Predictions for Cycle 21

Authors	Initial Minimum	Maximum	Final Minimum	$R_M(21)$
Schove (1955)	1978.5	1984.5	1989.5	145
Guo (1963)				<100
Jose (1965)	1977	1984	1990	(30)
Bell and Wolbach (1965)	1974.9		1985.7	
King-Hele (1966)		1978.5		110
Xanthakis (1967)				(80)
Bonov (1970)				37*–67*
Vasilyev and Kandaurova (1971)	1975	1980	1987	85*
Henkel (1971)				100
Gleissberg (present paper)	1975.2	1980.0	1986.3	56–96

*Some predictions concern $W_M(21)$, the highest annual mean of the sunspot-relative-numbers in Cycle 21, instead of $R_M(21)$; these values are marked with an asterisk.
References: Bell, B. and Wolbach, J. G.: 1965, *Icarus* **4,** 409; Bonov, A. D.: 1970, *Soln. Dann.* No. 7, 111; Guo, C.-S.: 1963, *Acta Astron. Sinica* **11,** 60; Henkel, H. R.: 1971, *Solar Phys.* **20,** 345; Jose, P. D.: 1965, *Astron. J.* **70,** 193; [King-Hele, D. G.: 1966, *Nature* **209,** 285; Schove, D. J.: 1955, *J. Geophys. Res.* **60,** 127; Vasilev, O. B. and Kandaurova, K. A.: 1971, *Soln. Dann.* No. 11, 109; Xanthakis, J.: 1967, *Nature* **215,** 1046.]
Source: Gleissberg, 1971, p. 245.

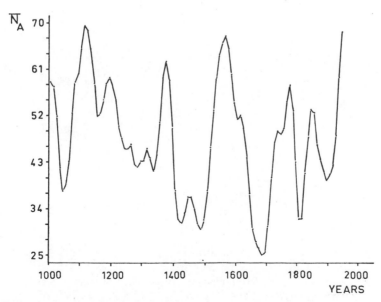

Figure 8 Auroral numbers (Paper 20 and cf. Paper 26) secularly smoothed. (*Reprinted from Gleissberg, W., 1971,* Solar Physics ***21:**242; copyright © 1971 by D. Reidel Publishing Company, Dordrecht, Holland.*)

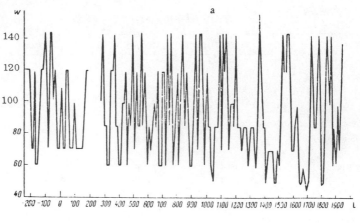

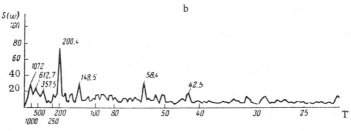

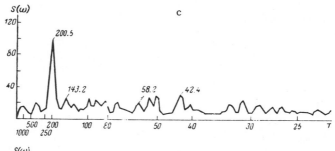

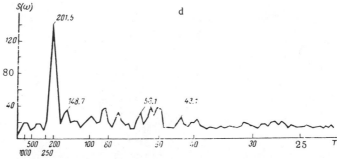

Figure 9 The unsmoothed Schove series and the spectral densities of the corresponding autocorrelation functions. (a) The autocorrelation function. (b) 214 B.C. to A.D. 1947. (c) A.D. 501 to A.D. 1947. (d) A.D. 850 to A.D. 1947. (*Reprinted from Zhukov, L. V., and Yu. S. Muzalevskii, 1969,* Soviet Astronomy **13:***475; copyright ©1969 by the American Institute of Physics.*)

16

century (see Paper 1 and Paper 36). Dendrochronology, nevertheless, promises to elucidate past sunspot cycles in unexpected ways inasmuch as the radiocarbon level of individual years can be determined and the ^{14}C flux is roughly inversely proportional to the sunspot activity.

Houtermans et al. (1967) demonstrated an approximately 200-year cycle in the radiocarbon record, at that time back only to A.D. 1000 (see Fig. 10 and especially Papers 22, 28, and 39).

Perhaps we can regard both the Maunder Minimum of 1645 to 1699 and the record solar activity of 1945 to 1983 as examples of what happens when the effects of the 80-year and the 200-year cycles reinforce one another. We show that longer and weaker cycles are to be expected in the future and in Part IV we thus turn to the effects of the several cycles—*short, medium,* and *long*—on our climate.

The relations of sunspot cycles with weather, prices, and history are often controversial. Most of what has been written in the past is without statistical justification (Pittock, 1978); more scientific studies such as those represented in the symposium volume *Sun and Climate* (see Schove, 1980b) show that solar-terrestrial relationships are not as simple as previously had been supposed.

The effect of the 11-year cycle on pressure has been authoritatively studied by Parker (1976; cf. Lamb, 1977), some of whose diagrams (Figs. 11 and 12) provide clues to winter and summer trends to be expected between 1987 and 1991.

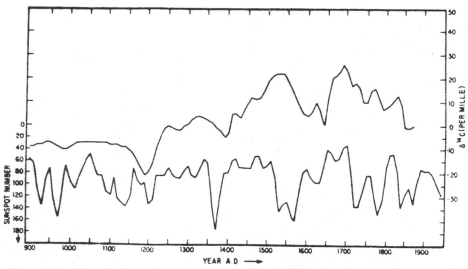

Figure 10 Radiocarbon in tree-rings (Δ^{14}C per mile) and sunspot number at maxima (on an inverted scale). Interpolated points at ten-year intervals. (*Cf.* Appendix C.) (*Reprinted from Houtermans, J. et al., 1967,* Radioactive Dating and Methods of Low-level Counting, *International Atomic Energy Agency, Vienna, p. 58; copyright ©1967 by the International Atomic Energy Agency.*)

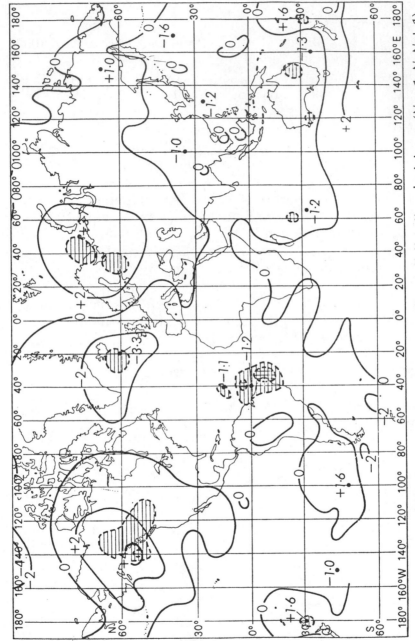

Figure 11 Difference between pressure at maximum ($X - 1$, X, $X + 1$) and minimum ($N - 1$, N, $N + 1$) phases in millibars, and significant areas by analysis of variance for Januarys, 1750–1958. Hatched area significant at five percent or beyond, cross-hatched at one percent or beyond. (*Reprinted by permission of the Controller of Her Brittanic Majesty's Stationery Office from Parker, B. N., 1976, Meteorol. Mag.* **105**:42; *copyright ©1976 by Her Majesty's Stationery Office.*)

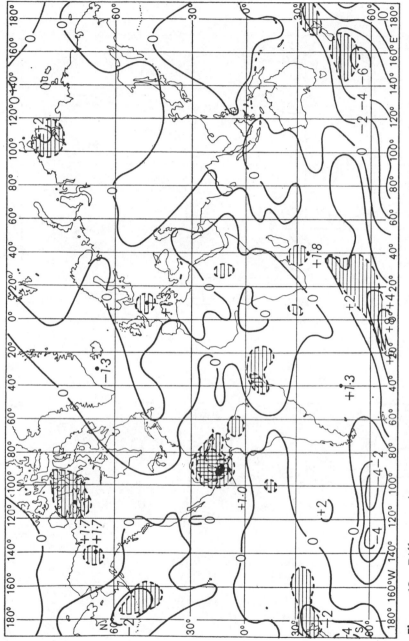

Figure 12 Difference between pressure at maximum $(X - 1, X, X + 1)$ and minimum $(N - 1, N, N + 1)$ phases in millibars, and significant areas by analysis of variance for Julys, 1750–1958. Hatched area significant at 5 percent or beyond, cross-hatched at 1 percent or beyond, and solid area at 0.1 percent. (*Reprinted by permission of the Controller of Her Brittanic Majesty's Stationery Office from Parker, B. N., 1976, Meterol. Mag. **105**:43; copyright ©1976 by Her Majesty's Stationery Office.*)

Other studies (Paper 35) show there is a tendency near strong sunspot maxima for low pressure and wetness in the Old World tropics and continental northeast winds in northwestern Europe, a *pattern* I associate with the *pressure parameter* (Papers 31, 32, 35, and 38).

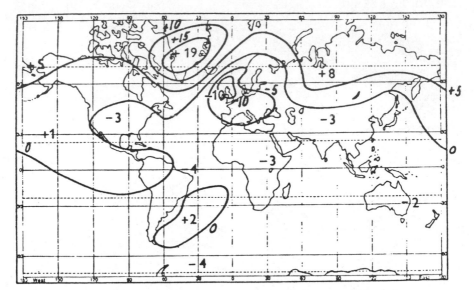

Figure 13   Change of mean winter pressure from sunspot minimum to the following major sunspot maximum (hundredths of an inch). (*Reprinted by permission from Willett, H. C., 1949,* Geog. Annaler **31**:301; *copyright ©1949 by Geografiska Annaler.*)

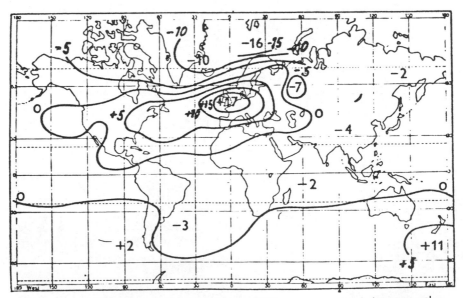

Figure 14   Change of mean winter pressure from sunspot minimum to the following minor sunspot maximum (hundredths of an inch). (*Reprinted by permission from Willett, H. C., 1949,* Geog. Annaler **31**:307; *copyright ©1949 by Geografiska Annaler.*)

Famines and revolutions (see Part VI) have been slightly more frequent in Europe at sunspot maxima, but drought famines in the Old World tropics are more frequent at sunspot minima.

Alternate sunspot cycles often have opposite climatic effects and the implied 22-year cycle has been studied by Willett (1949, 1961).

The next sunspot maximum will be an even one, 22 in the Zurich sequence, and therefore Willett's map (Fig. 14) may prove a better guide than Parker's (Fig. 11) to the pressure trends in winter in the coming period of sunspot increase (about 1987–1991). However, as Table 2 reveals, even maxima are not always *minor maxima*. The apparent influence of the double sunspot cycle in the other seasons is shown in Wagner's maps (1971, Fig. 5). The 22-year cycle has been important in both England (Fig. 15) and North America (Fig. 16) since the seventeenth century.

The 22-year cycle (23.6) in English temperature was effectively demonstrated by Bain (1976), who has since found (personal communication) that during the separate months, especially during the spring months, cycles of 20 to 25 years are always significant. The convincing results from the United States presented in Figure 16 have not been confirmed for the Middle Ages and, using ice-cores, Hibler finds (see Schove and Fairbridge, eds., 1983) the 20-year cycle back to A.D. 600 but notes that it is unreliable. Since 1850, when the 22-year cycle in solar amplitude commenced, meteorological effects of the 22-year cycle are to be expected near the neutral line in the world map of Paper 35; otherwise, the cause of this cycle is obscure.

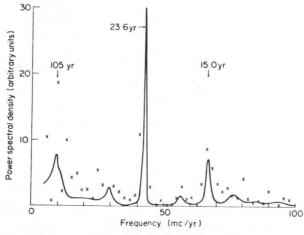

Figure 15 Power spectra for the annual mean temperatures in central England, 1659–1973. The periods corresponding to certain peaks are shown—by maximum entropy spectral analysis; x x by fast Fourier transform. (*Reprinted by permission from Bain, W. C., 1976, Royal Meteorol. Soc. Quart. Jour.* ***102:***465.)

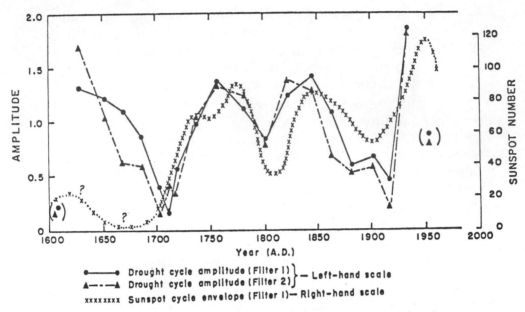

Figure 16 Chronology A.D. 1600–1962 of variations in amplitude of twenty-two-year drought rhythm based on each of two band-pass filters, and chronology of Hale sunspot envelope. Sunspot data for period 1600 to 1700 are tentative reconstructions from J. Eddy (Paper 22). (Reprinted *from Mitchell, J. M. et al., 1979*, Solar Terrestrial Influences on Weather and Climate, *D. Reidel, Dordrecht, Holland, p. 139; copyright ©1979 by D. Reidel Publishing Company.*)

Meanwhile, in the twentieth century there has been a progressive trend toward stronger solar maxima and a stronger solar wind, and some clues as to the effect of this 80-year maximum are given by the pressure trends between about 1890 and 1920; in *low latitudes* these trends continued into 1951 to 1980, a 30-year period designated as *c. 1965*.

Pressure has thus fallen in the Indian Ocean and risen in the Pacific Ocean, and similar *superoscillations* in earlier centuries (see list based on tree-rings and Nile statistics in Fig. 6 and page 88 of Schove, 1980*b*) often fit periods of great sunspot activity. In northwestern Europe the weather usually becomes cold and wet if the sunspot cycles suddenly become weak. These effects noted in Figures 17 and 18 may be partly due to the 200-year cycle (Papers 20, 28, and 38) revealed in Figure 19.

Even centuries such as the fourth, sixth, tenth, and twelfth have often had more sunspots and auroral activity, and such centuries are often slightly warmer than the intervening odd centuries. This suggests that the 200-year cycle found in solar activity is also found in global temperature. The so-called "Northward March of Civilization" has taken place primarily because of our increased control of indoor climate, but in some of the even centuries such as the later tenth, the increased warmth added to this northward movement.

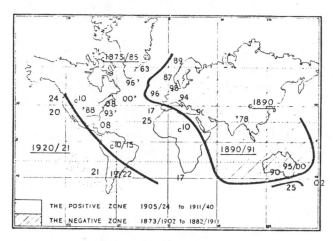

Figure 17 The P-max (thirty-year maximum of barometric pressure) between c. *1860* (= 60) and c. *1930* (= 30). (*Reprinted from Schove, D. J., 1961*, Geofisica Pura e Applicata **49**:257; copyright © *1961 by Birkhäuser Verlag.*)

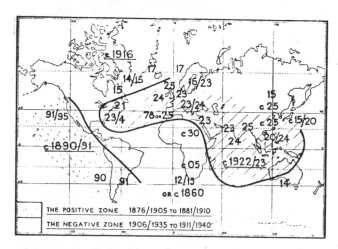

Figure 18 The P-min (pressure minimum) between c. *1860* and c. 1925 (= 1911/1940). (*Reprinted from Schove, D. J., 1961*, Geofisica Pura e Applicata **49**:257; copyright ©1961 by Birkhäuser Verlag.)

Geological applications of the study of sunspot cycles are discussed in Part VII where varves and tree-rings in combination promise to enlarge our horizons. At an archaeometry conference in 1964 I suggested that if the geological and archaeological dates were separated histograms of radiocarbon dates might reveal the dates of early "wiggles" on the radiocarbon curves. Such dates, where confirmed (such as 4730 b.p. and 4820 b.p. in De Jong et al., 1979), reveal that there are prehistoric prototypes of the Maunder Minimum, for example, 3350, about 3510, and about 3630 B.C. We include plausible

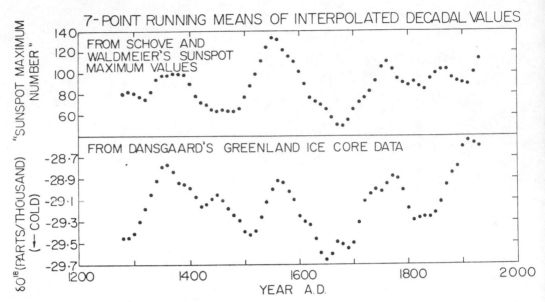

Figure 19 Sunspot maximum number and Greenland temperature index. (*Reprinted by permission from J. W. King,* Jeffreys Lecture to the Royal Astronomical Society, London, 1977.)

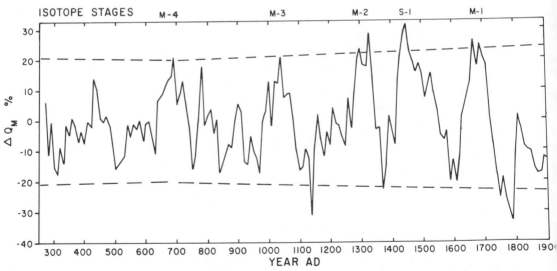

Figure 20 The post-300 A.D. ^{14}C production rate changes ΔQ_M relative to an average Q_M value (Stuiver and Quay, 1980). The ΔQ_M record was calculated from the atmospheric ^{14}C record through the use of a carbon reservoir model (Stuiver and Quay, 1980). The dashed line reflects the influence of Earth geomagnetic field changes on the magnitude of solar modulation of the cosmic ray flux. M-1 is the Maunder Minimum and M-2, M-3, and M-4 are earlier Solar Minima of similar character. (*Reprinted from Stuiver, M., and P. D. Quay, 1981,* Solar Physics **74**:480; *copyright © 1981 by D. Reidel Publishing Company, Dordrecht, Holland.*)

evidence for a 22-year weather cycle in pre-Holocene varves, but our papers emphasize that we need *very long* series of replicated varve counts and that Holocene varves should first be cross-dated with the absolute Bristlecone or Oak chronologies.

Overview

Our direct knowledge of sunspots is practically limited to the past 2000 years, and even in that period we cannot hope to fill the gaps in some centuries. Our knowledge of the 11-year sunspot cycles is being improved as additional aurorae come to light; there are still some periods where the evidence is insufficient to date the cycles and where the attempts of Bray (1980), Dai and Chen (1980), Wittmann (1978), and myself are at variance with one another. Some improvement will be possible as Stuiver (cf. Stuiver, 1982; Stuiver and Quay, 1980, 1981) and other members of the International Committee for the Calibration of the Radiocarbon Time Scale determine the radiocarbon content of tree-rings for each decade and each year. However, the hope that this would prove a physical method (Baxter and Walton, 1971; Tans et al., 1979) of dating 11-year cycles in the B.C. period has not yet happened. Our attention is now transferred to another isotope, Beryllium 10, being investigated in varves and ice-cores (see Schove and Fairbridge, eds., 1983).

The length of the cycle in the past two millennia (and probably in 3000–1000 B.C.) has been 11.1 years but we cannot yet prove that the period (since A.D. 300) of this "clock in the sun" is 11.09 years (as I believe) rather than 11.12 years (as I had supposed in 1948 and as many people still believe). We know that there is an 80-year cycle in phase but in amplitude this breaks up into cycles of 42 to 60, 90, and 133 to 148 years. There is also a 200-year cycle in both phase and amplitude. We still need to unravel the relationship between the 2-year and 3-year cycles and the 11-year cycle.

Solar-terrestrial relationships will become clearer when the International Quaternary Association (INQUA) section on the teleconnection of varves has produced long, precisely-dated series of proxy data to be tested for the phase of the 11-year cycle. We also need to separate two aspects of the 22-year cycle: those due to the tendency for odd cycles to be stronger and those due to the reversal of the sun's magnetism.

Current studies of *ice-cores, varves,* and *tree-rings* (see Schove and Fairbridge, eds., 1983) include spectral analyses of very long series of varves by Anderson and others (not yet published) that provide some new clues to sunspot cycles in the distant past. In a recent note in *Nature,* Williams (1981) claims that even in the Precambrian the 11-, 22-, and 80-year cycles were in existence in Australian varves. These varves are found in the Elatina formation

(32°25′ S, 137°59′ E). However, the photograph included in the article suggests that the layers may be microvarves and that there were about 11 in one year. Even if this proves to be the case, the discovery is of special interest as it would suggest that cycles of 2.2, about 8, and perhaps 13 years existed in an Ice Age 680 million years ago. Dr. Williams, the editor of *Megacycles* in this Benchmark series, is making further investigations (see Schove and Fairbridge, eds., 1983).

We are still in doubt about the future. The 80-year cycle suggests that sunspot cycles will become longer and the 200-year cycle suggests that they will become weaker; and certainly after the minimum expected in about 1987 to 1988 we should be able to provide an approximate date for the sunspot maximum expected in about 1991 to 1992.

We conclude the *Overview* with a translation of the earliest reference to sunspots (kindly supplied by Dr. J. C. H. Hsü) from a Chinese oracle bone: "Will the sun have marks? It really has marks" (Hsü, 1972). This *Late Shang fourth-period bone* reveals that speculation about sunspots goes back to at least the twelfth century B.C.

REFERENCES

Abetti, G., 1957, *The Sun,* Faber & Faber, London, 336p.

Bain, W. C., 1976, The Power Spectrum of Temperatures in Central England, *Royal Meteorol. Soc. Quart. Jour.* **102:**464–466.

Baxter, M. S., and A. Walton, 1971, Fluctuations of Atmospheric Carbon-14 Concentrations During the Past Century, *Royal Soc. (London) Proc.* **321A:**105–127.

Bray, J. R., 1980, A Sunspot-Auroral Solar Index from 522 B.C. to A.D. 1968, *New Zealand Jour. Sci.* **23:**99–106.

Bray, R. J., and R. E. Loughhead, 1964, *Sunspots,* Chapman & Hall, London, 304p. (Textbook of Sunspot Physics.)

Cole, T. W., 1973, Periodicities in Solar Activity, *Solar Physics* **30:**103–110.

Cousins, F. W., 1972, *The Solar System,* Baker, London, 300p. (Good popular account.)

Dai, N., and M. Chen, 1980, Chronology of Historical Aurora Data in China, Korea, and Japan, *Kejishiwenji* **6:**87–146. (In Chinese.)

Dall'Olmo, U., 1979, An Additional List of Aurorae from European Sources from A.D. 450 to A.D. 1466, *Jour. Geophys. Research* **84:**1525–1535.

De Jong, A. F. M., W. G. Mook, and B. Becker, 1979, Confirmation of the Suess Wiggles: 3700–3200 B.C., *Nature* **280:**48.

Gleissberg, W., 1971, The Probable Behavior of Sunspot Cycle 21, *Solar Physics* **21:**240–245.

Houtermans, J., H. E. Suess, and W. Munk, 1967, Effect of Industrial Fuel Combustion on the Carbon-14 Level of Atmospheric CO_2, in *Radioactive Dating and Methods of Low-Level Counting,* International Atomic Energy Agency, Vienna, pp. 57–68.

Hsü, J. C. H., 1972, *The Mendes Collection of Shang Dynasty Oracle Bones,* Royal Ontario Museum, Toronto, Canada.

Kane, R. P., 1978, Predicted Intensity of the Solar Maximum, *Nature* **274:**139–140.

Keimatsu, M., 1970–1976, A Chronology of Aurorae and Sunspots Observed in China, Korea, and Japan, *Kanazawa Univ. Ann. Sci.* **7:**1–10; **8:**1–16; **9:**1–36; **10:**1–32; **11:**1–36; **12:**1–40; **13:**1–32.

Kiepenheuer, K. O., 1953, Solar Activity, in *The Sun,* G. P. Kuiper, ed., University of Chicago Press, Chicago, pp. 322–465.

Lamb, H. H., 1977, *Climate: Present, Past and Future,* vol. 2, Methuen, London, 835p.

Link, F., 1962, Observations et catalogue des Aurorae Boreales apparues en Occident de -626 a 1600, *Geofys. Sb.* **10:**297–390.

Link, F., 1964, Observations et catalogue des Aurorae Boreales apparues en Occident de 1601 a 1700, *Geofys. Sb.* **12:**501–550.

Link, F., 1978, Solar Cycles Between 1540 and 1700, *Solar Physics* **59:**175–178.

Mitchell, J. M., C. W. Stockton, and D. M. Meko, 1979, Evidence of a 22-year Rhythm of Drought in the Western United States Related to the Half Solar Cycle Since the 17th Century, in *Solar Terrestrial Influences on Weather and Climate,* B. M. McCormac and T. A. Seliga, eds., Reidel, Dordrecht, Holland, pp. 125–143.

Newton, H. W., 1958, *The Sun,* Penguin, London, 208p.

Newton, R. R., 1972, *Medieval Chronicles and the Rotation of the Earth,* Johns Hopkins, Baltimore, 825p.

Parker, B. N., 1976, Global Pressure Variation and the 11-year Solar Cycle, *Meteorol. Mag.* **105:**33–44.

Pittock, A. B., 1978, A Critical Look at Long-term Sun-Weather Relationship, *Reviews Geophysics and Space Physics* **16:**400–420.

Schove, D. J., 1948, Sunspot Epochs, A.D. 188–1750, *Popular Astronomy* **56:**247–252.

Schove, D. J., 1961, Major Pressure Oscillation 1875/1960, *Geofisica Pura e Applicata* **49:**255–263.

Schove, D. J., 1979, Sunspot Turning Points and Aurorae Since A.D. 1510, *Solar Physics* **61:**423–432.

Schove, D. J., 1980a, Aurorae, Sunspots and Weather, Mainly Since A.D. 1200, in *Exploration of the Polar Upper Atmosphere,* C. S. Deehr and J. A. Holtet, eds., Reidel, Dordrecht, Holland, pp. 421–430.

Schove, D. J., 1980b, The 200-, 22- and 11-year Cycles and Long Series of Climatic Data, Mainly Since A.D. 200, in *Sun and Climate,* Centre National d'Etudes Spatiales, pp. 87–100.

Schove, D. J., in progress, *The Spectrum of Time.*

Schove, D. J., and R. W. Fairbridge, eds., 1983, *Ice-cores, Varves and Tree-rings,* Balkema, London.

Shapley, A. H., H. W. Koehl, and J. H. Allen, 1975, *Solar-Terrestrial Physics and Meteorology: A Working Document I,* Special Committee for Solar Terrestrial Physics, Washington, D.C., 142p.

Siscoe, G. L., 1980, Evidence in the Auroral Record for Secular Solar Variability, *Rev. Geophys. and Space Phys.* **18:**647–658.

Stuiver, M., 1982, A High-Precision Calibration and the A.D. Radiocarbon Time Scale, *Radiocarbon* **24:**1–26.

Stuiver, M., and P. D. Quay, 1980, Changes in Atmospheric Carbon-14 Attributed to a Variable Sun, *Science* **207:**11–19.

Stuiver, M., and P. D. Quay, 1981, A 1600-Year Long Record of Solar Change Derived from Atmospheric ^{14}C Levels, *Solar Physics* **74:**479–481.

Tans, P. P., A. F. M. de Jong, and W. G. Mook, 1979, Natural Atmospheric ^{14}C Variations and the Suess Effects, *Nature* **280:**826–828.

Wagner, A. J., 1971, Long-period Variations in Seasonal Sea-level Pressure Over the Northern Hemisphere, *Monthly Weather Rev.* **99:**49–66.

Waldmeier, M., 1961, *The Sunspot Activity in the Years 1610–1960,* Schulthess, Zurich, 171p.

Waldmeier, M., 1976, The Sunspot Activity in the Years 1961–1975, *Astronomische Mitt.* **346:**1–19.

Willett, H. C., 1949, Solar Variability as a Factor in the Fluctuations of Climate During Geological Time, *Geog. Annaler* **31:**295–315.

Willett, H. C., 1961, The Pattern of Solar Climatic Relationships. The Double Solar Cycle, *New York Acad. Sci. Annals* **95:**89–106.

Williams, G. E., 1981, Sunspot Periods in the Late Precambrian Glacial Climate and Solar-Planetary Relations, *Nature* **291:**624–628.

Wittmann, A., 1978, The Sunspot Cycle Before the Maunder Minimum, *Astronomy and Astrophysics* **66:**93–97.

Wolf, R., 1868, *Astronomische Mitt.* **24:**111. (Annual sunspot data since 1700.)

Zhukov, L. V., and Yu. S. Muzalevskii, 1969, A Correlation Spectral Analysis of the Periodicities in Solar Activity, *Soviet Astronomy* **13:**473–482.

Part I

EARLY SUNSPOTS

Editor's Comments
on Papers 1 Through 6

1   SCHOVE
Excerpts from *The Earliest Dated Sunspot*

2   SCHOVE
Excerpt from *Eclipses, Comets and the Spectrum of Time in Africa*

3   SCHOVE
Excerpts from *Sunspots, Aurorae and Blood Rain: The Spectrum of Time*

4   GOLDSTEIN
Excerpts from *Some Medieval Reports of Venus and Mercury Transits*

5   CLARK and STEPHENSON
Excerpts from *An Interpretation of the Pre-Telescopic Sunspot Records from the Orient*

6   GALILEO
Excerpts from *Letters on Sunspots*

Pretelescopic Spots to A.D. 1610

Papers 1 and 2 begin with a discussion of the earliest-dated sunspot. Spots on the sun were recorded before 1000 B.C. as Chinese oracle bones testify. The Greeks probably observed sunspots by the eighth century B.C. The earliest-dated sunspot has been generally regarded as that seen in China in 28 B.C. (Papers 3 and 5), but, arguing from synchronous references to aurorae, I suggested that a sign observed by the Chinese in the sun in 165 B.C. must also have been a sunspot (Paper 1). Bicknell (1968) suggested that as well as Halley's Comet a sunspot and an aurora may have been seen in 467 to 466 B.C. Western records of early sunspots (Papers 3 and 4) are hard to find and the sunspots usually cited (e.g., from von Humboldt, 1871; cf. Schove, 1951, pp. 134–135) are explained mostly by eclipses or other phenomena (cf. Schove and Fletcher, 1983). The pale sun recorded when Caesar died, included as a sunspot effect in some modern catalogs, may have been due to volcanic dust (Paper 3); in 1978 Hammer and colleagues reported that in the Greenland ice-cap there is a volcanic acid layer of the first century B.C.; Simkin and coworkers (1981) have found radiocarbon evidence for a New Zealand (Taupo) eruption about that century but a northern hemisphere eruption seems to be required. (The Newbury, Oregon eruption now appears to be later, e.g., A.D. 530s or 620s; see Schove and Fletcher, 1983.) A late Chinese commentary (cited in p. 398 of Paper 5) implies that sunspots could have been seen as well but this may be an attempt by the medieval Chinese to rationalize the 43 B.C. phenomenon.

The first clear Western record of a dated sunspot is that of 17–24 March A.D. 807, seen probably at Aachen. The original report (see Paper 4) has now been translated as follows: "On March 17 the star Mercury was seen as a small dark spot a little above the center of the sun, and it was seen by us for eight days. When it first entered and left the sun, we could not observe it well because of clouds" (Anon. in Scholz, trans., 1970, p. 87). This was no doubt the source used by Einhard (1974; cf. Paper 3) in his biography of the Emperor Charlemagne. The aurorae of this same year, 807, are cited in Paper 38.

The misinterpretation of sunspots as planets in transit across the face of the sun is partly due to the influence of Aristotle's belief that the sun, being a perfect body, could have no spots. The Islamic astronomers made the same mistake, and this accounts for the title of Paper 4.

The first clear Islamic record of a sunspot is that of 25 May 840, which seems to have been followed for three solar rotations ($3 \times 28.3 = 91$ days). This account, from a lost work (perhaps by the Benú Musā brothers) on comets, probably referred also to aurorae; thus the so-called comets before the death of Haroun-al-Rashid in 809 were presumably the auroral displays of the sunspot year 807. However, the genuine comets of December 840, July 841 to August 841, and December 841 to February 842 were no doubt recorded at Baghdad (cf. Schove and Fletcher, 1983); like the 840 sunspots these comets could well have been regarded as warnings of the death of Caliph Mu'tasim, Haroun's successor. Some recurrent auroras of 840 are mentioned later (see Paper 38), but we notice meanwhile that the 33-year difference between the two sunspot references leaves room for two intervening maxima, which can now be dated from auroral evidence as about 819 and about 827.

The sunspot recorded by Averroes is not dated but in view of the planetary interpretation Goldstein (Paper 4) suggests 15 May 1068. I find that this is two solar rotations ($28 \times 2 = 56$ days) before the Chinese aurora of July 10 (cf. Paper 17) so that Goldstein's date is plausible.

31

Sunspot sightings were relatively infrequent in both the eleventh and fifteenth centuries (cf. Figs. 4, 8, and 10) but strangely enough — apart from the Russian records quoted in Paper 1—our next Western record of a spot dates from about 1450 when two physicians, Guido Carrara and his son, Giovanni, saw a sunspot in Bergamo (Sarton, 1947). In one of Giovanni's books, *De constitutione mundi,* he speaks of his father's ideas as follows:

> At a certain time two drops of blood were seen in the sun and the populace was terrified. . . . However, my father, Guido of Carrara, whom I venture to describe as the outstanding figure of his age in every branch of letters, observed and compared the (positions of) planets and found that Venus and Mercury were the cause. (Sarton, 1947, p. 70)

This was possibly the sunspot recorded by the Chinese on 10 November 1449, although only Mercury was in front of the sun then.

In the sixteenth century sunspots must have been more frequent than recent writers have suggested, and two examples have been included in Paper 2. Kepler's spot was seen on Monday, May 18/28 in 1607 and once again misconstrued as "a singular phenomenon," or "Mercury in the sun" (cf. Paper 4), for Kepler had anticipated a transit and was watching the sun accordingly.

Our main source of pretelescopic spots lies in the Far Eastern historical records, and the first catalog of sunspots was that included in the *Chin Shu;* its astronomical section has been translated by Ho (see Paper 15) and reveals that the Chinese could easily see sunspots in the hazy weather between November and April in fourth-century China. The first extensive catalog worthy of the term *benchmark* was that published by Ma Tuan-lin in his thirteenth-century encyclopedia, the *Wen Hsien T'ung K'ao;* updated versions of this catalog became known in the West in the 1870s (see Needham, 1958, p. 435). However, we present here an authoritative modern version by Clark and Stephenson (Paper 5). This is more critical than the latest catalog from China (Yunnan Observatory, 1977), thus it omits the duplication of A.D. 180, the haze of 8 January 308 and the so-called spots of A.D. 502 and 505. Nevertheless, in rejecting certain later examples (in 1 December A.D. 807, 20 January A.D. 829, 6 May A.D. 832, 18 December A.D. 887, and 25 February A.D. 1129) I suggest that it is the brief text, not the diagnosis that is at fault. Other Chinese records considered by Keimatsu (1970-1976) are January 962 to February 962, 10 November 1449, 16 April 1569 to 11 August 1569, 8 May 1597, and 15 June 1597. Two additional records from Annam appear to come from local chronicles (c. 1275–1305 and c. 1577–1597) independent of the Chinese sources and read: "In the third month (17 March 1276 to 15 April 1276) there was a dark spot on the sun. It was as large as a hen's egg and it seemed to scintillate for a long time" (Ho, 1964, p. 141).

Sunspots were seen in Europe about the time of the battle of Muhlberg (24 April 1547). At sunrise on 3 January 1593 "there were seen for three days two dark spots shaped like crows inside the sun" (Ho, 1964, pp. 142–143).

Clark and Stephenson also provide an introduction to some of the wider implications of these early sunspot records, which are discussed later in this book (see especially Papers 22 and 28). Nevertheless, the auroral evidence suggests that extant sunspot representation in the Early T'ang in the eighth century and the Late Ming in the sixteenth century is inadequate.

Spots Through the Telescope, A.D. 1610 to 1699

The telescope was invented in the Netherlands by early 1609 and the news of the invention reached Galileo in Italy late in June 1609. About the same time the news must have reached Harriot in England, for his first telescopic observation of the moon is dated 26 July (Old Style or Julian calendar); however, 1609 was near the sunspot minimum and sunspots large enough to be noticed are not likely to have been seen before Harriot made his first drawing in 8/18 December 1610 (North, 1974, Fig. 4). Thomas Harriot had been mathematical tutor to Sir Walter Raleigh, and as a surveyor he had accompanied Sir Richard Grenville to Virginia. His subsequent series of numbered sunspot drawings, dated from 1/11 December 1611 to 18/28 January 1613 have been studied recently by Herr (1978; and cf. Wolf, 1858, pp. 135–139, 160), who has shown that the mean heliographic latitude of the spots was 16.7° and that the range was from 2° to 38°.

The chronology of the rediscovery of sunspots by Harriot, Fabricius, Galileo, and Scheiner has been discussed by both Shea (1972, pp. 49-53) and North (1974); Galileo's later claim to have seen sunspots in July 1610 to August 1610 is not now accepted. Harriot, however, never published his discoveries and the first printed account of telescopic sunspots was made in 1611 by Fabricius, a Frisian, whose work was dedicated on 13 June 1611 but whose observations appear to go back to March 1611. Fabricius used his observations, as the spots drifted across the solar disc, to verify the rotation of the sun.

Galileo's famous *Letters to Cosimo II of Florence* are, nevertheless, the obvious source for Paper 6. In the Autumn of 1611, Scheiner, a Jesuit lecturer at Ingolstadt, had written three letters to his friend Mark Welser in Augsburg informing him that he had discovered spots on the sun. Scheiner's dated observations extend from 21 October 1611 to 7 April 1612 (Wolf, 1862). Welser had the letters printed, one copy being sent to Galileo whose dated observations extend from 5 April 1612 to 21 August 1612. In his letters Galileo adopted a scientific

33

approach and effectively disposed of the belief that sunspots were planets in transit across the face of the sun. He pointed out that this was the mistake made by the Frankish chronicler in A.D. 807 and by Kepler in A.D. 1607. Excerpts from Galileo's letters of 1611 are now published and here we have attempted to select the most significant passages from his letter of 14 August 1612.

In the 1620s Scheiner continued to make sunspot observations and indeed his drawings of 1625 to 1626 are the basis (see Table 2 in the Introduction) for dating the sunspot maximum of that decade. His beautiful engravings were included in his *Rosa Ursina* (Scheiner, 1630; cf. Figs. 1.1 and 1.2 in Bray and Loughhead, 1964). Subsequently in 1642 to 1644 at Gdansk or Danzig, Hevelius made observations that have recently been used by Eddy and his collaborators (1977) to suggest that something peculiar was already happening to the sun; Eddy et al. claimed that in about 1643 the equatorial velocity was faster than normal and that the differential rotation of the zones was enhanced (but see Vitinsky, 1978). Certainly from 1645 until 1699, during the Maunder Minimum (see Part IV), sunspots were comparatively rare (see the chart in Link, 1978). A chronology of all the seventeenth century observations that Wolf could find was prepared by him in the 1860s as a manuscript of 11 pages in the Swiss Federal Observatory, Zurich (copy in G. E. Hale Collection, Hale Observatory Library, Pasadena, California).

REFERENCES

Bicknell, P. J., 1968, Did Anaxagoras Observe a Sunspot in 467 B.C.? *Isis* **59:**87–90.

Bray, R. J., and R. E. Loughhead, 1964, *Sunspots,* Chapman & Hall, London, 304p. (Textbook of Sunspot Physics.)

Eddy, J. A., 1977, The Case of the Missing Sunspots, *Sci. Am.* **236:**80–92.

Eddy, J. A., P. A. Gilman, and D. E. Trotter, 1977, Anomalous Solar Rotation in the Early 17th Century, *Science* **198:**824–829.

Einhard, 1974, *Two Lives of Charlemagne,* L. Thorpe, trans., Penguin Classics, London, 240p.

Hammer, C. U., H. B. Clausen, W. Dansgaard, N. Gunderstrup, S. J. Johnsen, and W. Reeh, 1978, Dating of Greenland Ice-Cores by Flow Models, Isotopes, Volcanic Debris, and Continental Dust, *Jour. Glaciology* **20:**3–26. (Cf. *Nature* **288:**230–235 (1980) and Schove and Fairbridge, eds., 1983.)

Herr, R. B., 1978, Solar Rotation Determined from Thomas Harriot's Sunspot Observations of 1611 to 1613, *Science* **202:**1079–1081.

Ho, P-Y., 1964, Natural Phenomena Recorded in the Dai-viet Su'ky Toanthu', *Am. Oriental Soc. Jour.* **84:**127–149.

Ho, P-Y., 1966, *The Astronomical Chapters of the Chin Shu,* Mouton, Paris, 271p.

Keimatsu, M., 1970–1976, A Chronology of Aurorae and Sunspots Observed in China, Korea, and Japan, *Kanazawa Univ. Ann. Sci.* **7:**1–10; **8:**1–16; **9:**1–36; **10:**1–32; **11:**1–36; **12:**1–40; **13:**1–32.

Link, F., 1978, Solar Cycles Between 1540 and 1700, *Solar Physics* **59:**175–178.

Needham, J., 1958, *Science and Civilisation in China,* vol. 3, Cambridge University Press, Cambridge, England, 877p.

North, J., 1974, Thomas Harriot and the First Telescopic Observation of Sunspots, in *Thomas Harriot,* J. W. Shirley, ed., Clarendon Press, Oxford, pp. 129–165.

Sarton, G., 1947, Early Observations of Sunspots, *Isis* **37:**69–71.

Scheiner, C., 1630, *Rosa Ursina sive Sol ex Admirando Facularum,* Apud Andream Phaeum Typographum Ducalem, Bracciani, Italy.

Scholz, B. W., trans., 1970, *Carolingian Chronicles: Royal Frankish Annals and Nithard's Histories,* The University of Michigan Press, Ann Arbor, 235p.

Schove, D. J., 1951, Sunspots, Aurorae and Blood Rain: The Spectrum of Time, *Isis* **42:**133–138. (See Paper 3.)

Schove, D. J., and R. W. Fairbridge, eds., 1983, *Ice-cores, Varves and Tree-rings,* Balkema, London.

Schove, D. J., and A. Fletcher, 1983, *Eclipse and Comet Chronology,* Boydell and Brewer, Woodbridge, England.

Shea, W. R., 1972, *Galileo's Intellectual Revolution,* Macmillan, London, 201p.

Simkin, T., L. Siebert, L. McClelland, D. Bridge, C. Newhall, and J. H. Latter, 1981, *Volcanoes of the World,* Hutchinson Ross Publishing Company, Stroudsburg, Pa., 233p.

Vitinsky, Y. I., 1978, Comments on the So-called Maunder Minimum, *Solar Physics* **57:**475–478.

von Humboldt, A., 1871, *Kosmos,* vol. 4, E. C. Otté and B. H. Paul, trans., Bell, London, 603p.

Wolf, R., 1858, Mitteilungen über die Sonnenflecken, *Astronomische Mitt.* **6:**127–143. (For sunspot formula, see pp. 133–134.)

Wolf, R., 1862, Mitteilungen über die Sonnenflecken, *Astronomische Mitt.* **13:**118.

Yunnan Observatory, 1977, A Recompilation of Our Country's Records of Sunspots Through the Ages and an Inquiry into Possible Periodicities in Their Activity, *Chinese Astronomy* **1:**347–359. (First published in Chinese in *Acta Astron. Sinica* **17:**217–227, 1976.)

1

Reprinted from pages 22, 22-23, 23-24, and 25 of *British Astron. Assoc. Jour.*
61:22-25 (1950)

The Earliest Dated Sunspot

By D. Justin Schove, B.Sc., F.R.Met.S.

Galileo was of course not the first to discover sunspots; they had been described frequently long before telescopes were invented. In China between December and April dry dusty winds from Northern Asia often bring a haze through which the spots can be seen without eyestrain. From the fourth century A.D. reports are fairly regular and help considerably in determining the dating of the 11-year sunspot cycle.* Distinctions are made between sunspots shaped like a flying-bird, those like an egg and those like an apple. No word or sign for sunspots has so far been identified in the ancient texts of the Near East, but there were doubtless many early observations mentioned in the original annals.

[*Editor's Note:* Material on European sunspots (A.D. 807 and A.D. 1607) has been omitted at this point (see Paper 3).]

Sunspots in Russia A.D. 1365—1371

Sunspots are of course easy to see in hazy weather or near sunset. In medieval Russia in 1365 and 1371 the smoke from forest fires made the Sun appear "bloody" and sunspots are clearly described in e.g. the Niconovsky chronicle (Vol. 11 of the St. Petersburg series of Russian Chronicles published in 1897), the extracts of which have recently been published in English by Vyssotsky (1949) as follows:—

"During this year (A.D. 1365) there was a sign in the sky. The sun was like blood and there were dark spots on it, and haziness lasted for half of the year. The heat was very intense; the forests, the marshes and the earth itself burned, the rivers dried up, some water-covered lowlands dried up completely, and there was terror, dread and sorrow among men."

"During this year (A.D. 1371) there was a sign in the sun. There were dark spots on the sun, as if nails were driven into it, and the murkiness was so great that it was impossible to see anything for more than seven feet. . . . Woods and forests were burning and the dry marshes began to burn, and the earth itself burned, and great fright and terror spread among men."

The latter extract now dispels doubts I expressed* as to veracity of the many Chinese sunspots ascribed to that year.

Sunspots in Japan and Korea.

A sunspot was recorded in Japan in December of A.D. 851 and in Korea on many occasions after A.D. 1151 March 21.

[*Editor's Note:* Material on Chinese sunspots (30 B.C.-A.D. 188) has been omitted at this point (see Paper 5).]

* *Cf.* Schove, 1948 b.
* Schove, 1947.

In my view there is however evidence for a much earlier sunspot seen in Spring 165 B.C. The older Chinese histories have not always survived, but the lost work of Wang Ch'i-ching, according to medieval authorities, stated that, in the time of the Emperor Wen, in the sun there was the Chinese character 主.‡

The contemporary significance of such a portent lay in its meaning of "prince" but to us it would appear that a sunspot was observed in Spring 165 B.C.

This interpretation is strengthened by the evidence of aurorae in the same context. In the annals of this date Pan Ku merely states:

"a yellow dragon appeared at Ch'eng-chi."*

Dragons are sometimes terrestrial animals and sometimes celestial phenomena. This one would appear to belong to the latter category inasmuch as it was associated with the sunspot and "a supernatural emanation in five colours . . . north-east of Ch'ang-an" and actually led to an imperial edict (see references, in Dubs' footnotes 2 and 3, to the Chinese annals of Szu-ma Ch'ien). It would seem that at least the emanation was a many-coloured aurora seen to the NE, probably (as frequently happens) the night after the sunspot had crossed the central meridian of the Sun.

Records of aurorae in Roman and other Chinese sources indicate sunspot maxima about the years 217 B.C., 206 B.C. and 195 B.C. but for the rest of the century (until 113 B.C.) the sources are too few and the dates too vague to determine whether the year 165 B.C. is likely to have been near a sunspot maximum.

Another method of determining the sunspot maxima about this time is based on the belief that tree-growth is greater and tree-rings wider about every 11 years in harmony with the sunspot cycle. As far as North Scandivanian conifers are concerned I could find† no real relationship and preliminary investigations of English trees show equally negative results. Nevertheless, the study of Italian trees by Buli (1949) suggests that in sunnier climates the relationship may be a real one. Assuming a cycle of 11 years in the 2500 year old sequoias of California is due to sunspots, Vercelli‡ infers that maxima would have occurred about the years 197 B.C., 187 B.C., 176 B.C., 164 B.C. If the same cycle is found in Roman timbers from archaeological sites this will be good evidence for my belief that the sign in the Sun of 166 B.C. was really a sunspot.

It may, by this means, be possible to determine the dates of the various sunspot maxima between the time of Theophrastus and A.D. 188 and thus to extend the table presented recently§ backwards for another six centuries. Meanwhile, I should like to express my thanks to the authors starred below for kindly sending me copies of the reprints named.

Early Far Eastern and Near Eastern records may yet provide examples of both aurorae and sunspots about 1000 B.C., which, with the help of new chronological techniques|| may soon be exactly dated.

‡ See Dubs, Vol. 1, 1938, p. 258, n. 2.
* Dubs, Vol. 1, 1938, p. 258.
† Schove, 1950.
‡ Vercelli, 1949, fig. 3, p. 10.
§ Schove, 1948 b.
|| *Cf.* Zeuner, 1950.

References.

Buli, U.	1949	Ricerche Climatiche sulle Pinete di Ravenna. Bologna.
Dubs, H. H.	1938 and 1944	History of the Former Han Dynasty by Pan Ku, with translation Vol. 1 (1938), 258. Vol. 2 (1944).
Schove, D. J.	1947	The Sunspot Cycle before 1750. *Terr. Mag.*, **52**, 233—237.
	1948a	Sunspots and Aurorae (500—250 B.C.). *Journ. Brit. Astr. Assn.*, London, **58**, 1948, 178—190 and 202—204.
	1948b	Sunspot Epochs A.D. 188—1610. *Popular Astronomy*, Northfield, Minn., U.S.A., **56**, 1948, 247—252.
	1950	Tree Rings and Summer Temperatures A.D. 1501—1930. *Scott. Geog. Mag.*, Edinburgh, **66**, 1950, 37—42.
Vercelli, F.	1949	Periodicita dendrologische e cicli solari. *Annali di Geofisica Roma*, **2**, 1949, 477—485.
Vyssotsky, A. N.	1949	Astronomical Records in the Russian Chronicles from 1000—1600 A.D. *Medd. fran Lunds Astro. Observatorium.* Scr. 2, Nr. 126, Historical Papers Nr. 22, Lund.
Zeuner, R. E.	1950	Dating the Past (2nd ed.), London.

[*Editor's Note:* The belief that tree-rings might help in detecting 11-year cycles proved false but a joint study with Dr. Bicknell of Chinese and classical aurorae fits Vercelli's date (see Paper 18).

A possible early sunspot is suggested by Bicknell (1968, Did Anaxagoras Observe a Sunspot in 467 B.C., *Isis* **59**:87–90). Anaxagoras may, like Humboldt, have regarded it as an opening in the sun, and he certainly seems to have predicted that a body would fall from the sun. Plutarch, in his "Life of Lysander" (Perrin, trans., 1950, Loeb classical series, 12, pp. 263, 266), quotes a lost work "On Religion" by Daimachus who described the 75-day comet in language that suggested aurorae in the same period: ". . . a fiery mass of vast size, as if a flaming cloud, not resting in one place but moving with intricate and irregular motions so that fiery fragments broken from it by its plunging and erratic course were carried flashing fire in all directions." The Parian Marble (ep. 57) of 264 B.C. confirms the date 467–466 B.C. for the meteorite and the Chinese sources confirm the comet.]

2

Reprinted from pages 96 and 98 of *British Astron. Assoc. Jour.* **78**:95–98 (1968)

ECLIPSES, COMETS AND THE SPECTRUM OF TIME IN AFRICA

D. J. SCHOVE

[*Editor's Note:* In the original, material precedes this excerpt.]

SUNSPOTS

Sunspots themselves are readily seen in the hazy weather associated with desert margins, although strangely enough no truly African early observations are known. The nearest instances are quoted below.

A Damascus chronicler of the sixteenth century refers, under A.H. 938 or 1531/2, to a lunar eclipse of 17 Ramadan (the only Ramadan eclipse about this date is 1530 April 12), and adds:

"The next morning [presumably 1530 April 13] in the dawn they saw, in the upper portion of the solar disc, two eyes and brows that gave it the likeness of a human face[20]."

The next instance comes from an English ship off the coast of West Africa. The log of the ship *Richard of Arundell* is quoted by Hakluyt[21] and the sunspot was seen in 1590 on December 7, 8 and 16 ". . . on the 7 at the going downe of the sunne, we saw a great black spot in the sunne, and the 8 day, both at rising and setting, we saw the like, which spot to our seeming was about the bigness of a shilling. . . ."

20 Ibn Guma, French translation of Ibn Guma, *c.* 1744, by H. Laoust, 1952, Paris, citing an earlier chronicler.
21 Hakluyt, 'Navigations', **ii**, 131, 1599 ed.

[*Editor's Note:* The day 17 is incorrect; the year may have been correct. The small lunar eclipse of 1 April 1531 could just have been noticed by astronomers who were expecting it. Auroral activity in 1531 (e.g., 13 April 1531 in China) now suggests that 2 April 1531 may well have been the date of the sunspot.

In page 95 of the original, African aurorae were noted in 879 (or possibly 878), 925, 1840, 1859, 1870, 1872, 1909, and 1921; sunspot maxima c. 1718, c. 1727, c. 1778, and c. 1837 were indicated as likely to have led to displays in North Africa.]

3

Reprinted from *Isis* **42**:133-134, 135-136 (1951)

Sunspots, Aurorae and Blood Rain: The Spectrum of Time

BY D. JUSTIN SCHOVE

INTRODUCTION

PROFESSOR SARTON has recently, in three separate places, discussed early observations of sunspots, aurorae and blood-rain.[1] Whereas there appears at first to be little connection, the three topics are all relevant to a single problem — the determination of the phase of the sunspot cycle in pre-telescopic times. It was this problem that led me to the project which I call "The Spectrum of Time."[2] The ambiguous interpretations of chroniclers and soothsayers confuse sunspots with planetary transits or eclipses, and confuse aurorae with comets or meteors. "Blood-rain" moreover is often, as will be shown in the second part of this paper, a confusion in itself and frequently, as in at least one of the cases mentioned in *Isis*, signifies an aurora. In view of these alternative explanations for the same celestial phenomena it was found necessary to include in one chronology natural events recorded in both East and West. When this was done many inexplicable "omens," by coincidence of dates, proved to be misinterpretations of northern lights.

The Spectrum of Time is a series of dates of natural phenomena which are more or less international. The time spectrum of a particular chronicle is a tabulation of the various portents and meteorological events. Often the dates, even in primary sources, are not known or incorrectly computed, and then the different annals are brought into chronological alignment by the recognition of certain "spectral lines," e.g. great comets, aurorae, meteoric displays, droughts, cold winters, pandemics and other phenomena.

EARLY SUNSPOTS

[*Editor's Note:* In the original, most of the so-called sunspots in Europe before 1607 are shown to refer to eclipses. That of 807 is a real sunspot.]

---

[1] Sunspots, in *Isis*, *37* (1947), 69–71, Query No. 111; aurorae, in *Introduction to the History of Science*, *3* (1948), 709–711; blood-rain, in *Isis*, *38* (1947), 96, Query No. 116.

[2] See my articles: Schove (1947), "The Sunspot Cycle before A.D. 1750," in *Terr. Mag. 52*, 1947, 233–238; Schove (1948a), "Sunspot Epochs A.D. 188–1610," in *Pop. Astro. 56*, 1948, 247–252; Schove (1948b), "Sunspots and Aurorae (500–250 B.C.)," in *Jnl. Brit. Astr. Assn. 58*, 1948, 178–190 and 202–204. See also my paper "Chronology of Natural Phenomena in East and West," to be published in the proceedings of the Sixth International Congress for the History of Science, held at Amsterdam in 1950.

The original source for the example of 807 has already been reproduced in *Isis* (1947, p. 69) where it was dated "807 or before 814." The latter alternative is implied by a current mistranslation of Einhard (or Eginhard), who referred to it as a portent of Charlemagne's death in 814. Actually Einhard did not mean, *During the last three years of his life* but, as apparent from the Latin, *"For three consecutive years, at the end of his life*, there were frequent eclipses, both solar and lunar and on the sun could be seen for seven days a spot of a dark colour." These eclipses can be calculated and appear in the "Spectrum of Time" as follows:

807: 11 February, Solar, France and England, etc.
809: 16 July, Solar, England, etc.
810: 20 June, Lunar, France, etc
810: 14 December, Lunar, France, etc.
810: 30 November, Solar, France, etc.

The unusual cluster of eclipses in the three years 807–810 proves that these are the years referred to, and it explains why in 810 the scholar Alcuin explained to Charlemagne the cause of eclipses. The "little black spot" seen for eight days "slightly below the centre of" the sun [5] in 807, 17–24 March, was probably the sunspot referred to, but that seen by the Chinese in 807 was seen in the "tenth moon" and there were intense aurorae seen in Europe in both 807 and 808. Solar activity then rapidly declined (the phenomena of 813 and 817 were probably barely-visible comets).

The table in Fritz suggests that sunspot activity was recorded in Pliny (II, 30, 99) for 44 B.C. Pliny really refers to 43 B.C. and speaks of "a whole year's continuous gloom." Plutarch mentions the paleness of the sun for a year after Caesar's death, but adds that for want of the sun's heat the fruits did not come to maturity. A writer Tibullus is quoted by Johnson [6] as saying "the misty year saw the darkened sun drive pale horses . . . ." Johnson does not quote Dio Cassius (XLV, 17, 5) who gives four phenomena: "A flash darted across from the east to the west and a new star was seen for several days. Then the light of the sun seemed to be diminished and even extinguished. . . . a plague spread over nearly all Italy. . . ."

Western writers have sometimes treated these statements as Roman extravagances in praise of Caesar, and indeed the great daylight comet of 44 B.C. was popularly interpreted (Dio Cassius XLV, 7, 1) "to mean that he had become immortal and

---

[4] Hermann Fritz, *Das Polarlicht*, 1881, Leipzig, p. 202.
[5] Cf. *Isis*, *37* (1947), 69.

[6] S. J. Johnson, *Historical & Future Eclipses*, 1895 ed., pp. 22–23.

received among the stars." However, reference to the "Spectrum of Time" shows similar observations made in China.[7]

The Han Shu quoted by Dubs refers to the Spring and Autumn frosts and the injury to the wheat harvest with resulting famine. It also says "In the fourth month (May/June) the color of the sun was pale blue and it cast no shadows; when it was exactly at the zenith it cast shadows (but) showed no brilliance. That summer was cold. In the ninth month (October), the sun, however, showed brilliance." An imperial edict in the subsequent Spring is quoted by Pan Ku as saying "the three luminaries have been veiled and indistinct."

In view of all the above references it seems certain that it was not sunspots that caused the diminution of the solar light (or heat). The great comet of the previous year was one of the most remarkable in history and it may not be unconnected, but the dust from a volcanic eruption is another possible hypothesis.

The same source (The Han Shu, quoted by Dubs, p. 384 note 5.6) does give a record of a real sunspot. This is that of 28 B.C., the earliest of those from Far-Eastern sources collected by S. Kanda of Tokyo.

In 28 B.C., on 22 February,

The sun came up red. In the second month (on April 4) the sun was red in the morning and when it went down it was also red. At night the moon was red. On (April 5) the sun came up red as blood without any brilliance. When the clepsydra marked four divisions and a half, (the sun) had some light, and lighted up the earth red and yellow. After breakfast (the sun) recovered (its natural light). . . . In the third month on (the day) yi-wei (probably a mistake for chi-wei, May 10), the sun came yellow with a black emanation as large as a cash right in the center of the sun.

Professor Dubs notes that this was "probably a sunspot" and Kanda translates baldly as "a black spot, as big as a coin." The redness was doubtless due to loess-dust.

[*Editor's Note:* Material has been omitted at this point.]

[7] Homer H. Dubs, *History of the Former Han Dynasty*, 2, 1944, p. 319 (note 7.10) and p. 320 (and note 8.5).

ERRATUM

Page 135, line 15 from the top, ". . . the scholar Alcuin . . ." should read ". . . the scholar Dicuil. . . ."

4

Reprinted from pages 50, 51–54, and 55–59 of *Centaurus* **14**:50–59 (1969)

Some Medieval Reports
of Venus and Mercury Transits

by

BERNARD R. GOLDSTEIN

Introduction

Medieval reports of transits arc generally taken as observations of sunspots because the descriptions, dates, and durations make more sense according to that interpretation. Indeed, I have no quarrel with those scholars who would use these reports as data for sunspot activity, but it must be also recognized that to understand the reports themselves, the medieval astronomical context has to be kept in mind. To be sure the possibility that a medieval report refers to an actual Venus transit cannot be excluded, for such an event, though rare, is visible to the naked eye[1].

[*Editor's Note:* In the original, observations of true transits[2, 3] and sunspots[3, 4] are given.]

1. The Transit Report of al-Kindī

Ibn al-Qifṭī (d. 1248) gives this report in Arabic.[5]

Ghars al-Naʿma Muḥammad b. al-Raʾīs Hilāl b. al-Muḥassin al-Ṣābī said in his book: I found in the handwriting of Jaʿfar b. al-Muktafī [d. 987][6] ... that in the year 225 during the caliphate of al-Muʿtaṣim there appeared a black spot (nukta) close to the middle of the sun. This took place on Tuesday, 19 Rajab 225 [May 25, 840], and when two days had gone from this date, i.e. after 21 Rajab, events (calamities) occurred. Al-Kindī mentioned that this spot lingered on the sun for 91 days and soon thereafter Muʿtaṣim died. Before the death of al-Muʿtaṣim two comets appeared, as some had before the death of al-Rashīd. Al-Kindī mentioned that this spot was due to the occulting of the sun by Venus, and their clinging together for this period... Up to here this (passage) is taken from the treatise of Ibn al-Muktafī.

In addition to Ibn al-Qifṭī's report, another source for this alleged transit is a statement by Averroes (who lived before Ibn al-Qifṭī). Averroes's account (see Section 3 below) is consistent with that of Ibn al-Qifṭī but no mention is made of the duration of the transit. If Ibn al-Qifṭī's report is based on an accurate transmission of al-

Kindī's interpretation, namely, that he considered the spot on the sun to be a Venus transit that lasted for 91 days (in fact a transit only lasts several hours), it does not speak well for al-Kindī's knowledge of astronomy. Moreover, on May 25, 840, Venus was near its greatest elongation from the sun.[7]

2. *The Transit Report of Avicenna*

Unlike the other transit reports considered in this paper, Avicenna's is not at all well known. I first came across Avicenna's report in looking at authors who mentioned al-Biṭrūjī, a twelfth century Spanish philosopher known for his attempt to replace Ptolemy's astronomical models.[8] In Yahuda b. Solomon Kohen's encyclopedia (originally in Arabic, but extant only in a Hebrew version by the author himself) there is a description of al-Biṭrūjī's theory preceded by a short introduction on the order of the planets.[9] Yahuda notes that al-Biṭrūjī put Venus above the sun and Mercury below it, whereas Jābir b. Aflaḥ[10] argued that both Mercury and Venus lie above the sun. But "Avicenna saw Venus appearing like a spot (*ketem*) in the midst of the sun and thus Venus lies below the sun."[11] The same observation by Avicenna is reported by Naṣīr al-Dīn al-Ṭūsī in his redaction of the Almagest at the beginning of Book IX,[12] where he also mentions a report by another author, Ṣaliḥ b. Muḥammad, whom I cannot identify.[13] The fact that the latter report mentions an interval of 20 years between two Venus transit observations automatically excludes a Venus transit for at least one of them.

> *I [Naṣīr al-Dīn] say: al-Shaikh al-Ra'īs abū 'Alī b. Sīnā [Avicenna] mentions in (one of) his books that he had seen Venus as a spot* (khāl aw shāma) *on the surface of the sun. Ṣaliḥ b. Muḥammad al-Zaynabī [?] al-Baghdādī mentioned, in his book which he called* The Almagest, *that al-Shaikh Abū 'Umrān, in Baghdad, and Muḥammad b. Abū Bakr al-Ḥakīm, in Farsīn in the neighborhood of Tūlak [?], saw the body of Venus on the disk* (qurṣ) *of the sun, twice, separated, by about 20 years.*

It was my hope that the original passage in Avicenna's own work would include a date from which one could decide if he had observed a transit, a sunspot at the time of conjunction, or a sunspot at an astronomically meaningless time.

I obtained a microfilm of an Arabic text from the Bibliothèque Na-
tionale, Paris, identified in the catalogue as Avicenna's Compendium of
the Almagest.[14] Avicenna here states (in his comments on the beginning
of Book IX concerning the order of the planetary spheres):[15] "I say that
I saw Venus as a spot on the surface of the sun," with the very words
cited by Naṣīr al-Dīn. Unfortunately, no date is given for the observation,
nor does he mention the other transits that Naṣīr al-Dīn cited. Since
Avicenna died in 1037 A.D., the only Venus transit in his lifetime took
place on May 24, 1032, and this transit may not have been visible where
he lived.[16]

3. *The Transit Report of Averroes*

The transit report associated with the name of Averroes is well known
because Copernicus took note of it in *De Revolutionibus*.[17] In 1893,
Steinschneider identified the work of Averroes in which this report occurs, preserved in a Hebrew translation of the original Arabic text.[18]
Averroes reports, in a comment on the arguments for the order of the
planets in Ptolemy's *Planetary Hypotheses* (*Šefer ha-sippūr*), that two
black spots (*ketamīm*) were seen on the sun at the time of Ibn Muʿādh
by Ibn Muʿādh's nephew (*ben aḥot*), and that by computation he, Averroes, found that Mercury and Venus were in conjunction with the sun
at that time. No date is given but Ibn Muʿādh is known to have observed
the solar eclipse of July 3, 1079.[19] Averroes goes on to mention that he
read in "some books" that a similar event was noted at the time of
al-Kindī, and when al-Kindī was asked about it, he reasoned that one
of the two planets was below the sun (i.e. Mercury or Venus was in our
line of sight).

It should be noted that the underlying report does not associate the
observation with a transit, but that in the interpretation of the report
Venus and Mercury are introduced. Presumably the widely held theory
of the incorruptability of the heavens led him to ignore the possibility
that the spots were on the body of the sun itself. Averroes implies, by
stating that he computed positions, that the original report included a
date. Since simultaneous transits of Mercury and Venus did not take
place, he either miscomputed, used faulty tables, or ignored the planetary
latitudes. It seems to me that the third possibility is most probable,
introduces the fewest assumptions, and is consistent with the text. If this

is the case, a possible, but by no means unique, date for Averroes's computation is May 15, 1068, when both Mercury and Venus were very nearly in conjunction with the sun.[20] In any event, Averroes has used an extant observation to prove that Mercury and Venus lie below the sun.

[*Editor's Note:* In the original, Levi ben Gerson[21, 22] in the fourteenth century is quoted. He doubted the planetary interpretation given by Averroes but gave no new evidence for either transits or sunspots.]

4. The Transit Report of Ibn Bājja

Quṭb al-Dīn al-Shirāzī (d. 1311), a student of Naṣīr al-Dīn al-Ṭūsī, wrote several important astronomical treatises, among them *Nihāyat al-idrāk*.[23] In the section on the order of the planets he cites an observation of a Venus transit by the Spanish philosopher Ibn Bājja (d. 1139) whose astronomical writings do not survive.[24] This passage was quoted by al-Birjandī (early sixteenth century) and ascribed to "the author of *Nihayāt al-idrāk*,"[25] and was also quoted by Alī al-Qūshjī (d. 1474/75) in his commentary on Quṭb al-Dīn's *al-Tuḥfa al-Shāhiya*.[26] This is what Quṭb al-Dīn wrote:[27]

> The scholar Abū Bakr b. (Ṣāʿigh)[28] known as Ibn Bājja[29] al-Andalusī mentioned in one of his books, "At sunrise one day I was standing on the roof of my house, and I saw two spots on the surface of the sun. I calculated the positions of Venus and Mercury at that time from the zij, and I found them both near the position of the sun. Therefore I concluded that the two spots were Venus and Mercury."

In this case the observation could not have been of a Venus transit, for none took place between 1040 and 1153, and Ibn Bājja died in 1139.

Conclusion

It is clear that we do not have a well authenticated medieval report of a transit; the only report for which the observation of a transit cannot be excluded immediately is that of Avicenna, whose report includes too little information.

NOTES

1. Cf. J. Meeus, "The Transits of Venus of Venus 3000 B.C. to A.D. 3000", *J. of the Brit. Astron. Assoc.*, *68* (1958), 98–108. Meeus writes (p. 99), "As transits of Venus over the Sun are visible with the naked eye, a table of such transits might be of some value to historians of astronomy." Of possible interest for the subsequent discussion are the four Venus transits between 800 and 1200 A.D.; Nov. 23–24, 910; May 24, 1032; May 22, 1040; and Nov. 23–24, 1153 (Meeus, p. 101).

2. B. Goldstein, "The Arabic Version of Ptolemy's *Planetary Hypotheses*," (*Trans. Amer. Phil. Soc. n.s. 57*, 1967), p. 6.

3. Kepler once thought he had observed a Mercury transit (using a camera obscura), which he expected on the basis of a computation prior to the event, and published a short treatise in 1609 announcing it (Cf. Kepler, *Gesammelte Werke* IV (1941), p. 92). However, Kepler later realized that what he had observed was in fact a sunspot and not a transit. For a discussion of this episode, cf. E. Rosen, *Kepler's Conversation with Galileo's Sidereal Messenger*, New York and London: Johnson Reprint Corp., 1965, pp. 97–99.

4. *Mon. Germ. Hist. Script.*, *1* (1826), 194, and *Annales Regni Francorum*, ed. F. Kurze, Hannover, 1895, p. 123: "A. 807... *Nam et stella Mercurii XVI. Kal. Aprilis* [March 17] *visa est in sole quasi parva macula, nigra tamen, paululum superius medio centro eiusdem sideris, quae a nobis octo dies conspicitur. Sed quando primo intravit vel exivit, nubibus impedientibus minime adnotare potuimus.*" Galileo was aware that this report was erroneous and stated that it described the observation of a sunspot (Cf. S. Drake, *Discoveries and Opinions of Galileo*, New York: Anchor Books, 1957, p. 117). To be sure, there was no Mercury transit at that time. Cf. G. Sarton, "Early observations of sunspots?" *Isis 37* (1947), 69–71, for this report and some others.

5. *Ta'rīkh al-ḥukamā*, ed. J. Lippert, Leipzig, 1903, p. 156. This passage may also be found in Casiri, *Bibliotheca Arabico-Hispana Escurialensis*, Madrid, 1760, vol. 1, p. 422–23.

6. Cf. H. Suter, *Die Mathematiker und Astronomen der Araber* (*Abhandlungen zur Geschichte der mathematischen Wissenschaften*, vol. 10, Leipzig, 1900), pp. 64–65, 212.

7. B. Tuckerman, *Planetary, Lunar, and Solar Positions A.D. 2 to A.D. 1649* (*Memoirs of the Amer. Phil. Soc.*, vol. 59, 1964), p. 438.

8. The Latin version of al-Bitrūjī's treatise is published: *De motibus celorum*, ed. F. J. Carmody, Berkeley and Los Angeles, 1952. I have prepared a critical edition of the original Arabic text with an English translation and notes, which will appear, I trust, in the not too distant future.

9. On Yahuda ben Solomon Kohen (thirteenth century, Spain), see M. Steinschneider, *Die Hebraeischen Uebersetzungen des Mittelalters*, Berlin, 1893, p. 1 ff. I have consulted the copy of this encyclopedia preserved in Ms. Bodleian *hebr. Michael* 551.

10. On Jābir b. Aflaḥ (twelfth century, Spain) see H. Suter, *op. cit.*, p. 119; Steinschneider, *op. cit.*, p. 543; and Delambre, *Histoire de l'astronomie du moyen âge*, Paris, 1819, pp. 179–85. The Latin version of Jābir's astronomical treatise was published by P. Apianus in 1534.

11. Ms. Bodleian *hebr. Michael* 551, fol. 160ᵃ.

12. On Naṣīr al-Dīn (thirteenth century, Iran) see E. S. Kennedy, *A Survey of Islamic Astronomical Tables* (*Trans. of the Amer. Phil. Soc.*, n.s. *46*, 1956), pp. 125, 161–62; A. Sayili, *The Observatory in Islam*, pp. 189 ff.; and Suter, *op. cit.*, pp. 146–53. For Naṣīr al-Dīn's redaction of the Almagest, I consulted Ms. British Museum *arab.* Reg. 16A. VIII; book IX begins on a folio numbered in two ways: 131ᵇ and 251ᵇ. This passage was noted by C. A. Nallino, *Scritti* V (1944), p. 82.

13. A. Sayili mentions observations of Venus transits by Ibn Sina, Muḥammad Ṣāliḥ ibn Muḥammad al-Baghdādī, and Muḥammad ibn Abī Bakr al-Ḥakīm, on the authority of the annotations by ʿAbd al-ʿAlī al-Birjandī (early sixteenth century) to Qāḍīzāda's commentary on the *Mulakhkhaṣ* of Chaghmīnī, published in Istanbul 1290 H., pp. 16, 40. Professor Sayili sent me a hand copy of the passage and I checked it with two manuscripts of this text of al-Birjandī: Princeton H. 504, fol. 84ᵃ–84ᵇ, and Yale A-307, fol. 33ᵃ. Al-Birjandī indicates that he is quoting from Naṣīr al-Dīn al-Ṭūsī's redaction of the Almagest (*taḥrīr al-majisṭī*). Cf. A. Sayili, "Islam and the Rise of the Seventeenth Century Science," *Türk Tarih Kurumu, Belleten, 22* (1958), 360.

14. Ms. Paris *arab.* 2484. The copy is dated 673 A.H. [1274/5] on fol. 143ᵇ, yet the catalogue (Paris, 1883–95) gives 683 A. H. Suter (*op. cit.*, p. 90) cites another copy of this work in the Bodleian Library on the basis of Uri's Catalogue of Arabic mss. there (1787, 1, 1012, ms. Marsh 621 dated 671 A.H. [1272/3]), and Brockelmann also accepted this identification (*Gesch. arab. Litt.*, 2nd ed., Leiden, 1943, G I 457, no. 70). This Bodleian text is definitely not the same work as that found in the Paris manuscript and there is no evidence in the manuscript to connect it with Avicenna. The Paris text states that it is Avicenna's Compendium of the Almagest (fol. 1ᵇ) and in general it follows the order of books and chapters in the Almagest. On the other hand, the Bodleian text gives neither a title nor an author (after a short invocation it begins: *qāla al-shaikh qaddasa allah rūḥahu*), and it does not follow the order of books and chapters in the Almagest. Avicenna believed that his observation of a Venus transit demonstrated that Venus lies below the sun, but the author of the Bodleian text states that Venus lies above the sun and Mercury below it (fol. 185ᵃ). The unusual theory of planetary sizes and distances as well as the many references to Ptolemy's *Planetary Hypotheses* make the Bodleian text of considerable interest. Dr. Noel Swerdlow of the University of Chicago is currently working with me on a study of this anonymous manuscript and we plan to publish our results shortly.

15. Ms. Paris, *arab.* 2484, fol. 97ᵃ–97ᵇ.

16. It is known that Avicenna lived in many different places in Iran and environs (cf. Suter, *op. cit.*, p. 86–90), but I do not know where he was on May 24, 1032. For this reason I asked Dr. Brian Marsden of the Smithsonian Astrophysical Observatory to consider whether the sun was above the horizon at the time of the transit anywhere in the region between Baghdad and India. His conclusion, communicated to me in a letter dated Nov. 8, 1967, was that on the basis of available modern astronomical data it remains an open question as to whether Avicenna (or anyone else in this region) could have seen the transit or not. Marsden's reasoning is presented below:

> "Using the equation in Meeus' article [see note 1], for the time of sunset in Baghdad on 1032 May 24, I get 16ʰ02ᵐUT. Not having any reliable eclipse data from the 11th century, we cannot be sure what the difference between

ET [Ephemeris Time] and UT [Universal Time] was, but it is my guess that Meeus' values for the UT time of the transit of Venus are correct to within about 15 minutes. The earliest possible time for the first contact would thus be 15^h21^m, and the latest possible time for the second contact (Venus fully on the solar disk) would be 16^h07^m. The longitudes of places where the sun is setting (upper limb of sun on refracted horizon) at these times at various latitudes are given in the following table:

UT Lat.	15^h21^m	16^h07^m
30°N	52°.8 E	41°.3 E
32	54.0	42.5
34	55.2	43.7
36	56.5	45.0
38	57.9	46.4
40	59.3	47.8

For a mountain site the longitude would be further east. Something of the transit *might* have been visible west of the longitudes in the first column (i.e. west of the line from Shiraz to Ashkhabad), and Venus almost definitely could have been seen completely on the disk west of the longitudes in the second column."

17. Copernicus, *De Revolutionibus*, I, 10, ed. F. Zeller and C. Zeller, Munich, 1949, p. 23. A note on this passage (p. 440–41) refers us to E. Zinner, *Entstehung und Ausbreitung der coppernicanischen Lehre*, Erlangen, 1943, p. 510, where it is argued that Averroes is a mistake for Aven Rodan ['Alī b. Riḍwān, eleventh century] and that Copernicus had misread the relevant passage in Pico della Mirandola. The evidence, however, leads to the conclusion that it is Zinner, and not Copernicus, who misread Pico. As Nallino points out (*Scritti*, V (1944), 82), Copernicus's source was probably Pico's *Disputationes in astrologiam* X, 4 (Opera omnia, Basel, 1572–73, vol. 1, p. 685): "*Averrois in paraphrasi magnae compositionis Ptolemaei dicit se quondam in Sole duas quasi maculas nigricantes annotasse, cumque numeros digessisset per id tempus, inventum Mercurium Solis radii oppositum.*" I consulted the 1495 edition (Bononiae, Benedictus Hectoris), and it agrees exactly with the Basel edition.
18. Steinschneider, *op. cit.*, p. 546 ff. I have consulted Ms. Paris *hebr.* 696, fol. 82ᵇ.
19. A report of his observation is preserved in Ms. Paris *hebr.* 1036, fol. 1–6. On Ibn Mu'ādh, cf. H. Hermelink, "Tabulae Jahen," *Archive for History of Exact Sciences 2* (1964), 108–12, and A. I. Sabra, "The Authorship of the *Liber de crepusculis*," *Isis 58* (1967), 77–85.
20. Tuckerman, *op. cit.*, p. 552. The longitude of the sun was 60.28°, of Mercury (in retrograde motion) was 59.59°, and of Venus (in direct motion) was 60.03°. The latitude of Mercury was not close to zero at this time. That Venus was in direct motion, i.e. at superior conjunction, and yet considered to lie between us and the sun, conforms to the Ptolemaic view that the sphere of Venus lies entirely below that of the sun.

21. On Levi ben Gerson (fourteenth century, southern France), see M. Steinschneider *Mathematik bei den Juden*, 2nd edition, Hildesheim, 1964, p. 129 ff.; and E. Renan, "Les écrivains juifs français du XIVᵉ siècle," *Histoire littéraire de la France*, Paris, 1893, vol. 31, p. 586 ff. For this report, I have consulted Ms. Paris *hebr.* 724, fol. 253ᵃ (*Milḥamōt Adonai*, Book V, part 1, ch. 133).

22. I wish to thank Mr. Jerome Bylebyl for identifying the corresponding passage in Galen's works: Galen, *Opera omnia* 20 vols., ed. K. G. Kühn, Leipzig, 1821–33, *De placitis Hippocratis et Platonis libri novem*, lib. II. cap. iv (vol. 5, p. 238–39):

> *Accidit hoc sane etiam in multis sacrificiis, quae sic de more celebrantur, et apparent animantia, corde jam aris imposito, non respirare tantum, aut clamare fortiter, sed etiam fugere, usque dum sanguinis profluvio commoriantur. Celerrime autem nimirum ex ipsis sanguis evacuatur, quatuor maximis vasis divulsis: sed quousque adhuc vivunt, et respirant, et clamant, et currunt.*

23. On Quṭb al-Dīn al-Shirāzī, see Suter, *op. cit.*, p. 158, and E. S. Kennedy, "Late Medieval Planetary Theory," *Isis 57* (1966), 365–78, especially 371 ff. For *Nihāyat al-idrāk* I have consulted British Museum Ms. Add. 7482.

24. On Ibn Bājja (also known as Avempace) as an astronomer, see Suter, *op. cit.*, pp. 116–17.

25. Cf. A. Sayili, *The Observatory in Islam*, Ankara, 1960, pp. 184–85. The texts cited in note 13 (above) also refer to the observation of Ibn Bājja.

26. On 'Alī al-Qūshjī, see Suter, *op. cit.*, p. 178. I am indebted to Professor Sayili for bringing this passage to my attention and sending me a copy of it based on Ms. Bursa, Hüseyin Celebi 750, fol. 70ᵇ.

27. British Museum Ms. Add. 7482, fol. 21ᵇ.

28. Unclear in the British Museum copy of *Nihāyat al-idrāk*, but quite clear in 'Alī Qūshjī's quotation of this passage.

29. In no manuscript is the *bā* pointed, and in some of them this letter could easily be taken for a *mīm*. Nevertheless, there can be no doubt that Ibn Bājja is meant.

[*Editor's Note:* Maya astronomical dates will be discussed by Schove in *Jour. History Astronomy* (October 1983) and in *(Estudios de Maya Cultura* (XIV, 1983). A possible Mercury transit is mentioned in *British Astron. Assoc. Jour.* (vol. 88, p. 104).]

5

Reprinted from pages 387, 388–402, 405, 406, 408, and 409–410 of *Royal Astron. Soc. Quart. Jour.* **19**:387–410 (1978)

An Interpretation of the Pre-Telescopic Sunspot Records from the Orient

David H.Clark

Royal Greenwich Observatory, Herstmonceux Castle, Hailsham, East Sussex

F.Richard Stephenson

Institute of Lunar and Planetary Sciences, University of Newcastle upon Tyne

(Received 1978 May 17)

SUMMARY

A catalogue of pre-telescopic sunspot records from the Orient is presented, based on a search of Chinese and Korean dynastic histories, and the completeness of this catalogue is analysed. The historical data are used to investigate possible major excursions in solar activity during the past two millennia. In addition to the now well-established 'Maunder Minimum' (AD 1645–1715), at least two other similar excursions are apparent in the naked-eye sunspot records – a 'Medieval Minor Minimum' (~AD 1280–1350) and the 'Spörer Minimum' (~AD 1400–1600). The reality of these excursions is supported by the atmospheric ^{14}C history preserved in tree rings. The implications of such long-term variability in solar activity are considered.

I INTRODUCTION

The mechanisms for the formation, maintenance and dissipation of solar activity, and its possible influence on terrestrial climate, are still only poorly understood, and these problems will only be resolved when the long-term solar behaviour is clearly established. In this paper we look at what information on the past history of the Sun can be obtained by interpreting the historical sunspot records from the Orient. In doing so, we recognize that sunspots themselves are not directly related to solar output or to most of the alternative indicators of solar variability – they are, however, a convenient index of almost all other activity on the Sun, and represent one of the few available sources of solar data prior to the establishment of Solar Physics with the advent of the telescope.

[*Editor's Note:* In the original, introductory material on early sunspots is included.]

The abundance of historical sunspot records from China may result, not only from the open acceptance of celestial imperfection, but also from the regular occurrence of conditions suitable for their detection. The geology of China renders it susceptible to dust-storms, particularly off the Gobi and Tarim plains. In addition, there is a high incidence of atmospheric haze associated with a continental climate. Many of the Chinese sunspot records mention reduced solar brightness commensurate with these conditions (e.g. AD 1145 – 'Within the Sun there was a black spot; the Sun had no brilliance.' AD 352 – 'The Sun was a dazzling red, like fire. Within it there was a 3-legged crow. Its shape was seen sharp and clear. After 5 days it ceased.'). Several records refer to spots being observable only at sunrise and sunset (e.g. AD 579 April 3 – 'When the Sun first rose and when it was about to set, within it on both occasions there was a black colour as large as a hen's egg, and after 4 days it was extinguished.'). Needham (1959) suggested the possibility that the Sun might have been viewed through pieces of semi-transparent jade, mica or smoky rock-crystal, but we have been unable to find any confirmation of this suggestion in the available contemporary literature. Indeed, evidence seems to exist to the contrary (e.g. the definite statements that spots were seen only at sunrise or sunset, through severe atmospheric haze, etc.).

The oriental descriptions of sunspots are particularly picturesque, often making an allusion to size or shape. Sunspots are described as 'like a plum', 'as large as a date', 'like a hen's egg', 'as large as a peach', etc. One can only speculate on the true sunspot size from such classification criteria, which were obviously far from consistent as the following example shows:

AD 1185 February 10 (China): 'Within the Sun there was produced a black spot as large as a date.'

AD 1185 February 11 (Korea): 'On the Sun there was a black spot as large as a pear.'

Both descriptions must refer to the same spot, although the choice of fruits of radically different size is somewhat puzzling.

Several simultaneously resolvable spots are often mentioned (e.g. AD 1112 May 2 – 'Within the Sun there were black spots, now 2, now 3, as large as chestnuts'), as well as what appear to be large groups, where the 'vapour' is said to have a particular shape (e.g. 'like a man', etc.). The duration of visibility is often given (e.g. AD 1137 – 'Within the Sun there was a black spot as large as a plum. After 10 days it gradually faded away.'). A particularly fascinating account dates from AD 188: 'The Sun was orange in colour and within it there was a black vapour like a flying magpie. After several *months* it gradually faded away.' This sunspot group was apparently tracked for several solar rotations, although no mention of actual disappearance and reappearance is included.

In Table I we have listed all the pre-telescopic sunspot observations we have found from oriental dynastic histories, encyclopedias, etc. We have used primary sources exclusively throughout our survey.

Previous catalogues have included those of Kanda (1933), and 'The Ancient Sunspot Records Research Group' of Yunnan Observatory (1976). All sightings reported in both these previous catalogues were independently found; however, in our interpretation several of the records in each are of atmospheric phenomena. Such spurious events are appended to our table, with the reasons for their exclusion from our catalogue. Our search revealed only a single new sunspot record not found by either the Kanda or Yunnan groups.

In Table I, each historical sunspot record is assigned a reference number; the relevant section of each historical record is given, with its estimated date. Records with the same date, from different sources but from the same country, have been given a common entry. Also included is the source of each record, usually a dynastic history, giving its name (e.g. *Han-shu*, history of the Former Han Dynasty, etc.), and the section containing the sighting (*chih*, 'Treatises'; *Pên-chi*, 'Basic Annals'; *Lieh-chuan*, 'Biographies'). Other sources of sunspot records include the *Wên-hsien-t'ung-k'ao* (Encyclopedia of Ma Tuan-lin), the *Koryŏ-sa* ('History of the Kingdom of Koryŏ', Korea), and the *Chŭngbo Munhŏn Pigo* (a Korean Encyclopedia).

TABLE I

Catalogue of pre-telescopic sunspot sightings from the Orient

Ref. No.	Place	Date	Description and source
(1)	C	28 BC May 10?	'The Sun appeared yellow and there was a black vapour as large as a coin at its centre.' (*Han-shu, chih*)
(2)	C	AD 20 March 17	'The Sun was very black. (Wang) Mang disliked it, and he issued an edict as follows: "Recently within the Sun there has been visible a shadow encroaching on it. A black vapour is an abnormality." All the people were alarmed at the strange omen.' (*Han-shu*, Biography of Wang Mang)
(3)	C	AD 187 March–April?	'There was a black vapour as large as a melon within the Sun.' (*Hou-han-shu, chih*)
(4)	C	AD 188 February–March?	'The Sun was orange in colour and within it there was a black vapour like a flying magpie. After several months it gradually faded away.' (*Hou-han-shu, chih*)
(5)	C	AD 268 November–December?	'Within the Sun there was a black spot.' (*T'ung-chih* – secondary source)
(6)	C	AD 299 February–March?	'Within the Sun there was something resembling a flying swallow. For several days this remained visible.' (*Chin-shu, chih* – this record is also reproduced in the *Sung-shu, chih*, but giving a duration of several *months* rather than *days*)
(7)	C	AD 301 January 19	'Within the Sun there was a black vapour.' (*Chin-shu, chih*)
(8)	C	AD 301 October 20	'Within the Sun there was a black spot.' (*Chin-shu, chih; Sung-shu, chih*)
(9)	C	AD 302 December–303 January?	'Within the Sun there was a black vapour.' (*Chin-shu, chih*)
(10)	C	AD 304 December–305 January?	'Within the Sun there was a black vapour; it divided the Sun.' (*Chin-shu, chih*)
(11)	C	AD 311 April 7	'Within the Sun there was something resembling a flying swallow.' (*Chin-shu, chih*)

Ref. No.	Place	Date	Description and source
(12)	C	AD 321 May 7	'Within the Sun there was a black spot.' (*Chin-shu, chih*; *Sung-shu, chih*)
(13)	C	AD 322 November 6	'Within the Sun there was a black spot.' (*Chin-shu, chih*; *Sung-shu, chih*)
(14)	C	AD 342 March 7	'Within the Sun there was again a black spot – [March 11], only then did it disappear.' (*Chin-shu, chih*; *Sung-shu, chih*)
(15)	C	AD 352–353?	'The Sun was a dazzling red, like fire. Within it there was a 3-legged crow. Its shape was seen sharp and clear. After 5 days it ceased.' (*Chin-shu, chih* – this record is also reproduced in the *Sung-shu, chih*, but giving a date corresponding to AD 371–372!)
(16)	C	AD 354 November 7	'Within the Sun there was a black spot as large as a hen's egg.' (*Chin-shu, chih*)
(17)	C	AD 355 April 4	'Within the Sun there were two black spots as large as peaches.' (*Chin-shu, chih*)
(18)	C	AD 359 November 7	'Within the Sun there was a black spot as large as a hen's egg.' (*Chin-shu, chih*)
(19)	C	AD 369 November 27?	'Within the Sun there was a black spot.' (*Chin-shu, chih*)
(20)	C	AD 370 March 29	'Within the Sun there was a black spot as large as a pear.' (*Chin-shu, chih*)
(21)	C	AD 373 January 28?	'Within the Sun there was a black spot.' (*Chin-shu, chih*)
(22)	C	AD 373 December 26	'Within the Sun there was a black spot as large as a pear.' (*Chin-shu, chih*)
(23)	C	AD 374 April 6	'Within the Sun there were two black spots as large as ducks' eggs.' (*Chin-shu, chih*)
(24)	C	AD 375 January 10	'Within the Sun there was a black spot as large as a hen's egg.' (*Chin-shu, chih*)
(25)	C	AD 388 April 2	'Within the Sun there were two black spots as large as pears.' (*Chin-shu, chih*)
(26)	C	AD 389 July 17	'Within the Sun there was another black spot as large as a pear.' (*Chin-shu, chih*)
(27)	C	AD 395 December 13	'Within the Sun there was another black spot.' (*Chin-shu, chih*)
(28)	C	AD 400 December 6	'Within the Sun there was a black spot.' (*Chin-shu, chih*)
(29)	C	AD 499 July 4?	'Within the Sun there was a black spot.' (*Wei-shu, chih*)
(30)	C	AD 500 January 29	'Within the Sun there was a black spot as large as a peach.' (*Wei-shu, chih*)

Ref. No.	Place	Date	Description and source
(31)	C	AD 500 January 30	'Within the Sun there were three black spots.' (*Nan-ch'i-shu, chih*)
(32)	C	AD 501 September 4	'The Sun was red and had no brightness; within it there was one black spot.' (*Wei-shu, chih*)
(33)	C	AD 502 February 8	'Within the Sun there was a black vapour like a goose's egg; (February 12?) it was seen again and there were two (additional) black vapours threading the Sun horizontally.' (*Wei-shu, chih*)
(34)	C	AD 502 March 26?	'Within the Sun there was a black vapour as large as a goose's egg.' (*Wei-shu, chih*)
(35)	C	AD 510 March 17	'Within the Sun there were two black vapours.' (*Wei-shu, chih*)
(36)	C	AD 511 December 16	'Within the Sun there were two black vapours as large as peaches.' (*Wei-shu, chih*)
(37)	C	AD 513 April 7?	'Within the Sun there was a black vapour.' (*Wei-shu, chih*)
(38)	C	AD 566 March 29	The Sun 'fought the light'; later it faded and in the Sun a crow was seen.' (*Pei-chou-shu, pên-chi*) 'The Sun had no brightness and a crow was seen.' (*Sui-shu, chih*)
(39)	C	AD 567 December 10	'When the Sun rose and set there was one black vapour as large as a cup within the Sun; (December 13) there was one more for 6 days; only then were they extinguished.' (*Pei-chou-shu, pên-chi; Sui-shu, chih*)
(40)	C	AD 578 December 25	At 'tea-time'. 'Within the Sun there was a black spot as large as a cup.' (*Sui-shu, chih*)
(41)	C	AD 579 April 3	'When the Sun first rose and when it was about to set, within it on both occasions there was a black colour as large as a hen's egg, and after 4 days it was extinguished.' (*Pei-chou-shu, pên-chi*)
(42)	C	AD 826 May 7	'Within the Sun there was a black vapour like a cup.' (*Hsin-t'ang-shu, chih*)
(43)	C	AD 826 May 24	'Within the Sun there was a black spot.' (*Hsin-t'ang-shu, chih*)
(44)	C	AD 832 April 21	'Within the Sun there was a black spot.' (*Hsin-t'ang-shu, chih*)
(45)	C	AD 837 December 22	'Within the Sun there was a black spot as large as a hen's egg. The Sun was red like ochre.' (*Hsin-t'ang-shu, chih*)
(46)	C	AD 841 December 30	'Within the Sun there was a black spot.' (*Hsin-t'ang-shu, chih*)

Ref. No.	Place	Date	Description and source
(47)	J	AD 851 December 22	'The Sun had no power. Within it there was a black dot as large as a plum.' (*Montohu Jitsurohu*)
(48)	C	AD 865 February–March?	'A white rainbow penetrated the Sun. Within the Sun there was a black vapour like a hen's egg.' (*Hsin-t'ang-shu, chih*)
(49)	C	AD 874–875?	'Within the Sun there was a black spot.' (*Hsin-t'ang-shu, chih*)
(50)	C	AD 875–876?	'Within the Sun there was something which was like a flying swallow.' (*Hsin-t'ang-shu, chih*)
(51)	C	AD 904 February 19	'Within the Sun there was seen the northern dipper.' (*Hsin-t'ang-shu, chih*)
(52)	C	AD 927 March 9	'Within the Sun there was a black vapour; its shape was like a hen's egg.' (*Wên-hsien-t'ung-k'ao*)
(53)	C	AD 974 March 3?	'Within the Sun there were two black spots.' (*Sung-shih, chih*)
(54)	C	AD 1077 March 7	'Within the Sun there was a black spot like a plum, until (March 21) when it dispersed.' (*Sung-shih, pên-chi* and *chih*)
(55)	NC	AD 1077 June 7	'Within the Sun there was a black spot.' (*Liao-shih, pên-chi*)
(56)	C	AD 1078 March 11	'Within the Sun there was a black spot like a plum, until (March 29), altogether 19 days, then it dispersed.' (*Sung-shih, chih*)
(57)	C	AD 1079 January 11	'Within the Sun there was a black spot as large as a plum, until (January 22), altogether 12 days, then it dispersed.' (*Sung-shih, chih*)
(58)	C	AD 1079 March 20	'Within the Sun there was a black spot as large as a plum until (March 29) when it dispersed.' (*Sung-shih, pên-chi* and *chih*)
(59)	C	AD 1104 December 11?	'Within the Sun there was a black spot as large as a date.' (*Sung-shih, chih*)
(60)	C	AD 1112 May 2	'Within the Sun there were black spots, now 2, now 3, as large as chestnuts.' (*Sung-chih, chih*)
(61)	C	AD 1118 December 18	'Within the Sun there was a black spot as large as a plum.' (*Sung-shih, chih*)
(62)	C	AD 1120 June 7	Within the Sun there was a black spot as large as a date.' (*Sung-shih, chih*)
(63)	C	AD 1122 January 10	'Within the Sun there was a black spot as large as a plum.' (*Sung-shih, pên-chi*)

Ref. No.	Place	Date	Description and source
(64)	C, NC	AD 1129 March 22	'Within the Sun there was a black spot, until (April 14) when it finally melted away.' (*Sung-shih, chih*) 'Within the Sun there was a black spot.' (*Kin-shih, chih*)
(65)	C	AD 1131 March 12	'Within the Sun there was a black spot for 4 days, then it finally melted away.' (*Sung-shih, pên-chî*) 'Within the Sun there was a black spot as large as a plum for 3 days; only then did it fade.' (*Sung-shih, chih*)
(66)	C	AD 1136 November 23	'Within the Sun there was a black spot as large as a plum until (November 27) when it finally melted away.' (*Sung-shih, chih*)
(67)	NC	AD 1136 November 27	'Within the Sun there were black spots; they were moving at an angle to one another.' (*Kin-shih, chih*)
(68)	C	AD 1137 March 1	'Within the Sun there was a black spot as large as a plum. After 10 days it melted away.' (*Sung-shih, chih*)
(69)	C	AD 1137 May 8	'Within the Sun there was a black spot until (May 22?) when it finally faded away.' (*Sung-shih, chih*) 'Within the Sun there was a black spot.' (*Sung-shih, pên-chi*)
(70)	C	AD 1138 March 16	'Within the Sun there was a black spot.' (*Sung-shih, pên-chi*)
(71)	C	AD 1138 November 26	'Within the Sun there was a black spot.' (*Sung-shih, pên-chi*)
(72)	C	AD 1139 March	'In this month within the Sun there was a black spot. After rather more than a month, only then did it fade away.' (*Sung-shih, pên-chi*)
(73)	C	AD 1139 November 20	'Within the Sun there was a black spot.' (*Sung-shih, pên-chi*)
(74)	C	AD 1145 July 23?	'Within the Sun there was a black vapour fading and reappearing.' (July 24) – 'Within the Sun there was a black spot; the Sun had no brilliance.' (*Sung-shih, chih*)
(75)	K	AD 1151 March 21	Within the Sun there was a black spot as large as a hen's egg. (March 31, April 1) – 'It was the same.' (*Koryŏ-sa*)
(76)	K	AD 1160 February 28	'Within the Sun there was a strange vapour for 3 days.' (*Koryŏ-sa*)
(77)	NC	AD 1160 September 26	'Within the Sun there was a black spot; its shape was like a man.' (*Kin-shih, chih*)
(78)	K	AD 1160 September 29	'Within the Sun there was a black spot.' (*Koryŏ-sa*)

57

Ref. No.	Place	Date	Description and source
(79)	K	AD 1171 October 20	'On the Sun there was a black spot as large as a peach.' (*Koryŏ-sa*)
(80)	K	AD 1171 November 16	Following Ref. (81). 'It was the same.' (*Koryŏ-sa*)
(81)	K	AD 1183 December 4	'On the Sun there was a black spot for two days.' (*Koryŏ-sa*)
(82)	C	AD 1185 February 10	'Within the Sun there was produced a black spot as large as a date.' (*Sung-shih, chih*)
(83)	K	AD 1185 February 11	'On the Sun there was a black spot as large as a pear.' (*Koryŏ-sa*)
(84)	C	AD 1185	(February 15) until (February 27) 'Within the Sun for the whole time there was a black spot.' (*Sung-shih, chih*) (February 15) 'Within the Sun there was a black spot.' (February 27) 'Within the Sun there was a black spot.' (*Sung-shih, pên-chi*)
(85)	K	AD 1185 March 27	Following Ref. (83) 'It was the same.' (*Koryŏ-sa*)
(86)	K	AD 1185 April 18	'On the Sun there was a black spot.' (April 19) 'It was the same.' (*Koryŏ-sa*)
(87)	K	AD 1185 November 14	Following Ref. (86) 'It was the same.' (*Koryŏ-sa*)
(88)	C	AD 1186 May 23	'Within the Sun there was produced a black spot as large as a date.' (*Sung-shih, chih*)
(89)	C	AD 1186 May 26	'Within the Sun there was a black spot.' (*Sung-shih, pên-chi*)
(90)	C	AD 1193 December 3	'Within the Sun there was a black spot, until (December 12) when it finally melted away.' (*Sung-shih, chih*) 'Within the Sun there was a black spot. (December 12) – Within the Sun there was a black spot; it faded away.' (*Sung-shih, pên-chi*)
(91)	K	AD 1200 September 19	'On the Sun there was a black spot as large as a plum.' (*Koryŏ-sa*)
(92)	C	AD 1200 September 21	'Within the Sun there was a black spot as large as a date until (September 26) when the Sun cleared.' (*Sung-shih, chih*) 'Within the Sun there was a black spot.' (*Sung-shih, pên-chi*)

58

Ref. No.	Place	Date	Description and source
(93)	C	AD 1201 January 9	Following Ref. (92) 'Another was produced until (January 29) when it finally melted away.' (*Sung-shih, chih*) 'Within the Sun there was another black spot; (January 29) it faded away.' (*Sung-shih, pên-chi*)
(94)	K	AD 1201 April 6	'Within the Sun there was a black spot as large as a plum.' (*Koryŏ-sa*)
(95)	K	AD 1202 August 23	'Within the Sun there was a black spot as large as a pear.' (*Koryŏ-sa*)
(96)	C	AD 1202 December 19	'Within the Sun there was produced a black spot as large as a date; (December 31) it finally melted away.' (*Sung-shih, chih*) 'Within the Sun there was produced a black spot.' (*Sung-shih, pên-chi*)
(97)	K	AD 1204 February 3	'Within the Sun there was a black spot as large as a plum, altogether for 3 days.' (*Koryŏ-sa*)
(98)	C	AD 1204 February 21	'Within the Sun there was a black spot as large as a date.' (*Sung-shih, chih*) 'Within the Sun there was a black spot.' (*Sung-shih, pên-chi*)
(99)	C	AD 1205 May 4	'Within the Sun there was a black spot.' (*Sung-shih, pên-chi; Sung-shih, chih*)
(100)	C	AD 1238 December 5	'Within the Sun there was a black spot.' (*Sung-shih, chih*)
(101)	K	AD 1258 September 15	'Within the Sun there was a black spot as large as a hen's egg. On the following day (September 16) there was another shaped like a man.' (*Koryŏ-sa*)
(102)	C	AD 1276 February 17	'Within the Sun there were black spots, rocking one another, like goose's eggs.' (*Sung-shih, pên-chi; Sung-shih, chih*)
(103)	K	AD 1278 August 31	'Within the Sun there was a black spot as large as a hen's egg.' (*Koryŏ-sa*)
(104)	K	AD 1356 April 4	'The Sun had no brilliance, and within it there was a black spot; (April 5) it was again like this.' (*Koryŏ-sa*)
(105)	K	AD 1361 March 16	'On the Sun there was a black spot.' (*Koryŏ-sa*)
(106)	K	AD 1362 October 5	'On the Sun there was a black spot.' (*Koryŏ-sa*)
(107)	C	AD 1370 January 1	'Within the Sun there was a black spot.' (*Ming-shih, chih*)

Ref. No.	Place	Date	Description and source
(108)	C	AD 1370 October 2	Following Ref. (107) 'It was the same.' (*Ming-shih, chih*)
(109)	C	AD 1370 October 21	Following Ref. (108) 'It was the same.' (*Ming-shih, chih*)
(110)	C	AD 1370 December 7	Following Ref. (109) 'It was the same.' (*Ming-shih, chih*)
(111)	K	AD 1371 January 2	'On the Sun there was a black spot.' (*Koryŏ-sa*)
(112)	C	AD 1370 December–1371 January	'In this month within the Sun, there were frequently black dots.' (*Ming-chao-t'ai-tien*)
(113)	C	AD 1371 March 31	'Within the Sun there was a black spot.' (*Ming-shih, chih*)
(114)	C	AD 1371 June 14	Following Ref. (113) 'It was the same.' (*Ming-shih, chih*)
(115)	C	AD 1371 November 6	Following Ref. (114) 'It was the same.' (*Ming-shih, chih*)
(116)	K	AD 1371 November 21	'Within the Sun there was a black spot.' (*Koryŏ-sa*)
(117)	C	AD 1372 February 6	'Within the Sun there was a black spot.' (*Ming-shih, chih*)
(118)	C	AD 1372 April 3	Following Ref. (117) 'It was the same.' (*Ming-shih, chih*)
(119)	K	AD 1372 May 8	'On the Sun there was a black spot.' (*Koryŏ-sa*)
(120)	C	AD 1372 June 19	'Within the Sun there was a black spot.' (*Ming-shih, chih*)
(121)	C	AD 1372 August 25	Following Ref. (120) 'It was the same.' (*Ming-shih, chih*)
(122)	K	AD 1373 April 26	'On the Sun there was a black spot for 2 days. (*Koryŏ-sa*)
(123)	K	AD 1373 October 23	'On the Sun there was a black spot.' (*Koryŏ-sa*)
(124)	C	AD 1373 November 15	'Within the Sun there was a black spot.' (*Ming-shih, chih*)
(125)	C	AD 1374 March 27–March 31	'It was the same.' (*Ming-shih, chih*)
(126)	K	AD 1375 March 20	'On the Sun there was a black spot. (March 21) it was the same.' (*Koryŏ-sa*)
(127)	C	AD 1375 March 23	'Within the Sun there was a black spot.' (*Ming-shih, chih*)
(128)	C	AD 1375 October 21	Following Ref. (127) 'It was the same.' (*Ming-shih, chih*)
(129)	C	AD 1376 January 19	Following Ref. (128) 'It was the same.' (*Ming-shih, chih*)
(130)	C	AD 1381 March 22	Following Ref. (129) 'It was the same, until (March 25).' (*Ming-shih, chih*)
(131)	K	AD 1381 March 23	'On the Sun there was a black spot.' (*Koryŏ-sa*)

Ref. No.	Place	Date	Description and source
(132)	K	AD 1382 March 9	'On the Sun there was a black spot as large as a hen's egg; altogether for 3 days.' (*Koryŏ-sa*)
(133)	C	AD 1382 March 21	'Within the Sun there was a black spot.' (*Ming-shih, chih*)
(134)	C	AD 1383 January 10	'Within the Sun there was a black spot.' (*Ming-shih, chih*)
(135)	K	AD 1387 April 15	'On the Sun there was a black spot.' (*Koryŏ-sa*)
(136)	K	AD 1402 November 15	'Within the Sun there was a black dot.' (*Chŭngbo Munhŏn Pigo*)
(137)	K	AD 1520 March 9	'Within the Sun there were black vapours rocking one another.' (*Chŭngbo Munhŏn Pigo*)
(138)	K	AD 1603 April 16	'The Sun was red and without brilliance, and there were three small dots of black vapour, in shape like a large cash. From the outside they were attached to the Sun. They seemed to separate and join together.' (*Chŭngbo Munhŏn Pigo*)
(139)	K	AD 1604 October 25	'Within the Sun there was a black spot as large as a hen's egg.' (*Chŭngbo Munhŏn Pigo*)

Place references: C = China
NC = Northern China
K = Korea
J = Japan

APPENDIX TO CATALOGUE

Records rejected as spurious

1 *Yunnan catalogue.* BC 43 May–June. 'The Sun's colour was pale blue and there were no shadows (on the ground). At noon exactly there were shadows but no brightness.' (*Han-shu, chih.*) This is clearly not a reference to sunspots, but a late commentary on the passage, which the Yunnan group cites, states: 'The Sun was dark, and located lopsidedly (there was something) as large as a crossbow pellet.' This may well be an allusion to a sunspot, but we cannot say how reliable the secondary source is.

AD 180 February–March. 'The Sun's colour was orange and within it there was a black vapour like a flying magpie. After several months it gradually faded away.' (*T'ung-chih.*) This is clearly a duplication of the AD 188 entry; it is verbatim, and the month is identical. The *T'ung-chih* is a secondary source of doubtful reliability.

AD 1597 June 15. 'The Sun's light rotated and rocked; soon it became a black disc.' This is clearly not an allusion to a sunspot. The Yunnan group regard it as 'dubious'.

2 *Kanda's catalogue.* AD 505 January 4. 'A black vapour penetrated the Sun.' (*Wei-shu, chih.*) The use of the term *kuan* ('penetrated') seems to suggest something entering the Sun from outside – more probably an atmospheric effect than a sunspot. The term *chung* ('within') was almost always used to describe the position of a true spot on the Sun.

AD 807 December 1. 'Beside the Sun there was a black vapour like a man in shape, kneeling and in both hands holding up a dish towards the Sun with a vapour in the dish like a man's head.' (*Hsin-t'ang-shu, chih.*)
This highly imaginative account presumably refers to some kind of atmospheric phenomenon.

829 January 20. 'There was a black vapour in the vicinity of the Sun, as if fighting (it).' (*Hsin-t'ang-shu, chih.*) The use of the rather vague term *yu* ('in the vicinity of') prevents us confidently interpreting the black vapour as a sunspot.

832 May 6. 'A black vapour obscured out the Sun.' (*Hsin-t'ang-shu, chih.*) Like the 829 observation, the nature of this event is uncertain.

887 December 18. 'Below the Sun there was a black vapour.' (*Hsin-t'ang-shu, chih.*) Here is a clear reference to some phenomenon outside the Sun.

1129 February 25. 'When the Sun first rose two black vapours shaped like a man clasped the sides of the Sun and did not disperse until the hour *szu*.' (*Sung-shih, chih.*) Again an interpretation in terms of an atmospheric phenomenon seems more plausible.

The data are depicted graphically in Fig. 1. Not all of these are independent events, the same sunspot activity sometimes being sighted and recorded in China and Korea – such dual sightings are bracketed in the figure. The bulk of the data are from China, with its richly documented astronomical history, although after AD 1150 there is an abundance of records from Korea. Kanda (1933) first drew attention to the single historical record from Japan in AD 851.

3 THE COMPLETENESS OF THE HISTORICAL SAMPLE

It is apparent from Fig. 1 that the occurrence of historical sunspot records appears to be highly fragmentary and sporadic. Both Eddy (1977a) and Bray (1974) have suggested that the observations were apparently merely accidental, Bray claiming that their occurrences were gleaned from casual mention in the diaries of the literati. We cannot agree with this interpretation. It is certainly true that no well-documented evidence exists that a regular watch (perhaps at sunset) was kept for sunspots throughout oriental history. However, during almost the whole of Chinese history since Han times, the acknowledged systematic and careful approach to observational astronomy and its compendious documentation, the abundance of records of various other day-time phenomena, and the deeply held belief that celestial phenomena were precursors of terrestrial events (see discussion in Clark & Stephenson 1977) make it unlikely that only sunspots were omitted from regular patrol. Despite this belief, it must be acknowledged that at no period were more than 30 sightings recorded per century – a regular search might have been expected to have revealed a somewhat greater number if sunspot activity was at present-day levels.

Clearly it is important to check the above interpretation before trying to use the historical sunspot records as indicators of long-term solar variability. We have attempted to do this by comparing the frequency of sunspot records for particular dynasties with the frequency of records of other celestial phenomena. This study reveals a number of difficulties inherent in an interpretation of the historical sunspot data, as the following example shows. Fig. 2 illustrates the statistics of selected astronomical observations made during, and for a few decades prior to, the Chin dynasty in China (AD 265–420). The number of solar eclipse records (important for calendrical purposes) may give a somewhat false indication of the completeness of astronomical observations, since a significant fraction of the reports (at least a quarter) were predicted but not observed. Indeed, at certain times eclipse predictions often

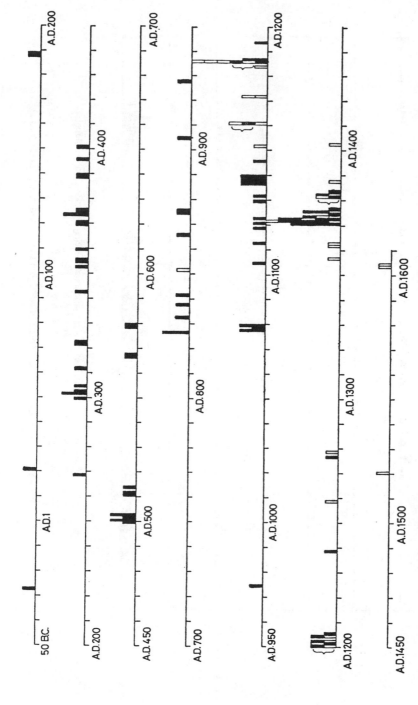

Fig. 1. The pre-telescopic sunspot records from the orient. Each 'bar' represents a single sighting – solid bars, from China; open bars, from Korea or Japan.

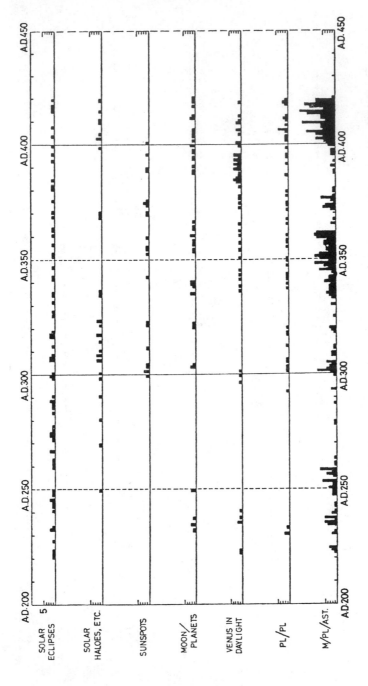

MOON/PLANETS ≡ occultations and conjunctions involving the moon and a planet.
PL/PL ≡ conjunctions of two or more planets.
M,PL/AST ≡ conjunctions of the moon or a planet with asterism or star.

FIG. 2. The statistics of various astronomical observations from the Chin dynasty (AD 265–420).

64

appear to be included in the astronomical treatises merely to give the impression of completeness – although it must be emphasized that such predicted and non-observed events represent the minority of eclipse records. There are, however, certain enigmatic gaps in the records of other phenomena, particularly in all the records prior to AD 300, and in the sunspot records after AD 400. The almost total lack of astronomical records except for eclipses and occasional conjunctions prior to AD 300 calls into question the completeness of the documentation of celestial events of astrological significance at this time. By contrast, the lack of any sunspot records between AD 400 and 420 probably represents a total lack of any sightings in this short period, since here there is an abundance of records of all other phenomena.

Bielenstein (1950) has suggested that, while with very few exceptions records were not actually falsified, they were often left incomplete – the degree of incompleteness being related to the popularity of a particular reign. The Emperor was believed to be the 'Son of Heaven', and could rule only as long as he governed wisely and justly. A departure from the way of virtue was expected to result in Heaven withdrawing its mandate – and this might be indicated by the occurrence of certain celestial phenomena. If 'warnings from Heaven' seemed not to be required, celestial events might not be memorialized. Certainly there is substantial evidence to support Bielenstein's view that 'popularity' meant popularity with the high officials of court who dictated what should be recorded by the historiographers, rather than with the mass of the people. An abundance of records of astrological significance with their accompanying admonitions and prognostications are often found to precede the overthrow of a dynasty (see, e.g. Fig. 2). Indeed, Needham (1959) has suggested that there was a stylized pattern of what was supposed to take place (by way of celestial portents) at the catastrophic close of a dynasty.

From the beginning of the Han (206 BC) onwards there are systematic astronomical records in all the dynastic histories. However, political and social pressures clearly influenced the recording of phenomena of astrological interest, and this must include sunspots. Periods of scarcity of historical sunspot records might result from protracted minima in sunspot activity, similar to the Maunder Minimum – equally, they may merely result from lack of interest. It is clearly important to try to establish whether sunspots were observed or not at these times – and if they were, why they were not memorialized. We will return to this problem in the next section.

Fig. 1 indicates that during the period AD 1100 to 1400 there were more recorded sunspot sightings than during any other comparable period of the pre-telescopic era. In China, the Sung Dynasty (AD 960 to 1279) was the focal point of scientific and technological activity. According to Needham (1959), there is some cause to question the diligence of Sung astronomers prior to AD 1070, but reforms instigated at this time assured the high accuracy of subsequent astronomical descriptions. The abundance of sunspot records from China between about AD 1050 and 1250 is therefore hardly surprising. What is surprising is the sudden onset of interest in Korea, emphasized in Fig. 3. The similarity in style of descriptions of sunspots suggests that the Korean interest had its origins in China.

[*Editor's Note:* Material including Figs. 3, 4, 5, and 7 has been omitted. The authors note gaps in sunspot records as follows: A.D. 400–500, A.D. 600–800, A.D. 980–1090, A.D. 1280–1350, and A.D. 1400–1600.]

On the basis of the above discussion, only gaps (2), (4) and (5) appear worthy of further investigation as possible protracted minima in solar activity. Figs 4 to 6 plot the occurrence of atmospheric phenomena from the astronomical treatises of the *Chiu-t'ang-shu* (History of the T'ang dynasty), the *Yüan-shih*, and the *Ming-shih* respectively. Atmospheric phenomena were chosen on the basis of our experience with a more complete analysis of the Chin astronomical data (Fig. 2); such phenomena as haloes, parhelia, dimming of the Sun, etc., give some measure of the degree of interest in day-time events of astrological significance, and also sunspots are often listed amongst these records.

Because of the completeness of the atmospheric records, it is tempting to interpret gaps (4) and (5), and possibly also gap (2) subject to the reservations expressed previously, as solar-activity protracted minima, similar to the now well-established Maunder Minimum of 1645–1715 – this interpretation will be given further credence in the following section.

[*Editor's Note:* The material omitted at this point relates to a comparison with Radiocarbon history (see our Appendix C).]

6 CONCLUSION

Eddy (1976, 1977a), by confirming the reality of the Maunder Minimum in sunspot activity, has finally broken the shackles which have bound solar physicists for two centuries to a Sun behaving with persistent regularity – a concept no less restrictive to scientific advancement than the Aristotelian dogma suffered by medieval astronomers.

Major excursions from the quasi-regular 11-year cycle of solar activity, similar to the Maunder Minimum, are preserved in the ^{14}C history; the incidence of naked-eye sunspot sightings provides 'circumstantial evidence' for such excursions. A detailed reanalysis of the only other obvious indicator of past solar variability, namely the historical auroral records, would now seem to be worth while.

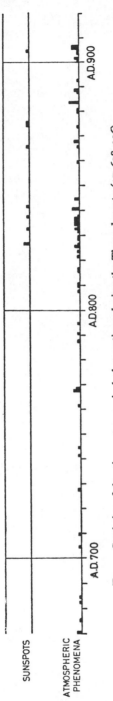

SUNSPOTS

ATMOSPHERIC
PHENOMENA

A.D.700

A.D.800

A.D.900

FIG. 4. Statistics of day-time astronomical observations during the T'ang dynasty (AD 618–906).

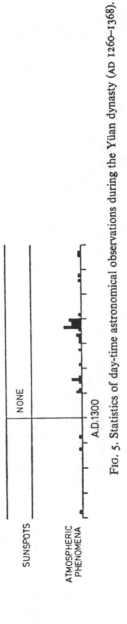

SUNSPOTS

NONE

ATMOSPHERIC
PHENOMENA

A.D.1300

FIG. 5. Statistics of day-time astronomical observations during the Yüan dynasty (AD 1260–1368).

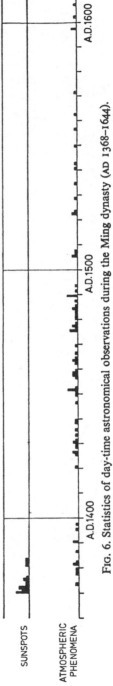

SUNSPOTS

ATMOSPHERIC
PHENOMENA

A.D.1400

A.D.1500

A.D.1600

FIG. 6. Statistics of day-time astronomical observations during the Ming dynasty (AD 1368–1644).

67

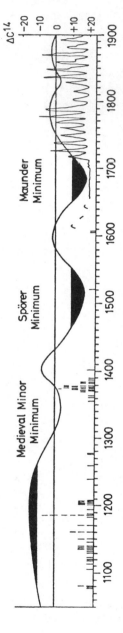

Fig. 8. Smoothed ^{14}C history (from Eddy 1976). The post-telescopic mean sunspot number is plotted, and the pre-telescopic sunspot records from our catalogue shown as bars.

The implications of long-term solar variability implied by the work of Eddy, and by the present study, must have a profound impact on solar–terrestrial physics. As Eddy (1977a) has suggested, long-term changes in solar activity may be accompanied by equally ponderous changes in the radiative or particulate output of the Sun – either of which could have a major impact on the Earth's climatic and biological history.

REFERENCES

Bielenstein, H., 1950. *Bull. Mus. Far. E. Antiq.*, **22**, 127.
Bray, J. R., 1974. In *Scientific, Historical, and Political Essays in Honour of Dirk J.Struik*, eds Cohen, R.S. *et al.*, Boston Studies in the Philosophy of Science, Vol. 15, D.Reidel, Dordrecht, Holland.
Clark, D.H. & Stephenson, F.R., 1977. *The Historical Supernovae*, Pergamon, Oxford.
Dubs, H.H., 1938. *Osiris*, **5**, 499.
Eddy, J.A., 1976. *Science*, **192**, 1189.
Eddy, J.A., 1977a. In *The Solar Output and its Variation*, Colorado Associated University Press.
Eddy, J.A., 1977b. *The Sun since the Bronze Age*, invited paper presented at the AGU/International Symposium on Solar-Terrestrial Physics.
Goldstein, B., 1969. *Centaurus*, **14**, 49.
Kandu, Shigeru, 1933. *Proc. Imp. Acad., Japan*, **9**, 293.
Maunder, E.W., 1890. *Mon. Not. R. astr. Soc.*, **50**, 251.
Needham, J., 1959. *Science and Civilisation in China*, Vol. 3, Cambridge University Press.
Schove, D.J., 1955. *J. geophys. Res.*, **60**, 167.
Spörer, F.W.G., 1887. *Astr. Gesellschaft Vrtljschr (Leipzig)*, **22**, 323.
Spörer, F.W.G., 1889. *Bull. Astronom.*, **6**, 60.
The Ancient Sunspot Records Research Group, 1976. *Acta. Astr. Sinica*, **17**, 217.

[*Editor's Note:* In a supplementary "Note on Sunspot Records from China," J. H. C. Wang (1980, E. O. Pepin et al., eds., *The Ancient Sun: Record in the Earth, Moon, and Meteorites,* Pergamon, Oxford, England) suggests corrections as follows: 17 April 513 (not 7) and 27 March 1138 (not 16).

Seventeenth century records for China given (in Chinese) by Z. Xu and Y. Jiang (1979, *Nanking Univ. Annals Phys. Sci.* **2**:31) are as follows: clusters in 1613–1618, 1621–1631, and 1635–1643, and then single spots in 1647, 25 October 1650, 30 April 1655, 26 January 1656 to 23 April 1656, 30 February 1665, and 16 March 1684 to 18 March 1684.]

6

Reprinted by permission of Doubleday & Company, Inc. from pages 106–107,
116–117, and 119 of *Discoveries and Opinions of Galileo*, S. Drake, ed. and trans.,
Doubleday, New York, 1957, 306p.

LETTERS ON SUNSPOTS

G. Galileo

[*Editor's Note:* In the original, material precedes these excerpts.]

I therefore repeat and more positively confirm to Your
Excellency that the dark spots seen in the solar disk by
means of the telescope are not at all distant from its surface,
but are either contiguous to it or separated by an interval
so small as to be quite imperceptible. Nor are they stars
or other permanent bodies, but some are always being produced and others dissolved. They vary in duration from one
or two days to thirty or forty. For the most part they are of
most irregular shape, and their shapes continually change,
some quickly and violently, others more slowly and moderately. They also vary in darkness, appearing sometimes
to condense and sometimes to spread out and rarefy. In
addition to changing shape, some of them divide into three
or four, and often several unite into one; this happens less
near the edge of the sun's disk than in its central parts.
Besides all these disordered movements they have in common a general uniform motion across the face of the sun in
parallel lines. From special characteristics of this motion one
may learn that the sun is absolutely spherical, that it rotates from west to east around its own center, carries the
spots along with it in parallel circles, and completes an entire revolution in about one lunar month. Also worth noting
is the fact that the spots always fall in one zone of the solar
body, lying between the two circles which bound the dec-
linations of the planets—that is, they fall within 28° or 29°
of the sun's equator.

The different densities and degrees of darkness of the
spots, their changes of shape, and their collecting and
separating are evident directly to our sight, without any
need of reasoning, as a glance at the diagrams which I am
enclosing will show. But that the spots are contiguous to the
sun and are carried around by its rotation can only be deduced and concluded by reasoning from certain particular
events which our observations yield.

First, to see twenty or thirty spots at a time move with
one common movement is a strong reason for believing that
each does not go wandering about by itself, in the manner
of the planets going around the sun. In order to explain
this, let us define the poles in the solar globe and its circles
of longitude and latitude as we do in the celestial sphere.
If the sun is spherical and rotates, there will be two points
at rest called the poles, and all other points on its surface
will describe parallel circles which are larger or smaller

according to their distance from the poles. The largest of all will be the central circle, equally distant from the two poles. The dimension of the spots along these circles will be called their breadth, and by their length we shall mean their dimension extending toward the poles and determined by a line perpendicular to that which determines their breadth.

These terms defined, let us consider the specific events observed in the sunspots from which one may arrive at a knowledge of their positions and movements. To begin with, the spots at their first appearance and final disappearance near the edges of the sun generally seem to have very little breadth, but to have the same length that they show in the central parts of the sun's disk. Those who understand what is meant by foreshortening on a spherical surface will see this to be a manifest argument that the sun is a globe, that the spots are close to its surface, and that as they are carried on that surface toward the center they will always grow in breadth while preserving the same length.

[*Editor's Note:* In the original, Galileo explains his method of drawing the diagram.]

I have since been much impressed by the courtesy of nature, which thousands of years ago arranged a means by which we might come to notice these spots, and through them to discover things of greater consequence. For without any instruments, from any little hole through which sunlight passes, there emerges an image of the sun with its spots, and at a distance this becomes stamped upon any surface opposite the hole. It is true that these spots are not nearly as sharp as those seen through the telescope, but the majority of them may nevertheless be seen. If in church some day Your Excellency sees the light of the sun falling upon the pavement at a distance from some broken windowpane, you may catch this light upon a flat white sheet of paper, and there you will perceive the spots. I might add that nature has been so kind that for our instruction she has sometimes marked the sun with a spot so large and dark as to be seen merely by the naked eye, though the false and inveterate idea that the heavenly bodies are devoid of all mutation or alteration has made people believe that such a spot was the planet Mercury coming between us and the sun, to the disgrace of past astronomers.

[*Editor's Note:* Material has been omitted at this point.]

I kiss Your Excellency's hand reverently, and pray God for your happiness.

From Florence, August 14, 1612.

Your Illustrious Excellency's very devoted servitor,

GALILEO GALILEI L.

Part II

THE ELEVEN-YEAR CYCLE

Editor's Comments
on Papers 7 Through 11

7   SCHWABE
     Excerpt from *Solar Observations During 1843*

8   WOLF
     Excerpts from *Universal Sunspot Numbers: Sunspot
     Observations in the Second Part of the Year 1850*

9   WOLF
     Excerpt from *Sunspot Epochs Since A.D. 1610: The Periodic
     Return of Sunspot Minima*

10   MAUNDER
     Excerpt from *Note on the Distribution of Sun-spots in
     Heliographic Latitude, 1874 to 1902*

11   SCHOVE
     Excerpts from *Sunspot Cycles*

    The Maunder Minimum ended about 1700 and sunspot cycles soon became both regular and strong, but astronomers (except for Horrebow from his 1761–1769 observations) were not expecting a periodicity and they noticed (e.g., LaLande [1771] 1966) only fluctuations that they took to be irregular. As we shall see in Part III even de Mairan (1754, part of which is reprinted as Paper 12), with his many tabulations of annual displays, failed to recognize an auroral cycle.

    About 1800 the sunspot cycles were again weak (see Figs. 3 and 8) and irregular (Deehr and Holtet eds., 1980) causing William Herschel (1738–1822), who had come to England from Germany, to miss the discovery in his famous article in 1801 (Paper 29), and the breakthrough came only with Schwabe's little note of 1843 (Paper 7).

    In the Middle Ages sunspots (see Paper 4) had been mistaken for planetary transits, and it was because Schwabe was motivated by the belief in a new planet within the orbit of Mercury that he watched the sun regularly and conscientiously for so many years. Heinrich Schwabe (1789–1875), of Dessau in Germany, had studied science at the University of Berlin and had returned as an apothecary to Dessau where he began his daily watch of the sun in 1826. Eventually, in 1843, his tabulations suggested to him that there was a sunspot period of about 10 years. His idea was rejected until it was taken up by von Humboldt in the third German (1851) edition of his *Kosmos* (see von Humboldt, 1871), so that in 1857 when Schwabe was awarded the Gold Medal of the Royal Astronomical Society, its president could say: "Twelve years he spent to satisfy himself: six more years to satisfy [astronomers] and still thirteen more years to convince mankind" (Johnson, 1858) Schwabe had meanwhile accepted Wolf's view that the average length of the cycle was 11 years and not 10 years, and his tabulations were eventually extended to 1868 (reproduced in Chambers, 1889, p. 26) and beyond.

Sunspot Numbers

The next breakthrough in quantification was the concept of universal sunspot numbers by Wolf (1816–1893) in 1851 (Paper 8). The concept itself is often dated 1848, and Wolf himself recalled (1877, p. 656) that the idea had come to him in 1847. His paper shows that in 1850 he was already combining the number of sunspot groups with the number of sunspots to yield monthly numbers for 1849 and 1850 of 10.8 and 7.4 respectively. At that time Wolf was lecturing in Berne but in 1855 he was appointed director of the new Zurich Observatory, where the daily determination of the relative sunspot number R was continued with collaboration from other observatories until 1980. Also, Wolf published (1858, pp. 133–134) the formula $R = k(10g + f)$, where g is the number of spot groups and f is the total number of spots (whether in groups or not). For Wolf the constant k was 1, but each observer still has to determine his own value, which depends on his "personal equation," the instrument used, and so forth by comparison with the "official" international sunspot numbers. Wolf's formula may seem arbitrary but sunspot areas, published since 1874 by the Greenwich Observatory (cf. Norton, 1973), are found to be proportional to the *Wolf numbers*. Indeed, various other methods tried (cf. Ruf, 1977; Chernosky and Hagan, 1958; Vitinsky, 1965, p. 5; Kuklin, 1976; and Meeus, 1977), have not rendered any change necessary.

Wolf did more than just invent sunspot numbers. He worked backwards through history to 1610 in the light of Schwabe's discovery so that already in 1852 he could claim that the mean length of the sunspot period was 11.1, a more accurate value than is often quoted today (cf. Table 1 in Part IV). Paper 9 describes the results of his project up to that time. The monograph Wolf cited never appeared in print but most of the results came out eventually in his journal, which became the *Astronomische Mitteilungen*. In the same year (Wolf,

1852) he considered the evidence back to 1607, grouping his material into 11-year epochs, 1600.00–1611.11 and so on, and recognizing the Scheiner maximum as 1626.0 ± 1.0 and the Hevelius minimum as 1645.0 ± 1.0 (cf. Table 2). He frequently revised his work in the light of new finds such as Harriot's manuscripts (Wolf, 1858, pp. 135–137). Full tables of the epochs of maxima and minima appeared in English in *Royal Astronomical Society Monthly Notices* (in 1861 the table was extended to 1860) and *Royal Astronomical Society Memoirs* (in 1877 the table was extended to 1870). Despite recent doubts cast on Wolf's seventeenth century dates, a minor alteration (1804.2–1805.2) by his successor Wolfer (1902), and Baur's (1964; cf. Parker, 1972) modified definitions, Wolf's dates have been confirmed by auroral evidence (Schove, 1979), roughly within the probable errors he had specified. Wolf published daily data back to 1828, monthly back to 1749, and annually back to 1700. His numbers for 1700 to 1748 (Wolf, 1868) are more accurate than often supposed, but they have never been revised to overlap the post 1749 values and my own several attempts to do this from 1955 to 1981 (Table 1 in Introduction) are provisional. The concept of *smoothed* monthly numbers was introduced later by Wolf (1879); the smoothed value for July 1981 is thus defined as the mean of January 1981 to December 1981 and February 1981 to January 1982.

Magnetism

About 1850 Lamont (1805–1879), a Scotsman at Munich, found that the varying daily range of the magnetic needle was subject to a cycle, which he at first regarded as 10½ years. Once he had announced his discovery in 1851, the connection with the similar auroral cycle was likewise noted; the auroral evidence, which is now an important clue to past sunspot cycles, will be considered in Part III. The auroral connection found by Wolf (Paper 9) was the inspiration of the work by Fritz (1873 and 1881; cf. Schroeder, 1981). The magnetic effects at sunspot minimum, measured by what is known as the *aa-index* (cf. Dicke, 1978; Kane, 1978; Brown, 1976), are used today to predict the intensity of the next sunspot maximum, and Mörner (1978*a,* and 1978*b*) claims that a response to the magnetic cycle can be detected in the annual sediments known as varves (cf. Part VII).

The Law of Zones

The next breakthrough was the discovery by Carrington (1826–1875), an English amateur, of the so-called *latitude drift* or *law of zones*. Carrington recorded (1858) the zone leap as the equatorial spots became extinct and new sunspots developed in higher heliographic latitudes. This was confirmed by Wolf in 1858 and occurs at each sunspot minimum.

Carrington's discovery was followed up by Spoerer from Berlin (1822–1896), who worked from 1860 at his small observatory at Anklam in Pomerania, and after 1873 at Potsdam Observatory in what is now East Germany. Sometimes the law of latitudes is known as Spoerer's Law, for in his 1889 paper (see Paper 21 and Spoerer, 1887) he confirmed that the law had been in operation in earlier cycles back to 1621. Solar rotations are, however, still numbered after Carrington, that of October 1979 being Carrington rotation number 1687.

The best Benchmark paper in this field is, however, that of an Englishman, Maunder (1904), who expressed the results graphically in his famous Butterfly Diagram (Paper 10). A more detailed history of this period is provided in Kiepenheuer (1953) and in general works such as those by Cousins (1972) and Newton (1958).

Continued comparisons over different cycles have since led to the following conclusions. Gnevyshev and Gnevysheva in 1949 (see Vitinsky, 1965, pp. 19, 122, 126) tabulated the mean latitudes in relation to the year at sunspot minimum: minimum year $(N) = 7°$ and $28°$; $N + 2 = 21°$; $N + 4 = 16°$; $N + 6 = 12°$; $N + 8 = 10°$; and $N + 11 = 7°$. Waldmeier (1935, 1939) had, however, shown that the magnitude (R) of the cycle introduced modification. The initial latitude of the high spots thus varies from $26°$ to $30°$ and the mean latitude at maximum varies from $12\frac{1}{2}°$ to $18°$ as R_M varies from 60 to 140. This effect is clear in Kopecký's chart (1973, Fig. 3 p. 117) of large spots and has proved helpful in the estimation of seventeenth-century sunspot numbers (Table 1 in Introduction).

Current research on the Butterfly Diagram has shown that there is no latitude drift in the strict sense, as Gleissberg and Damboldt's (1979) review makes clear.

Recent Results

Smoothed sunspot cycles (Fig. 3) appear to have a single peak but Gnevyshev (1967, 1977) realized that the actual values show a double peak and the gap in large sunspots between the two peaks is now known as the Gnevyshev gap, clearly shown in the diagrams of Gleissberg (our Fig. 1), Gnevyshev (1967, 1977), Wittmann (1978), and Kopecký and Kotrc (1976).

Soviet scientists play an important role in solar research, as the works of Vitinsky (1965) and Rubashev (1964) make clear. Greenwich,

Boulder, Brussels, and Tokyo are world centers cooperating in the preparation of *International Sunspot Numbers.*

Some ideas on solar cycles and their effects were noted in an encyclopedia article, which is excerpted in Paper 11. This serves as an introduction to the concept of longer cycles (see Part III), reflected in a trend of the asymmetry of the individual cycles themselves (cf. Waldmeier, 1961).

Current work on solar activity is discussed in *Solar Physics, Nature,* and in the International Astronomical Union publications (e.g. Bumba and Kleczek, eds., 1976). The latter includes a review by Kuklin (1976) in which he describes Soviet and Czech discoveries and speculations.

REFERENCES

Baur, F., 1964, New Establishments of Epochs of Sunspot Maxima and Minima, *Meteorol. Abhandlungen* **50**(3):1–38. (In German.)

Brown, G. M., 1976, What Determines Sunspot Maximum, *Royal Astron. Soc. Monthly Notices* **174:**185–189.

Bumba, V., and J. Kleczek, eds., 1976, *Basic Mechanisms of Solar Activity,* Reidel, Dordrecht, Holland, 481p.

Carrington, C., 1858, The Distribution of the Solar Spots in Latitude Since the Beginning of the Year 1854, *Royal Astron. Soc. Monthly Notices* **19:**1–3.

Chambers, G. F., 1889, *A Handbook of Descriptive and Practical Astronomy,* Clarendon Press, Oxford, England, 676p.

Chernosky, E. J., and M. P. Hagan, 1958, The Zurich Sunspot Number and Its Variations for 1700–1957, *Jour. Geophys. Research* **63:**775–788.

Cousins, F. W., 1972, *The Solar System,* Baker, London, 300p. (Good popular account.)

Deehr, C. S., and J. A. Holtet, eds., 1980, *Exploration of the Polar Upper Atmosphere,* Reidel, Dordrecht, Holland, 498p.

de Mairan, J. J. d'O, 1754, *Traité Physique et Historique de l'Aurorae Boreale,* 2nd ed., Académie Royale des Sciences, Paris, 592p. (1st ed., 1733.)

Dicke, R. H., 1978, Is There a Chronometer Hidden Deep in the Sun? *Nature* **276:**676–680.

Fritz, H., 1873, *Verzeichniss Beobachteter Polarlichter,* C. Gerold's Sohn, Wien, Austria, 265p.

Fritz, H., 1881, *Das Polarlicht,* Brockhaus, Leipzig, 348p.

Gleissberg, W., and T. Damboldt, 1979, Reflections on the Maunder Minimum of Sunspots, *British Astron. Assoc. Jour.* **89:**440–449.

Gnevyshev, M. N., 1967, On the Eleven-Year Cycle of Solar Activity, *Solar Physics* **1:**107–120.

Gnevyshev, M. N., 1977, Essential Features of the Eleven-Year Solar Cycle, *Solar Physics* **51:**175–183.

Johnson, M. J., 1858, On Presenting the Gold Medal of the Society to M. Schwabe, *Roy. Astron. Soc. Mem.* **26:**196–205.

Kane, R. P., 1978, Predicted Intensity of the Solar Maximum, *Nature* **274:**139–140.

Kiepenheuer, K. O., 1953, Solar Activity, in *The Sun,* G. P. Kuiper, ed., The University of Chicago Press, Chicago, pp. 322–465.

Kopecký, M., 1973, The Periodicity of Large Sunspot Groups, *Astron. Inst. Czechoslovakia Bull.* **24:**113–118.

Kopecký, M., and P. Kotrc, 1976, Some Regularities of the Occurrence of Large Sunspot Groups, *Contr. Astron. Observ.* **6:**243–247.

Kuklin, G. V., 1976, Cyclical and Secular Variations of Solar Activity, in *Basic Mechanisms of Solar Activity,* V. Bumba and J. Kleczek, eds., Reidel, Dordrecht, Holland, pp. 147–190.

LaLande, J., 1966, *L'Astronomie,* vol. 3, Johnson Reprint Corporation, New York, pp. 286–287. (1st ed., 1771.)

Maunder, E. W., 1904, The Duration of Sunspots, *Royal Astron. Soc. Monthly Notices* **64:**759–762.

Meeus, J., 1977, On an "Unexpected Anomaly" in the Annual Distribution of the Maxima of the Eleven-Year Sunspot Cycle, *Jour. Interdisciplinary Cycle Research* **8:**230–233.

Mörner, N-A., 1978a, *Paleomagnetism and Varved Clays (Abstract),* Israel Association of Science 10th Congress, Jerusalem.

Mörner, N-A., 1978b, Annual and Interannual Magnetic Variations in Varved Clay, *Jour. Interdisciplinary Cycle Research* **9:**229–241.

Newton, H. W., 1958, *The Sun,* Penguin, London, 208p.

Norton, A. P., 1973, *Norton's Star Atlas,* Gall & Inglis, London, 116p and 8 plates.

Parker, B. N., 1972, Keeping Up with Sunspots, *Weather* **27:**247–251.

Rubashev, B. M., 1964, *Problemy solnechnoy aktivnosti,* Nauka, Moscow and Leningrad, 362p. (*Problems of Solar Activity,* translated in NASA TTF 244.)

Ruf, K., 1977, American and Zurich Sunspot Numbers, *Jour. Interdisciplinary Cycle Research* **8:**215–217.

Schove, D. J., 1979, Sunspot Turning Points and Aurorae Since A.D. 1510, *Solar Physics* **61:**423–432.

Schroeder, W., 1981, Hermann Fritz, Wegbereiter der Polarlichtforschung, *Naturforsch. Gesell. Zurich Vierteljahrsschr.* **126:**199–204.

Spoerer, F. W. G., 1887, Ueber die Periodicität der Sonnenflecken seit dem Jahre 1618, *Astronomische Gesell. Vierteljahrsschr.* **22:**323–329.

Vitinsky, Y. I., 1965, *Solar Activity Forecasting,* Israel Program for Scientific Translations, Jerusalem, 129p. (*Prognozy solnechnoi aktivnosti,* Akad. Nauk Izd. SSSR, Leningrad, 1962, 151p.) (Standard work.)

von Humboldt, A., 1871, *Kosmos,* vol. 3, E. C. Otté and B. H. Paul, trans., Bell, London, 603p.

Waldmeier, M., 1935, Neue Eigenschaften der sonnenflecken Kurve, *Astronomische Mitt.* **133:**105–130.

Waldmeier, M., 1939, Sunspot Activity, *Astronomische Mitt.* **138:**470.

Waldmeier, M., 1961, *The Sunspot Activity in the Years 1610–1960,* Schulthess, Zurich, 171p.

Wittmann, A., 1978, The Sunspot Cycle Before the Maunder Minimum, *Astronomy and Astrophysics* **66:**93–97.

Wolf, R., 1852, Bestimmung der Epochen für das Minimum und Maximum der Sonnenflecken-Bildung, *Naturf. Gesell. Bern Mitt.*

Wolf, R., 1858, Mitteilungen über die Sonnenflecken, No. 6, *Astronomische Mitt.,* pp. 127–143. (For sunspot formula, see pp. 133–134.)

Wolf, R., 1868, Mitteilungen über die Sonnenflecken, No. 24, *Astronomische Mitt.,* p. 111. (Annual data since 1700.)

Wolf, R., 1877, *Geschichte der Astronomie,* Oldenbourg, Munich, 815p.

Wolf, R., 1879, Mitteilungen über die Sonnenflecken, No. 50, *Astronomische Mitt.,* p. 29. (Tables since 1749.)

Wolfer, A., 1902, Revisions of Wolf's Sunspot Relative Numbers, *Monthly Weather Rev.* **30:**171-176.

7

Solar Observations During 1843[†]

H. SCHWABE

In Dessau

[*Editor's Note:* In the original, material precedes this excerpt.]

From my earlier observations, which I have reported every year
in this journal, it appears that there is a certain periodicity in the
appearance of sunspots and this theory seems more and more
probable from the results of this year. Although I described the
numbers of groups from 1826 to 1837 in the 15th volume of
Astronomische Nachrichten, no. 350, p. 246, I should now like to
add a complete report of all my observations of sunspots up to
the present, in which I have indicated the number of days of
observation and the days when there were no spots to be seen,
as well as the number of groups. The number of groups alone
does not give enough information to judge the period as I am
convinced that with very great accumulations of sunspots the
number of groups is estimated too low, and during a period when
there are few spots the estimate of the number of groups is rather
too large. In the first instance, several groups often merge together
into a single one and in the second, one group may divide into two
by the separating off of some spots. For this reason, I am sure
the reader will excuse the repetition of facts given in earlier
reports.

[†] *Astron. Nach.* **20,** No. 495.

Year.	No. of Clusters.	Days when no Spots were Observed.	Observation Days.
1826	118	22	277
1827	161	2	273
1828	225	0	282
1829	199	0	244
1830	190	1	217
1831	149	3	239
1832	84	49	270
1833	33	139	267
1834	51	120	273
1835	173	18	244
1836	272	0	200
1837	333	0	168
1838	282	0	202
1839	162	0	205
1840	152	3	263
1841	102	15	283
1842	68	64	307
1843	34	149	324

If one compares the number of groups with the number of days when no spots are visible, one will find that sunspots have a period of about 10 years, and that for five years of this period they appear so frequently that during that time there are very few or no days when no spots at all are visible.

The future will tell whether this period persists, whether the minimum activity of the sun in producing spots lasts one or two years and whether this phenomenon takes longer to build up or longer to decline.

On 10th and 11th April and on 10th May the sun was so particularly clear in a slightly overcast sky that the darkening at its limb showed up very clearly.

Although I kept a careful lookout for the so-called "light-flashes" round the sun I only saw them on 6th May at midday when a great number shot across the field of vision of the telescope, in a direction different from that of the wind and clouds.

[*Editor's Note:* In the original, some further 1843 observations follow this excerpt.]

8

UNIVERSAL SUNSPOT NUMBERS: SUNSPOT OBSERVATIONS IN THE SECOND PART OF THE YEAR 1850

R. Wolf

These excerpts were translated expressly for this Benchmark volume by Th. Damboldt from Naturf. Gesell. Bern. Mitt. 1:89–95 (1851). Copyright ©1979 by Th. Damboldt. The tables that appear in this translation have been reprinted from the original.

The state of the solar surface was observed as often as possible. Observations were made on the following: 28 days in July, 29 days in August, 26 days in September, 18 days in October, 16 days in November, and 11 days in December, which comes to a total of 128 days. The results are given in the following tables (columns **C, D, E**) for the numbers of spot groups, spots, and more or less conspicuous faculae and scales (Fackeln und Schuppen). The tables show five columns for each month: **A** indicates the cloudiness insofar as it affected the observations: (1) denotes an unobstructed sun, (2) observations of the sun through clouds and (3) the sun not visible at all. **B** indicates the instrument used for the observations, namely (1) a four-legged Fraunhofer with a magnification of 64, which was used as often as possible, and (2) a portable telescope used during unfavorable days and during excursions. **C** gives the number of spot groups. **D** gives the number of spots in all groups. **E** applies to faculae and scales, where (1) denotes normal and (2) abnormal frequency and intensity.

Sonnenflecken–Beobachtungen A. 1850.

	Juli.					August.					September.			
	A	B	C	D	E	A	B	C	D	E	A	B	C	D	E
1	1	2	2	3	–	2	1	3	7	–	1	1	6	21	1
2	3	–	–	–	–	3	–	–	–	–	1	1	8	46	2
3	1	2	2	4	–	1	1	5	27	2	1	1	6	45	2
4	1	2	2	4	–	1	1	5	21	2	1	1	6	47	1
5	1	–	–	–	–	1	2	3	5	–	1	1	5	36	1
6	1	2	0	0	–	1	2	4	7	–	1	1	5	54	1
7	2	2	1	2	–	3	–	–	–	–	1	1	5	56	1
8	2	2	1	1	–	1	1	4	25	2	1	1	5	48	1
9	2	2	2	4	–	1	1	4	20	2	1	1	4	46	1
10	3	–	–	–	–	2	2	1	3	–	1	1	6	46	2
11	2	2	1	2	–	1	1	4	19	1	1	1	5	24	1
12	1	2	6	16	–	2	1	3	17	1	1	1	8	38	1
13	2	1	2	13	–	2	2	1	3	–	1	1	5	28	1
14	1	1	6	50	1	2	2	1	2	–	1	1	4	18	1
15	1	1	7	35	1	1	1	3	21	1	1	1	4	36	1
16	1	1	8	30	1	1	1	2	5	1	1	1	3	22	1
17	1	1	6	17	1	2	2	1	5	–	1	1	4	31	1
18	2	1	3	10	1	1	1	4	31	1	1	1	5	42	1
19	2	1	2	5	–	1	1	4	42	2	1	1	6	54	2
20	2	1	2	8	1	2	2	2	7	–	1	1	5	37	1
21	1	1	1	3	1	1	1	4	36	1	2	2	3	10	–
22	1	1	1	3	1	2	2	3	7	–	2	–	–	–	–
23	1	1	0	0	1	2	1	4	28	1	1	2	4	10	–
24	2	1	0	0	1	2	2	2	4	–	1	2	3	16	–
25	1	1	0	0	1	1	1	3	18	1	3	–	–	–	–
26	2	2	0	0	–	1	1	6	26	2	2	–	–	–	–
27	2	1	1	3	1	2	1	2	2	–	2	2	2	5	–
28	1	1	1	3	1	1	1	6	21	2	3	–	–	–	–
29	2	1	0	0	–	2	1	4	14	2	2	2	4	5	–
30	1	1	3	7	2	2	2	1	3	–	2	2	4	5	–
31	2	1	3	8	2	1	1	6	22	2					

Sonnenflecken–Beobachtungen A. 1850.

	October.					November.					December.				
	A	B	C	D	E	A	B	C	D	E	A	B	C	D	E
1	2	2	3	5	–	2	2	1	4	–	3	–	–	–	–
2	3	–	–	–	–	3	–	–	–	–	3	–	–	–	–
3	3	–	–	–	–	1	1	3	25	1	1	1	2	10	1
4	2	2	3	4	–	1	1	3	23	1	3	–	–	–	–
5	3	–	–	–	–	1	1	2	10	1	3	–	–	–	–
6	3	–	–	–	–	1	1	0	0	1	1	1	2	5	1
7	1	2	3	10	–	1	1	1	2	1	1	1	4	14	2
8	2	2	2	2	–	1	1	1	1	1	1	1	3	16	2
9	1	2	2	3	–	2	1	1	1	–	3	–	–	–	–
10	3	–	–	–	–	1	1	2	7	1	3	–	–	–	–
11	2	2	0	0	–	1	1	2	8	1	3	–	–	–	–
12	3	–	–	–	–	3	–	–	–	–	3	–	–	–	–
13	1	1	4	24	2	2	2	1	1	–	1	1	5	27	4
14	1	1	5	29	2	3	–	–	–	–	1	1	4	40	1
15	1	1	5	41	1	1	1	5	19	1	3	–	–	–	–
16	1	1	6	55	1	1	1	4	17	1	3	–	–	–	–
17	1	1	5	42	1	3	–	–	–	–	2	2	0	0	–
18	1	1	6	54	1	3	–	–	–	–	3	–	–	–	–
19	1	1	6	62	1	3	–	–	–	–	3	–	–	–	–
20	2	1	4	40	–	3	–	–	–	–	3	–	–	–	–
21	3	–	–	–	–	3	–	–	–	–	3	–	–	–	–
22	3	–	–	–	–	2	1	1	4	–	3	–	–	–	–
23	3	–	–	–	–	1	1	6	48	2	3	–	–	–	–
24	3	–	–	–	–	1	1	6	62	1	3	–	–	–	–
25	3	–	–	–	–	3	–	–	–	–	2	1	2	7	–
26	3	–	–	–	–	3	–	–	–	–	1	1	4	18	1
27	3	–	–	–	–	3	–	–	–	–	1	1	4	8	1
28	1	1	5	36	2	3	–	–	–	–	1	2	3	5	–
29	2	2	1	5	–	3	–	–	–	–	3	–	–	–	–
30	1	2	2	7	–	3	–	–	–	–	3	–	–	–	–
31	1	1	3	38	1						3	–	–	–	–

The comments given in the previous communication about sunspots are, in general, confirmed by the observations of this half year. These latter observations call for only a few remarks:

1. In 1848 and 1849 sunspots could be seen every time the sun was observed—mostly in a considerable number—and this was true also in the first half of 1850; however, there were no spots on 9 occasions in the second half of 1850.

   [*Editor's Note:* Details of dates are omitted here.]

3. On 31 August there were 6 groups with 22 spots; of these only one group consisting of one spot was located north of the equator. Generally the northern hemisphere seemed to be less spotted than the southern hemisphere.

   [*Editor's Note:* Material has been omitted at this point.]

6. It would be highly desirable to check not only whether greater spots and groups develop in the same zone on the sun but whether certain regions within this zone favor the formation of spots.
7. Some phenomena seem to promote the view that the formation of sunspots arises from the sun's interior, so to speak, as if gases come from the interior to the surface and form bubbles that explode after sufficient growth.
8. Sunspot observations of 1849 and 1850 (unfortunately observations of 1848 were not made according to the same system) are grouped together in such a way that the observed number of spot groups is increased by 1/10 of the relevant number of spots during the days when observations were made with the large telescope).[1] From these numbers the mean sunspot number for each month is found, resulting in the following summary of sunspots during the two-year period:[2] 1849 = 10, 8 and 1850 = 7, 4.

   [*Editor's Note:* Monthly and seasonal totals are omitted here.]

From this summary a gradual decrease in sunspot numbers seems to follow with good reliability—a decrease that coincides with the opinion of Schwabe. He states in a letter dated 31 December 1848 (*Astron. Nach.*, 667): "If a sunspot periodicity of 10 years should prove correct—and I have observed sunspots for 23 years—from 1849 on a decrease of 5 years duration and then an increase in their frequency up to 1858 should take place."

NOTES

1. In my opinion the total area of all the spots would be the most suitable measure for the spot reading, but as the short time did not allow me to make the necessary measurements and estimates, I believe that with the previous summary I obtained an adequate substitute.
2. The sun was observed on 552 days within the two years; however, only on 258 days could I observe the unobstructed sun with the larger telescope, and these values alone are the basis for this summary.

9

SUNSPOT EPOCHS SINCE A.D. 1610: THE PERIODIC RETURN OF SUNSPOT MINIMA

R. Wolf

This excerpt was translated expressly for this Benchmark volume by D. J. Schove, St. David's College, England, from Acad. Sci. Comptes Rendus 35:704–705 (1852). Copyright © 1979 by D. J. Schove.

The Academy of Sciences (at Paris) has expressed interest in the relationship, established in my last communication, between sunspots and terrestrial magnetism. Since then I have continued to study these phenomena and used up some 400 notebooks in order to include all observations of sunspots since their discovery. This has led to a monograph that I shall finish shortly; its content seems to be important enough to present here the gist of it in a few words. My monograph is divided into six parts as follows.

In the first chapter, sixteen different epochs (each) established by a minimum and a maximum of sunspots, are the basis for my demonstration that the mean sunspot cycle must be fixed at 11.111 ± 0.038 years so that nine periods equal one century exactly.

In the second chapter, I shall show that in each century the years 0.00, 11.11, 22.22, 33.33, 44.44, 55.56, 66.67, 77.78, and 88.89 correspond to the sunspot minima. The interval between minimum and maximum is variable: its average is five years.

The third chapter will contain a listing of all sunspot observations from the time of Fabricius and Scheiner up to that of Schwabe, arranged throughout to fit my period. The agreement is surprising.

The fourth chapter will establish remarkable analogies between solar spots and variable stars, so that a close connection between these singular phenomena may be supposed to exist.

In the fifth chapter, I shall show that my period of 11.111 years coincides still more precisely with the variations of magnetic declination than the period of 10⅓ years established by M. Lamont. The magnetic variations agree with the sunspots, not only in their regular changes but also in all the little irregularities, and I consider that this last observation will suffice to have proved definitely the important relationship.

The sixth chapter will discuss a comparison between the solar cycle and the meteorological indications contained in a Zurich chronicle covering the years 1000 to 1800. From this, in accord with the ideas of William Herschel, the years when the spots are most numerous are also in general drier and more fertile than the rest; the others on the contrary are more moist and thundery. Aurorae and earthquakes, indicated in this chronicle, increase in a striking way in sunspot years.

EDITOR'S NOTES

Chapter I: Fully published in scattered articles in Wolf's *Astronomische Mitt.* (e.g., 1859, p. vi ff).

Chapter II: Now known to be approximately true back to A.D. 800 (Schove, D. J., 1948, *Popular Astronomy* **54:**247–252), written before I had seen Wolf's remarkable work.

Chapter III: Never published.

Chapter IV: Still discussed but no obvious parallel.

Chapter V: Continued at Zurich until 1980.

Chapter VI: True for aurorae (see our Part III); otherwise still debated (see our Part VI).

10

Reprinted from page 760 and Figure 8 of *Royal Astron. Soc. Monthly Notices*
64:747–761 (1904)

NOTE ON THE DISTRIBUTION OF SUN-SPOTS IN HELIOGRAPHIC LATITUDE, 1874 TO 1902

E. W. Maunder

[*Editor's Note:* In the original, material precedes and follows this excerpt.]

Fig. 8 (Plate 16) displays for the entire period 1877–1902 the same features shown in fig. 7 for the four years 1878–1881. But in this case, in order to bring the facts within reasonable compass, instead of the individual spot-centres being represented, the spot-distribution for each synodic rotation of the sun has been given. Wherever a spot-centre has fallen on one or more days during a given rotation in a particular degree of latitude a line has been drawn across that particular degree. The diagram, therefore, like fig. 7, represents distribution in latitude, but takes no account of area. In assigning a spot-centre to any particular degree of latitude all latitudes from, say, 6°·5 to 7°·4 have been reckoned as 7°, and similarly with others.

An examination of the diagram brings out Spoerer's Law with remarkable clearness. There is but a single movement in either hemisphere, the general trend of which is downward from immediately after minimum to immediately before minimum. During the years of great spot-activity almost every degree of latitude is affected. The interruptions are perfectly irregular and sporadic. The indications of subordinate zones, which would be shown, if present, by continuous barren tracks, through the middle of the vertical lines, are quite wanting. Nothing can be more striking than the contrast between the very fair approach to continuity of the distribution lines, during the greater part of the solar cycle, and the definiteness with which they are limited, both on the side of the equator and on that of high latitudes, by barren belts descending towards the equator. The general form of spot-distribution in both cycles is that of a hollow wedge.

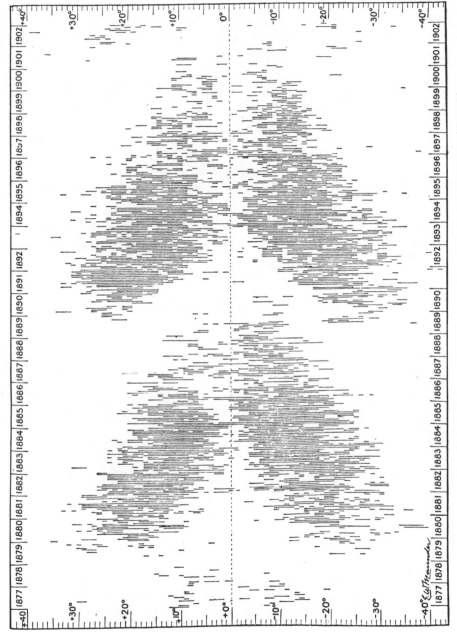

FIG. 8.—DISTRIBUTION OF SPOT-CENTRES IN LATITUDE, ROTATION BY ROTATION, 1877–1902.

91

SUNSPOT CYCLES

D. J. Schove

(1) Sunspot Cycles since 1610 A.D.

(a) **Introduction.** Sunspot numbers (see *Sunspots and Solar Activity*) vary considerably from year to year and from cycle to cycle. Values for the years 1944–1964 are illustrated in Fig. 1, which is so arranged as to emphasize the 11-year cycle. The columns are arranged according to the (Schove) remainder when the last two digits are divided by 11. It will be noted in Fig. 1 that the remainder at the maximum of the last two cycles has been about 3. This is less than normal: the sunspot cycles are at present running early.

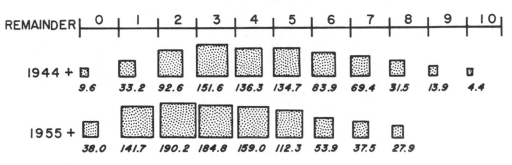

Fig. 1. Annual sunspot numbers, 1944–1964.

[*Editor's Note:* Material including Figures 2 and 3 has been omitted at this point. The cross-references in this article refer to other entries in *The Encyclopedia of Atmospheric Sciences and Astrogeology.*]

(d) **Asymmetry.** This has been investigated by Gleissberg (cf. 1965). The period of rise ($T = M - m$) is shorter than the period of fall ($U = m - M$), notably in strong cycles. As an empirical approximation since 1850 we may write

$$T = 14 - 5 \log R_M$$

where R_M is the maximum smoothed sunspot number in question. The index of asymmetry

$$A = \frac{U - T}{U + T}$$

A secular trend may be noted (cf. Fig. 3).

(e) **The Butterfly Diagram.** Sunspots in the early phase of a cycle, as in 1944 or 1966, occur in high solar latitudes; in the later phases, as in 1964, they occur near the solar equator. The "Butterfly Diagram" (Fig. 2) shows the pattern relating spot latitude to the phase of the solar cycle.

(f) **The 27-day Cycle.** A sunspot or other active solar region may reappear with successive solar rotations. In the case of a medium-latitude spot near the beginning of a cycle, the period is 28 days, but an equatorial spot near the end of a cycle reappears after 27 days. The same rotation period can be recognized, for example, in the twelfth century.

(2) Solar Cycles since 500 B.C.

(a) **The 11.1-year Solar Cycle.** The mean length of the solar cycle in the present century has been 10 years, although the long-term value is 11 years. Indeed, as in Fig. 1, since 800 A.D., years such as 1944 with the last two digits divisible by 11 are approximately years of few spots and the (Schove-) "remainder", after such division, defines the phase of the maximum (cf. Fig. 1). The mean length of the cycle over the past two millenia has been very close to 11.1 years. This is illustrated in Fig. 3.

(b) **The 22.2-year Magnetic (Hale) Cycle.** The magnetism of the leading spots is reversed in alternate cycles, so that the true solar cycle is 22.2 not 11.1 years. Indeed, during the period 1850–1930 alternate maxima were also relatively stronger than in neighboring cycles. Moving 10-year means of sunspot numbers include maxima (*s*-max) at, for example, 1731, 1782, 1834, 1849, 1868, 1894/5, 1921 and 1954, and minima (*s*-min) at 1749, 1803,

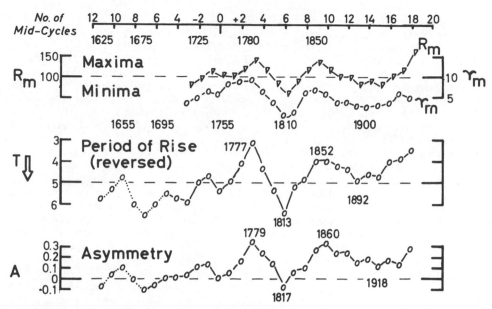

Fig. 4. The "80-year" cycle: fluctuations in amplitude (R_M), time of rise (T), and asymmetry (A) in the solar cycles since A.D. 1610 (from Schove, 1965). Each point is the mean for three cycles, plotted at the time of the central cycle (e.g., No. 1 = *ca.* 1760).

R_M = sunspot number at maximum
r_m = sunspot number at minimum
T = time of rise from previous minimum to maximum (note inverted scale)
A = index of asymmetry, as defined by Gleissberg

1853, 1878, 1910 and 1930; no regular 22-year cycle appears in such data. A feeble tendency to stronger alternate maxima can nevertheless be traced back to medieval times. If we assume that the author's calculation of the earlier cycles is broadly correct back to 800 A.D., the more active cycles were those with leading spots with N-polarity in the northern solar hemisphere.

(c) The 80-year "Gleissberg" Cycle. Longer cycles of 55–100 years frequently occur in the *strength* of solar cycles and the tendency to cycles of about 80 years in the *length* of the solar cycle has been demonstrated by Gleissberg (1965). In the past $2\frac{1}{2}$ centuries, both maxima and minima have followed a cycle of slightly over 90 years. Overlapping 30 year means of sunspot numbers indicate the principal maxima (S-max) at c. *1729* (i.e., 1715/44), c. *1790*, c. *1864* and probably c. *1955±5*, and the principal minima (S-min) at c. *1704*, c. *1760*, c. *1825* and c. *1903/4* and c. *1914/15*.

S-max show some tendency to cluster near 30-year periods at years of form A.D. ($78q + 48$) where *q* is an integer. Maxima of the 80-year cycle are dated, judging from Gleissberg (1965), A.D., c. 360,

440, 510, 575, 665, 755, 840, 920, 990, 1115, 1185, 1250, 1305, 1375, 1455, 1545, 1605, 1715, 1765, 1840 and (?) 1955.

[*Editor's Note:* Material has been omitted at this point.]

(d) Radiocarbon. The varying initial radiocarbon activity of different centuries (depending on variable cosmic flux) lends support to this suggestion. Indeed, Damon and Long find the 200-year cycle with an amplitude of 1% extending back through the last three millenia.

(4) Solar-climate Relations

(a) Apparent Mechanism. The observed solar-climate relations are weak and unreliable but they may be interpreted (according to the present author) by assuming:

(i) Solar activity leads to an eruption of air from the domain of the Upper High at 200 millibars. This leads to a lowering of barometric pressure throughout the Indian Ocean region.

(ii) The geographical extent of the area affected increases with the strength of the solar cycle in question. The climatic effects are greatest in medium cycles with sunspot numbers (R_M) between 80 and 140. Such cycles are characterized by an increase in the Pressure Parameter, i.e., a lowering of pressure in the "Domain of the Upper High" from the Rockies and the Andes in the west to Tibet and the Tasman Sea in the east.

(b) The 11-year Cycle. The climatic effects of a sunspot maximum are primarily those associated with an increase in the pressure parameter, and when this fails to occur, other correlations fail likewise. On the other hand, volcanic dust also can produce a similar increase in the pressure parameter together with the associated "solar" correlations. Moreover, in general, weak solar cycles misfire and strong solar cycles produce their greatest effect before the maximum is reached. In any case, the length of the solar cycle varies from 9–16 years. No 11-year periodicity appears therefore in meteorological timeseries.

(c) The 22-year "Hale" Cycle. The feeble tendency noted above for alternate cycles to be stronger appears to be responsible for a weak cycle of 20–24 years in the pressure parameter and associated climatic fluctuations. This cycle appears to be real in some long tree-ring (see *Meteorological Cycles)* and varve (see Vol. I, *Sunspots and Sedimentation)* series, and may have been in existence in Devonian times.

(d) The 80-year "Gleissberg" Cycle. This cycle is not sufficiently regular to attain much significance in meteorological series, and in any case, some cycles of about this length in climatic data are probably not of solar origin. There is thus a "superoscillation" reflected in 30-year means of the Pressure Parameter, with maximum pressure in the Indian Ocean (*P*-max or *PP*-min) about 1880–1905 (c. *1890*) and maximum pressure in the Pacific Ocean about 1905–1940 (*PP*-max c. *1920*). More significant is a temperature cycle with warm phases in the Arctic c. *1845* and c. *1930*. However, warm phases also occurred in Europe at c. *1185*, c. *1255/85*, c. *1375*, c. *1470* and c. *1520* and if a cycle is to be involved its mean length would seem to be either c. 93 or 83 years. Tree-ring series sometimes show cycles of about 80 years, but it is not known whether the phase keeps in step with the irregularities of the 80-year solar cycle. The

Nile Flood does show some tendency to keep in phase over the last 1300 years.

(e) The 205-year Cycle. Tree-rings in Northern Finland reflect summer temperature in individual years, and Sirén has found that they respond to a two-century cycle. "Missing tree-rings," caused by severe droughts, in the American Southwest, are more frequent in odd centuries. These appear to be climatic reflections of the two-century solar cycles.

(f) Harmonics. A cycle of 2.2 years ("Southern Oscillation") is the only cycle greater than the year to attain much significance in meteorological time series; amplitude is so high in the equatorial stratosphere that there is a regular alternation of west and east winds (see *Equatorial Oscillation*). It has been claimed that the length of this curious cycle is one-fifth of the solar cycle. Wave trains have also been observed in the pressure parameter at certain periods (e.g., 2.7, 3.0, 3.4, 3.75, 4.85, 6.9...), and there is a cycle in the Greenland Upper Trough (7.8 years). Whether or not these are the effects of persistent "harmonics" is not yet decided.

(g) Possible Long Cycles. Other long climatic cycles that have some claim to significance (see *Cycles and Periodicities*) are one of 1265 years in the postglacial period, one of 40,000 or 80,000 years in the Würm glacial (and perhaps the Carboniferous), and one of 30 million years in the Mesozoic. There is, finally, an orogenic revolution cycle of about 400 million years, which may perhaps be linked in some way with the Galactic Cycle (see *Time and Astronomic Cycles*). We cannot yet tell whether these cycles have any solar associations. Many other long-period cycles have been claimed, but no significance seems to be attached to them. Nevertheless, powerful computer techniques applied to ice cores and varved sediments are expected to lead to a more scientific and critical approach to long-period cycles than has hitherto been possible, and the new methods of dating palaeoclimatic phases will make it possible to put even the solar theories to the test.

References

Gleissberg, W., 1965, "The eighty-year cycle in auroral frequency numbers," *J. Brit. Astr. Assoc.*, **75**, 227–231.

*Schove, D. J., 1965, "Solar Cycles and Equatorial Climates," *Geol. Rundschau*, **54**, 448–477.

ERRATUM

Line 10 of **(d) Asymmetry,** "... (cf. Fig. 3)" should read "... (cf. Fig. 4)." Line 9 of **(d) The 80-year "Gleissberg" Cycle,** "... 1880–1905 ..." should read "... 1876–1905 ..."

Part III

EARLY AURORAE

Editor's Comments
on Papers 12 Through 19

12   **De MAIRAN**
      Excerpt from *Auroral Fluctuations*

13   **SCHOVE**
      Excerpts from *Sunspots and Aurorae, 500–250 B.C.*

14   **SCHOVE**
      The Spectrum of Time

15   **SCHOVE and HO**
      Excerpts from *Chinese Records of Sunspots and Aurorae in the Fourth Century A.D.*

16   **SCHOVE**
      Excerpts from *Visions in North-West Europe (A.D. 400–600) and Dated Auroral Displays*

17   **SCHOVE and HO**
      Excerpts from *Chinese Aurorae: I, A.D. 1048–1070*

18   **STOTHERS**
      Excerpt from *Ancient Aurorae*

19   **SCHOVE**
      Auroral Numbers Since 500 B.C.

We begin with two views of the aurora seen in Lent A.D. 793: "In this year dire portents appeared over Northumbria and sorely frightened the people. They consisted of immense whirlwinds and flashes of lightning, and fiery dragons were seen flying in the air. A great famine . . ." (Whitelock, 1955, p. 167). From Alcuin's letter the account states that "What portends this bloody rain . . . we saw fall menacingly on the north side from the summit of the roof, though the sky was serene" (Whitelock, 1955, p. 776).

Auroras, rather than the all-too-sporadic sunspots, are at present our principal clue to the 11-year cycles before 1610. The word *aurora* means *dawn* and the phrase *like the dawn* was occasionally used as a simile for horizon displays in the Middle Ages (A.D. 584, 962, 1173), *northern dawn* (aurora boreale) being adopted by Galileo or Guidicci in 1619. As a term for *Northern Lights* it came into general use only in the eighteenth century; until then each display was often regarded as a unique phenomenon or even a religious experience. For the Spectrum of Time project (Schove, 1960, 1961; see also Paper 14) it proved necessary to cull the information from religious sources such as records of Buddhist ceremonies in Japan (A.D. 762–767) and historical sources such as the Chinese dynastic histories. The records collected were utilized in the writer's studies (Papers 19 and 20), but auroral descriptions are so numerous and lengthy that here we have had to limit ourselves to short excerpts.

During the Maunder Minimum of the seventeenth century (cf. Figs. 5, 8, and 10) displays were often seen in the British Isles (cf. Schove, 1953, 1979) despite statements sometimes made to the contrary. However, in the eighteenth century displays became more vivid and frequent and the first auroral catalog in the West was that of De Mairan (1754), whose comments (Paper 12) reveal that he noticed the connection with sunspots and that he may have known the half cycle was 5 or 6 years. The *solar air* of Descartes is indeed what we now term the *solar wind* (cf. Deehr and Holtet, eds., 1980; Bruzek and Durrant, eds., 1977, Chapter 13). Later in the century, Pilgram, searching for meteorological periodicities, took the first step toward the autocorrelation function in his table of recurrences.

Pilgram (1788) thus gave a list of *Nordlicht* years from A.D. 394 to 1784 and studied the intervals between them. He found peaks about 9 and 19 years (partly explained by the shortness of active cycles), and then clusters especially at 200 years but also at about 100-, 270-, 303-, 400-, and 532-year intervals. However, the period that he selected, which was of special interest, was that of 47 to 54 years. His results often correspond to peaks in the autocorrelation curve of sunspots found by the Yunnan group (Yunnan Observatory, 1977, Figs. 1 and 2) or in dates between sunspot maxima found by myself (Paper 19). We now appreciate that Pilgram's clusters are the intervals between integral multiples—4, 5, 9, 24, 27, 36, and 48—of a sunspot cycle 11.1 years in length, and perhaps two, three, and four times a longer period of about 133 years.

Early auroral catalogs such as those by Frobesius (1739) and Schoening (1760) often give fuller information for the early period than the better-known works of Fritz (1873, 1881), who followed up Wolf's work on sunspots with parallel studies of the aurora (cf. Schroeder, 1979).

The next few excerpts reflect the East-West collaboration (see Schove 1949, 1960, 1961) of the Spectrum of Time (SOT) project. Schove's 1948 work (excerpted in Paper 13) on the period 500 B.C. to 250 B.C. is largely superseded by the work of Stothers (Paper 18), but the diagrams show the method and the variability of the historical sources, a variability that must be taken into account in assessing when and whether the sunspot cycle was particularly weak. A synthesis of the classical and Chinese evidence is in progress.

In the fourth century A.D. the Chinese auroral observations, with sunspot records now available in Paper 5, made it possible to sketch the sunspot curve and the diagram has been included in Paper 15. Paper 16 (see also Schove, 1950) shows that in the West the "visions of saints" can be utilized in a similar way; those seen when Saint Columba died (7 June 597) were probably auroral.

In the late eleventh century aurorae were not numerous (Figs. 4 and 8 in our Introduction) but Chinese observations were very thorough and some conclusions from our study of Sung aurorae are included in Paper 17. Professor Ho has been successful in finding many displays that have not even been included in the lists of Keimatsu (1970–1976) or of Dai and Chen (1980) who have written the best printed catalog of Far Eastern aurorae. The principal Western catalog is that by Link (1962, 1964). Some additional medieval displays are cited in Dall'Olmo (1979) and in Papers 16 and 38 (see also Figs. 4 and 5 in our Introduction). The years and sources of the Greco-Roman displays have been listed by Stothers (Paper 18); in another paper Stothers (1979) found possible solutions with alternative periodicities of 15.0, 8.7, or 11.1 years (although from a small set of selected data he preferred 11.5 years) by assuming that the standard deviation if divided by the correct period should be at a minimum.

In all these catalogs, as the compilers point out, there are inevitably ambiguous cases; some quotations in the lists appear also in *comet* catalogs (e.g., Barrett, 1978) and many so-called aurorae especially in Keimatsu (1970–1976; see also Papers 13 and 14) are meteoric. Separate studies of other phenomena using the Spectrum of Time collections (cf. Papers 13 and 14) have suggested the exclusion of examples believed to be: *comets* (e.g., 5 B.C., A.D. 442, 467, and a *supernova* in 1182) (cf. Schove, 1975); *shooting stars* (e.g., 687 B.C., A.D. 745, 765, 934, 1093–1094) (cf. Schove, 1972); *incorrectly dated secondary reports* (e.g., 402, 434, 580, 911, 1003, 1005, 1114, 1197). A chronology of these and other natural phenomena is being published separately (Schove, in progress) as part of the Spectrum of Time project. A further analysis by Schove of the geographical relation of aurorae and sunspot cycles will appear in *Annales Geophysicae* (Zurich, 1983).

Decadal indices of auroral activity are estimated in Paper 19 (cf. Fig. 8 in our Introduction). Botley (1957, 1967) has described and listed some dated examples of *aurora tropicalis,* and displays in lower latitudes are very useful in pinpointing the dates of the maxima. Indeed, further records of strange visions (see Eather, 1980) are likely to add to our knowledge of past sunspot cycles.

REFERENCES

Barrett, A. A., 1978, Observations of Comets in Greek and Roman Sources Before A.D. 410, *Royal Astron. Soc. Canada Jour.* **72:**81–106.

Botley, C. M., 1957, Some Great Tropical Aurorae, *British Astron. Assoc. Jour.* **67:**188–191. (Since A.D. 1700.)

Botley, C. M., 1967, Unusual Auroral Periods, *British Astron. Assoc. Jour.* **77:**328–330. (Since A.D. 300.)

Bruzek, S., and C. J. Durrant, 1977, *Illustrated Glossary for Solar and Solar-Terrestrial Physics,* Reidel, Dordrecht, Holland, 204p.

Dai, N., and M. Chen, 1980, Chronology of Historical Aurora Data in China, Korea, and Japan, *Kejishiwenji* **6:**87–146. (In Chinese.)

Dall'Olmo, U., 1979, An Additional List of Aurorae from European Sources from A.D. 450 to A.D. 1466, *Jour. Geophys. Research* **84:**1525–1535.

Deehr, C. S., and J. A. Holtet, eds., 1980, *Exploration of the Polar Upper Atmosphere,* Reidel, Dordrecht, Holland, 498p.

De Mairan, J. J. d'O, 1754, *Traité Physique et Historique de l'Aurorae Boreale,* 2nd ed., Académie Royale des Sciences, Paris, 592p. (1st ed., 1733.)

Eather, R. H., 1980, Majestic Lights: the Aurora in Science, History, and the Arts, American Geophysical Union, Washington, D.C., 323p.

Fritz, H., 1873, *Verzeichniss beobachteter Polarlichter,* C. Gerold's Sohn, Wien, Austria, 265p.

Fritz, H., 1881, *Das Polarlicht,* Brockhaus, Leipzig, 348p.

Frobesius, J. N., 1739, *Nova et antique luminis atque aurorae borealis spectacula,* C. F. Weygand, Helmstadt, Germany, 70p.

Keimatsu, M., 1970–1976, A Chronology of Aurorae and Sunspots Observed in China, Korea, and Japan, *Kanazawa Univ. Ann. Sci.* **7:**1–10; **8:**1–16; **9:**1–36; **10:**1–32; **11:**1–36; **12:**1–40; **13:**1–32.

Link, F., 1962, Observations et catalogue des Aurorae Boreales apparues en Occident de -626 a 1600, *Geofys. Sbornik* **10:**297–390.

Link, F., 1964, Observations et catalogue des Aurores Boreales apparues en Occident de 1601 a 1700, *Travaux de l'institut geophysique de l'academie Tchecoslovaque des Sciences* **212:**501–550.

Pilgram, A., 1788, *Wetterkunde,* vols. 1 and 2, Wien, Austria.

Schoening, G., 1760, Nordlyset Aelde, *Skrift Kjobenh.* Selsk 8, Copenhagen.

Schove, D. J., 1948, Sunspots and Aurorae, 500–250 B.C., *British Astron. Assoc. Jour.* **58:**178–190, 202–204.

Schove, D. J., 1949, Chinese "Raininess" Through the Centuries, *Meteorol. Mag.* **78:**11–16.

Schove, D. J., 1950, Visions in North-West Europe (A.D. 400–600) and Dated Auroral Displays, *British Archaeol. Assoc. Jour.,* ser. 3, **13:**34–49.

Schove, D. J., 1953, London Aurorae of A.D. 1661, *British Astron. Assoc. Jour.* **63:**266–270, 321–325. (*Cf.* **62:**38–41, 63–66.)

Schove, D. J., 1960, Chronology of Natural Phenomena in East and West, *Archives Internat. d'Histoire des Sciences* **52–53:**263–268.

Schove, D. J., 1961, The Spectrum of Time, 1420–1540, *British Astron. Assoc. Jour.* **71:**320–322. (*Cf.* Paper 14.)

Schove, D. J., 1972, The Leonids: Who Saw Them First? *Sky and Telescope* **43:**156–157.

Schove, D. J., 1974, Chronology and Historical Geography of Famine, Plague and Other Pandemics, *Internat. Congress History and Medicine (23rd) Proc.,* pp. 1265–1272.

Schove, D. J., 1975, Comet Chronology in Numbers, A.D. 200–1882, *British Astron. Assoc. Jour.* **5:**40, 407.

Schove, D. J., 1979, Sunspot Turning Points and Aurorae Since A.D. 1510, *Solar Physics* **61:**423–432.

Schove, D. J., in progress, *The Spectrum of Time.*

Schroeder, W., 1979, Auroral Frequency in the Seventeenth and Eighteenth Centuries and the Maunder Minimum, *Jour. Atmos. and Terrest. Physics* **41:**445–446.

Siscoe, G. L., 1980, Evidence in the Auroral Record for Secular Solar Variability, *Rev. Geophysics and Space Physics* **18:**647–658.

Stothers, R., 1979, Solar Activity Cycle During Classical Antiquity, *Astronomy and Astrophysics* **77:**121–127.

Whitelock, D., trans., 1955, *English Historical Documents I, c. 500–1042,* Eyre & Spottiswood, London, 867p.

Yunnan Observatory, 1977, A Recompilation of Our Country's Records of Sunspots Through the Ages and an Inquiry into Possible Periodicities in Their Activity, *Chinese Astronomy* **1:**347–359. (First published in Chinese in *Acta Astron. Sinica* **17:**217–227, 1976.)

12

AURORAL FLUCTUATIONS

J. J. d'O. De Mairan

*This excerpt was translated expressly for this Benchmark
volume by D. J. Schove, St. David's College, England, from
Traité Physique et Historique de L'Aurore Boréale, 2nd ed.,
Academie Royale des Sciences, Paris, 1754, p. 264.
Copyright ©1979 by D. J. Schove.*

Descartes did not omit to consider a solar atmosphere, consisting of very
rarified matter, which he called *air,* comparable to the air that surrounds
our earth; he supposed it to extend (from the sun) to the sphere of Mercury
and even beyond, and ascribed to it a similar origin. He believed that
sunspots in dissipating added much new material to this solar air. [The
observation of prominences could have provided Descartes with evidence
for this.]

What appears to favor this idea is that for the last 5 or 6 years [i.e., the
maximum half of the 11-year cycle] when the aurorae, as often happens,
have become very frequent, and, according to our hypothesis, [should be
associated with] great extensions of this *air,* sunspots have likewise been
numerous.

[*Editor's Note:* De Mairan proceeds to emphasize the frequency of aurorae
and sunspots in the early seventeenth century until the 1620s, and their
rarity in the period 1650 to 1670.]

13

1948 by The British Astronomical Association
Reprinted from pages 178–180, 185, and 190 of *British Astron. Assoc. Jour.*
58:178–190 (1948)

SUNSPOTS AND AURORAE, 500–250 B.C.

D. J. Schove

The Earliest Examples

The first record of a sunspot, according to the usual view, occurs in the Chinese archives in A.D. 188 and the first aurora in Roman sources of A.D. 194. Nevertheless, there exists the little known work of Fritz (1873-96) who begins his catalogue of auroræ with the year 503 B.C. but whose authorities are not very satisfactory. For dates more recent than about 220 B.C., however, new evidence has now come to light from both Chinese and Roman records. Professor Homer H. Dubs is now publishing (1938 +) his translation of Pan Ku's "History of the Former Han Dynasty" and Dr. F. B. Krauss "An Interpretation of the Omens, Portents and Prodigies recorded by Livy, Tacitus and Suetonius". The discussion of the so-called Auroræ in Fritz forms the basis of the first part of this paper; the new evidence from the other two works is the basis of the second part.

Written history was confined in early times to latitudes between 45° and 30° N. In such latitudes displays of auroræ (northern lights) are unusual. Commentators on Roman and Chinese histories do not therefore expect to find auroræ in the records. Lights seen in the sky have been otherwise interpreted even by astronomers; Kugler believed that the "fire" seen in the heavens by the Romans was meteoric dust accompanied by glowing meteorites and Eberhard (1933) suggested volcanoes or dust-storms as the origin of similar phenomena in China. Professor Dubs specifically noted a Chinese observation of a sunspot but made no mention of northern lights; he suggested to me however, that any displays seen would have been called "emanations" and has kindly allowed me to use his card-index for the investigation.

A recent map of auroral frequency 1700—1942 (*Nat. Geog. Mag.* 1947, p. 676) indicates that auroræ occur about once in ten years in the Mediterranean and hardly ever occur in China. Nevertheless, isolated examples occur of almost world-wide auroræ. Examples have occurred in 1859, 1870, 1872, 1909, 1921 and 1938. In the display of 1938 January 25, fire-engines even in S.E. Europe were called out to deal with supposed "conflagrations" beyond the horizon; in 1901 and 1921 polar lights (Botley, 1947) "belied their name with a vengeance by being visible at Singapore, 70 miles only from the equator".

The "Diagnosis" and dating of Aurorae 500—250 B.C.

Chronologically, there are two well-marked periods corresponding to the two parts of this paper. In the earlier period, 500—250 B.C., there are no records of sunspots and the auroræ are difficult to "diagnose" and to date. After 250 B.C. datable examples of both sunspots and auroræ can be found.

Aurorae in early records cannot always be distinguished from comets, meteoric and volcanic phenomena. An attempt has accordingly been made in this paper to include all these events and to sort them out chronologically, giving the reference to the original source.

102

The difficulty of dating remains. The dates of eclipses can be checked by calculation. Chinese history from 720 to 481 B.C. can be confirmed in this way; Greek history is likewise confirmed by frequent eclipses after 480 B.C. and occasional earlier ones. Roman dating can be checked this way only since 217 B.C., the earlier eclipse of the poet Ennius (*fl. c.* 200 B.C.) quoted by Cicero being variously identified as that of 400 B.C., 388 B.C. or 288 B.C.

The difficulties of Roman chronology are discussed in the "Cambridge Ancient History" (Vol. 7, Chapter 10). Cary (1935, p. 48 and 44) summarizes the position as follows:—

> "The expulsion of the Tarquins . . . dated 509 . . . should be assigned to 507; the capture of Rome by the Gauls should be post-dated from 390 to 387 and the Licinian Rogation from 367 to 362". The dates "stand fairly well in accord with the authentic chronology of the Greek historians. The discrepancy never exceeds eight years, and at points is narrowed down to two or three years . . . the traditional dates have been retained because . . . none of the modern substitutes have found general acceptance".

The Roman plague of 433 B.C. appears to synchronize with the great plague of 430 B.C. in the Greek world; Altheim claims that this supplies "a certain confirmation" of Roman dating, but there are earlier Roman plagues ascribed to 464—459 and 454—443 which have no recorded counterparts in the Greek world. Accordingly, if the astronomical phenomena to be discussed below can be identified or dated, it will prove significant historically.

[*Editor's Note:* An early version of the Spectrum of Time for 500–250 B.C. has been omitted. A better list of aurorae is now given by Stothers (Paper 18) who shows that the phenomena of c. 500 B.C. and c. 333 B.C. should be deleted from Fig. 1.

Other early alleged aurorae are as follows: 649 B.C. (China, probable), c. 627 B.C. (visions of Jeremiah 1:13), c. August 594 B.C. (vision of Ezekiel 1:1), February 519 B.C. (vision of Zacharius), 494 B.C. (China), 481–480 B.C. (Pliny's horn-shaped phenomenon I prefer to identify with the Chinese comet), 467 B.C. (Anaxagoras), c. 460 B.C. and 457 B.C. (provisionally corrected dates of Livy's aurorae). The remainders of the years given are 8, 8, 7, 5, 7, 9 or 10, 12, 8 and 11 (see Paper 20). This persistent tendency for remainders near 8 suggests that most of the phenomena are genuine aurorae. The 494 B.C. event comes from the partly fabricated *Bamboo Books* but these are now believed to have incorporated a genuine lost source for such phenomena.]

Bibliography.

Altheim, F.	(1938)	"A History of Roman Religion." (Translation)
Botley, Miss C. M.	(1947)	"Polar Lights." (Tunbridge Wells)
Cary, M.	(1935)	"History of Rome." (London)
Chambers, G. F.	(1888)	"Handbook of Astronomy," 4th edition.
	(1909)	"The Story of the Comets.'
Chavannes, E.	(1899)	"Memoires Historiques," Vol. III.
Chizhevsky, A.	(1924)	"Physical Factors of the Historical Process." (In Russian). (Kaluga)
Diels, ed. W. Kranz	(1934-35)	"Fragmente der Vorsokratiker," 5th edition. (Berlin)
Dubs, H. H.	(1928)	"The Works of Hsuntze."
Dubs, H. H.	(1938+)	Translation of Pan Ku's "History of the Former Han Dynasty," 3 vols.
Eberhard, W.	(1933)	"Beit. z. Kosmologischen Spekulation der Chinesen der Han Zeit." (Berlin)
Farrington, B.	(1939)	"Science and Politics in the Ancient World." (London)
Freeman, Kathleen	(1946)	"The Pre-Socratic Philosophers." (Oxford)
Fritz, H.	(1873)	"Verzeichniss beobachter Polarlichter." (Vienna)
	(1881)	"Das Polarlicht." (Leipzig)
	(1896)	"The periods of solar and terrestrial phenomena." (Zurich)
		English translation in *Monthly Weather Review* (1928), **56**, 401—407.
Galle, J. G.	(1894)	"Verzeichniss der Elemente der Bisher Berechneten Cometenbahner."
Gartlein, C. W.	(1947)	*Nat. Geog. Mag.*, **92**, 673—704.
Ginzel, Dr. F. K.	(1899)	"Spezielle Kanon der Sonnen und Mondfinsternisse." (Berlin)
Heath, Sir T. L.	(1932)	"Greek Astronomy." (London)
Hellman, Dr. C. Doris	(1944)	"The Comet of 1577." (N. York)
Hoang, P.	(1925)	"Varietés Sinologiques," No. 56. (Shanghai)
Krauss, F. B.	(1930)	"An Interpretation of the Omens, Portents and Prodigies recorded by Livy, Tacitus and Suetonius." (Philadelphia)
Lycosthenes, Conr.	(1557)	"Prodigorum ac ostentorum chronicon" (in Latin) Basle. Brit. Mus. 581 i. 5.
Nihusius, B.	(16??)	"De divinatione sacra et profana." (This is the work as quoted by Fritz, but I can find no other record of it).
Obsequens, Julius	(1552)	"De Prodigiis." 1st printed edition at Basle edited by Lycosthenes. The 1720 Leiden edition (B.M. reference 955 *c.* 19) was used in this paper.
Plummer, H. C.	(1942)	"Halley's Comet and its Importance," *Nature*, **150**, p. 249.
Schove, D. J.	(1947)	*Terr. Mag.*, **52**, 233—238. "The Sunspot Cycle Before 1750".
	(1948)	*Popular Astronomy* May. "Sunspot Epochs 188—1610."
Stiftar, V. Th.	(19??)	in *Hermes*, **7**, 57—59, 151—153, 342—346, 364—369, 395—399, 420—428. "The Prodigies in Titus Livius" (in Russian and not consulted).
Williams, John	(1871)	"Observations on Comets (in China) from 611 B.C. to A.D. 1640." (London)

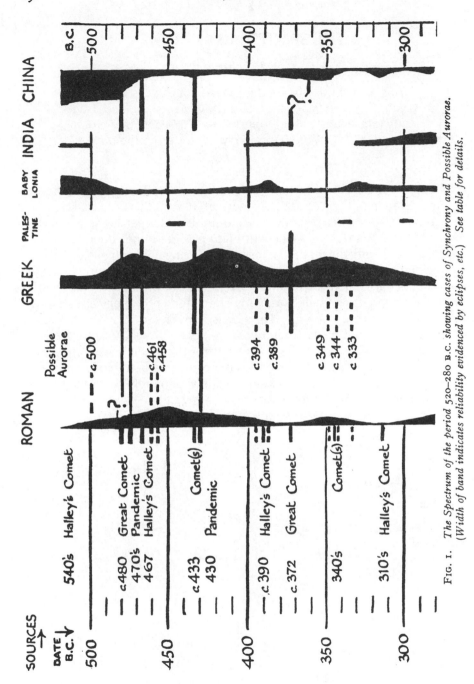

Fig. 1. *The Spectrum of the period 520–280 B.C. showing cases of Synchrony and Possible Aurorae.*
(Width of band indicates reliability evidenced by eclipses, etc.)   See table for details.

14

THE SPECTRUM OF TIME

D. J. Schove

This article was abridged expressly for this Benchmark volume by D. J. Schove, St. David's College, England, from Croydon Astron. Assoc. Jour. 9:1-6 (1967). The figure that appears in this article has been reprinted from the original.

The Arts and the Sciences meet in the Spectrum of Time, and those who are interested in both often ask how they can help in this project. The research has been described in various journals and is partly explained by Figure 1, but this article has been written to provide some concrete suggestions. Briefly the method consists of reading any early historical source and extracting from it all those dated or dateable astronomical and meteorological events and portents—even the extraordinary or impossible ones—and submitting them to as many checks on chronology as can be made before the results are typed and classified. The method was fully explained at international conferences (see Schove, 1960). The separate records are often obscure in the original context; a marvellous portent in one chronicle might seem to historians to be any one of the following: a comet, a fireball, an aurora, a meteoric display, or merely distant lightning. The comparison of information obtained from East and West usually enables us to determine what the phenomena really was and its true date.

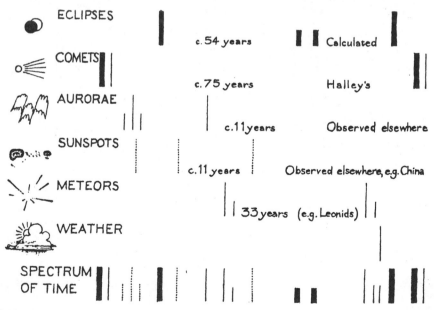

Figure 1. Astronomical and meteorological events "add up" to the Spectrum of Time.

In this way we can construct what I have termed (Schove, 1948, pp. 202–204) the *Spectrum of Time*—a calendar of those world events that were actually observed in both the East and the West.

The classification of the various observations is made as follows: *Astronomical Events*—eclipses (finsternisse), comets, shooting stars or meteors (etoiles filantes), aurorae (nordlicht) and sunspots (sonnenflecken), novae, occultations, and so forth; *Other Events*—meteorological, earthquakes (erdbeben), famines (hungersnot, disette, carestia), plagues and epidemics (pest), locusts, and biological phenomena; *Natural Records*—treering dating (dendrochronology), clay-varve dating, fluorine and carbon dating.

Most early sources in England have been well combed, although those interested might help typing or translating the Latin notebooks prepared by Britton (1937). However, even without a knowledge of languages, one can still participate by investigating some of the little known early English and Scottish diaries listed in Matthew's (1950) book. There are numerous manuscript sources in the Public Record Office but for this it is necessary to be able to read early scripts. Whether we are linguists or not we must have the imagination to interpret the words of the diarist or chronicler into the meteorology of the sky.

[*Editor's Note:* Material has been omitted at this point. I am continuing a further discussion of the matter above in *The Spectrum of Time,* which is now in progress.]

The more familiar writers have now been fairly well studied including most early Irish and pre-900 Continental sources. Byzantine sources—for those who read Greek—still hold much material. Syrian and Arabic sources have been used but the very full Chinese sources have not yet been exhausted and there are diaries and chronicles in various obscure languages for those who have the ability to translate them. Even Hispanic sources still contain unused information about the weather in various parts of the world in the period 1500–1800.

Rules of the Game

The following rules are suggested for those who are willing to participate:

1. Do not choose the earliest sources in your language: unless the language or source is obscure, these have probably been used already. On the other hand, chronicles usually cease to be interesting for comets after 1610 and for aurorae after 1715.
2. Find a contemporary chronicle or diary and read it back to front! The best bits are usually in the second half, which will have been kept up to date "with the flow of time." The first part is often a rehash or summary of some other chronicler's work.

3. Inform the author of the time range of the chronicler under investigation. In return he can provide the true dates of comets, and so forth, and the dates of other items to watch for—the *spectrum* of time in question.
4. If possible, a copy of the full quotation in the original language is useful as well as the English translation (which should be a literal and full one).
5. If you can copy type in Latin there are various sources, including the manuscript collections, which have been kindly presented.
6. If your only language is English and you can copy type or write clearly there is still material in English that can be traced through typescript indexes available at the British Astronomical Association for the *British Astron. Assoc. Jour.* and for the *Observatory.* Certain diaries in Scotland and England need full investigation.

The sunspot cycle in the fifteenth century thus needs to be investigated thoroughly by considering the evidence for aurorae in various chronicles. Baltic chronicles include the detailed Polish sources listed by Walawender (1932, 1935), which are very good on the second half of the century (in Latin) and are in the British Museum. The city chronicles of the Hanseatic towns are in German and Latin and are often listed in Weikinn's (1958) catalog of floods, and a few of these chronicles are available at the Institute of Historical Research, London. Unfortunately the aurora at that time was not called by any particular name and it is necessary to record the date of all sorts of phenomena—fires, armies, weapons, battles in the sky, rosy glows in the northern horizon, and various other peculiar visions as some of the Spectrum of Time articles (see References) make clear. Link's recent catalogs (1962, 1964) are mainly in Latin but they do not become full until after about 1520. Meanwhile, there are thought to be further Far Eastern records in the Ming encyclopedias and provincial Chinese histories. (In addition to the many Chinese local histories there is the "veritable daily record" kept at Peking. The Chinese Academy of Sciences is planning to publish its results shortly.)

REFERENCES

Britton, C. E., 1937, *A Meteorological Chronology to A.D.* 1450, His Majesty's Stationery Office, London.

Link, F., 1962, Geofys. Sb. **10:**297–390.

Link, F., 1964, Geofys. Sb. **12:**501–550.

Matthews, W., 1950, *A Bibliography of British Diaries,* University of California Press, Berkeley, 300p.

Schove, D. J., 1948, Sunspots and Aurorae, 500–250 B.C. *British Astron. Assoc. Jour.* **58:**178–190, 202–204.

Schove, D. J., 1960, Chronology of Natural Phenomena in East and West, *Archives Internat. d'Histoire des Sciences* **52–53:**263–268.

Walawender, A., 1932, 1935, *Kronika Klesk Elementarnych w Polsce w latach 1450–1586,* 2 vols., London School of Economics Library.

Weikinn, C., 1958, *Quellentexte zur Witterungsgeschichte Europa von der Zeitwende bis zum Jahre 1850,* Akademie-Verlag, Berlin, 531p.

15

Reprinted from pages 105–107 and 110–111 of *Am. Orient. Soc. Jour.* **87**:105–112 (1967)

CHINESE RECORDS OF SUNSPOTS AND AURORAE IN THE FOURTH CENTURY A.D.

D. J. Schove and P.-Y. Ho

[*Editor's Note:* In the original, material precedes and follows these excerpts including a glossary of Chinese characters. Reference is made to uncertain auroral evidence from visions seen near the Mediterranean. Additional Chinese aurorae are now dated 1 January 303, 20 June 305, and June 306.

The following acknowledgements appear on page 112 in the original: "Our thanks are due to Dr. S. Kanda, Mr. D. Leslie, Dr. Wang Ling, and Dr. Joseph Needham, for help with sources and translations."]

However, the principal source for the reliable evidence is the *Chin Shu*,[a] the official history of the Chin Dynasty in China. A translation of the Astronomical Chapters had been made by one of us (Ho, 1966), and a copy of the typescript kindly presented to the other author by Professor Ho provided the basis for the present article.[4]

VERY STRONG SOLAR MAXIMUM A.D. 302

The first sunspot record relevant to the fourth century is that of A.D. 299, and it is the first known spot belonging to the "very strong" cycle with a maximum in 302 or 301.

MANY SUNSPOTS

299 Feb/Mar Ho (1966) p. 161

During the first month (17 Feb. to 18 Mar. A.D. 299) of the 9th year (of the Yüan K'ang[b] reign-period of the Emperor Hui-Ti)[c] sunspots resembling a number of swallows in flight appeared in the sun. They were seen for a few days before they (finally) disappeared. This phenomenon was interpreted by Wang Yin[d] (One of the Emperor's official historiographers. See note by Ho, 1966, p. 161) as a presage of the abdication and death of the two Emperors Min (-Ti)[e] and Huai (-Ti).[f]

---

[4] Ho Peng-Yoke, *The Astronomical Chapters of the Chin Shu* (Paris, 1966).

This record comes from Chapter XII: Section on Astronomical Changes (cf. Ho, 1966, p. 160) (c) Haloes and Sunspots. The next two items in that section, on 7 Jan. and 5 Nov. 300, appear to relate to haloes and dust-haze respectively.

SUNSPOT(S)

301 Jan. 19:   Ho (1966) p. 161

On a *kêng-hsu* day in the twelfth month (of 300) dark vapour (i. e. sunspots) appeared in the sun. According to the *I Chuan*[g] of Ching Fang, these abnormalities arise from failure to appease the heavens, and hence dark vapours (sunspots) appeared in the sun.

It should be noted that there is no record for a sunspot in the solar year 300.

SUNSPOT(S)

301 Oct. 20   Ho (1966) p. 161

On a *chia-shen* day in the ninth month of the first year of the Yung-Ning[h] reign-period *hei tzu*[1] (sunspots) appeared. According to the prognostication of Ching Fang's *I* ......sunspots appear whenever government officials fail to prevent the emperor from taking the wrong course, and allow it to come to the knowledge of the people, thus bringing about public disapproval. Sunspots are also said to occur whenever government officers smirch the good name of the emperor.

SUNSPOT(S)

302 Dec. (Before 4 Jan. 303)   Ho (1966) p. 161

In the eleventh month (6 Dec. to 4 Jan. of the 1st year of the T'ai-An[j] reign-period dark vapour (*hei ch'i*[k]) appeared.

AURORA (or METEORITE)

303 Sep. 25:   Ho (1966), p. 149

In the earlier part of the same section on Astronomical Changes, Part (a), the heading of which is translated by Ho as " Auroras and Thunder-like Noises" we read:

On a *kêng-wu* day in the eighth month of the 2nd year of the T'ai-An reign-period the heavens divided at the centre into two parts, accompanied three times by sounds resembling thunder (aurora borealis and " swish of the aurora" suggests Ho p. 149). This was a sign that the emperor had failed in his duties and that insolent Ministers had seized power.

The mention of a thunder-like noise, if it is really associated, would make the meteoritic interpretation more likely, but other meteorites of this decade are correctly relegated to the section (i) entitled Meteors and Meteor Streams (303 Dec. 6; 304 Sep. 15; 305 Nov. etc. See Ho, p. 245 ff.), and it seems that the sounds may well have been rumours. Phenomena are not as fully described in the fourth century as in the eleventh (cf. Schove

and Ho, 1958) and, even in the comet section there is a group of seven minor "comets" from 299 to 305 Sep. which may include auroras. (Section [h] is entitled "Ominous Stars and Guest Stars," and translated by Ho as "Comets and Novae." See Ho, 1966, p. 232 ff. cf. summary in Ho, 1964, pp. 158-9.)

AURORA (?)

303 Dec. 27   Ho (1966), p. 251

On the night of a *jên-yin* day in the twelfth month of the 1st year of the Yung-Hsing[l] reign-period a red vapour (aurora) appeared across the heavens accompanied by a crashing sound.

The description of what was seen appears to have been an aurora not a meteor. In this case the sound was presumably either subjective or a subsequent addition, but the meteoric possibility remains.

AURORA

304   Chapter *Wu Hsing Chih*[m]

In the official Chinese Histories there are chapters *Wu Hsing Chih* ("Five Element Chapters") which include obscure portents such as the aurora. This chapter has not been included in Ho's translation, but was investigated direct by DJS with the help of Dr. Wang Ling. A reference in 304 to "At night a light like fire" is almost certainly a reference to an aurora.

SUNSPOTS

304 Dec/305 Jan: Ho (1966), p. 162

Sunspots must still have been numerous in 304, as the next solar record, the last in this cycle, reads:

In the eleventh month (14 Dec/11 Jan) of the first year of the Yung Hsing reign-period the sun was divided by dark vapour.

A solar portent lasting from 4th to the 6th June 306 follows in the text, and may refer either to red sunsets or to the effect of dust-haze drought (Ho, p. 162).

AURORA

305 Nov. 21:   Ho (1966), p. 251

In the section (j) Extraordinary Clouds and Vapours, we read:

On a *ting-ch'ou* day in the tenth month of the second year a red vapour appeared in the north stretching across the heavens from west to east......(Ho, p. 251).

This is clearly an aurora, although it would be rather late in the cycle for an auroral arc with the present magnetic dip. The capital of the Western Chin was Lo-yang[n] (Approx. 34° 43' N and 113°

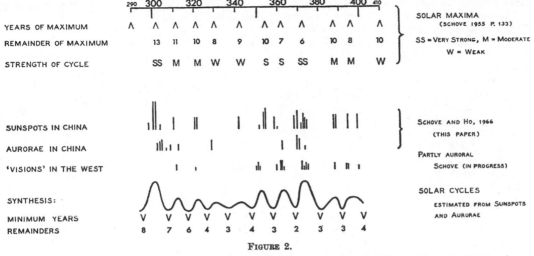

FIGURE 2.

Pattern of the Solar Cycle in the 4th Century A. D. inferred mainly from the Chinese evidence in this paper.

30′ E) in modern Honan. (Ho, 1966, p. v. and cf. p. xiv for the political confusion of A. D. 291-306).

[Editor's Note: Material discussing the spots and aurorae of A.D. 309 to A.D. 395 has been omitted at this point (see Paper 5). New radiocarbon evidence confirms the trends of Figure 2.]

Chinese aurorae have also been discussed by Needham,[9] who explains why it is difficult to assemble the aurorae scattered in Chinese sources. The phenomenon has various forms, and it was not clearly recognized as an entity, so that it "went by many names, e.g. *ch'ih ch'i* (red vapour), *pei chi kuang* (north polar light) etc., etc. They are classified in encyclopaedias in different places, appearing in the Five Element Chapters of Dynastic Histories (e.g. A. D. 882), as portents associated with the lives of Emperors, in divinatory literature (e.g. the lost *Yao Chan* or "Divination by Weird Wonders," temp. Han dynasty but quoted A. D. 718 by *K'ai-Yüan Chan Ching*).

The less scientific literature of double rainbows (cf. A. D. 1070 Needham, op. cit., p. 473), earthshine (*Ching-hsing* or *Te-hsing*, cf. Needham. III. 422/3) could be investigated. A check of the dates of this latter phenomenon (e. g. cf. *Shih Chi*, ch. 27, p. 330, *Chin Shu*, ch. 12, p. 4a, *Sung Shih*, ch. 56, p. 21a) against the known auroral maxima might reveal that a significant proportion were auroral.

Haloes were usually distinguished with scientific exactness by the Chinese (cf. Needham, III, p. 476), but blood rain and red rain in Europe is often auroral and date lists of red rain (e. g. in the *T'u Shu Chi Ch'eng, Shu cheng tien,* ch. 143) might be similarly checked.

The auspicious birds noted by Schafer (1963) [10] were checked in this way, and eight appearances coincide with the dates of sunspot maxima given by Schove .(1955). Therefore, it seems possible that one or two of the (nocturnal) birds were really auroral; Schafer points out how some of the

[9] J. Needham, " Science and Civilization in China," III (1959), 482-3.

[10] E. H. Schafer, " The Auspices of Tang," *Jour. Amer. Oriental Soc.*, 83/2 (1963), 197-225.

birds (cf. the phoenix) were regarded as magical and he notes that some were red or flame-coloured. The description of *luan* and *feng*, and the phenomena of 517 B.C., 80 B.C., 55 B.C., A.D. 238, 502, 615, 647, 676, 903, sound promising. So far, however, no precise date has been found in Schafer's list that synchronizes with a specific aurora and many auspicious birds have been shown by Schafer to refer to particular species. (Schafer, 1963, Index, p. 225).

Sunspots are much more readily recognized (cf. Needham, III, p. 434/6), but, as this paper shows, the list of Kanda is not complete, and the T'ang occurrences before 807 have not been traced in either the *T'ang Shu* or the *Chiu T'ang Shu*.

Hail and thunder were once thought to be associated with solar cycles and were even used in an early attempt (by Fritz, cf. Schove, 1955) to date solar maxima. The relationships between weather and sunspots are now known to be too complex to justify the use of weather (or e.g. locusts) as evidence.[11]

The climatic peculiarity of the 4th century appears to be drought: the drought: flood ratio was twice as great as that of any other century according to Chu K'o-Chen (Needham, III, Table 39, p. 473). The droughtiness of the mid-fourth century, at least in the 350's, is confirmed by more detailed investigations.[12] Migrations from the Steppe to the Sown were perhaps associated with this dry phase.

Radio-carbon and pollen evidence is however required before we can generalize about the climate as a whole.

The solar maxima, whatever our uncertainties about the climate, are dated clearly from the Chinese evidence above. Indeed, since this paper was written the writer finds that Kanda, using much of the above information had previously come to similar conclusions as to the dates of maxima (302/6, 311/3, 322, Missed, 342, 354, 359, 372, 388 and 398), although Kanda's dates are merely means and he did not estimate dates of

minima.[13] In other centuries Kanda's conclusions are often incorrect.

TABLE I

SUNSPOT MAXIMA AND MINIMA

MAXIMUM	"REMAINDER"	STRENGTH Strong Moderate or Weak	MINIMUM	AURORAL NUMBERS
c. 290	13	M or S	c. 296 (298)	50
302	13	SS	307	70
(312)	12	M	(316)	60
321	10	M	(325)	50
(c. 331)	9	W	(335)	30
c. 342	9	WW	(349)	40
(353)	9	S	(357)	60
c. 362	7	(M)	(367)	60
(371)	5	SS	(c. 381)	70
c. 387	10	M	(392)	50
(c. 397)	(9)	M	c. 404	60
c. 410	10	W		

In conclusion, the table above summarizes the results. Slight revisions from the 1955 table (Schove, 1955, p. 133) have been made, but these exceed one year only in the case of the 298 minimum which was originally dated 296. These revisions are made less from new observational evidence (which is still very scanty) than from new knowledge of the shape of sunspot cycles since 1850 (cf. Gleissberg in *Naturwissenschaft*, 1960, fig. 2, p. 2). The estimated "auroral numbers" per decade (of form 01-10) are derived from Schove, 1962.[14] There is no clear evidence of the 22.2 year cycle in this century, but Gleissberg's "80-year cycle" appears to have peaks c. 300 and c. 370 which are maxima as far as intensity is concerned and minima of length.[15] The fourth century, like every even century since, represents a distinct maximum (say c. 340) of the "200-year" cycle. Sunspots and aurorae from Far Eastern sources are proving very useful in throwing light on these various supercycles and further information would be greatly appreciated, especially for the third, thirteenth and fifteenth centuries A.D. and for periods 590-640, 770-805, 870-900, 1090-1120 and 1300-1350.

[11] Cf. D. J. Schove, "Solar Cycles and Equatorial Climates," *Geologische Rundschau*, Bd. 54, H. 1 (1965), 448-477.

[12] Cf. Schove, "Chinese 'Raininess' through the centuries," *Meteorological Magazine* 78 (1949), 11-16, citing the work of Yao; also the *Chin Shu*, ch. 27, p. 6b, 106-116.

[13] S. Kanda, *Proc. Imp. Acad. Japan*, 9 (1933), 293.

[14] D. J. Schove, "Auroral Numbers since A.D. 300," *Jour. Brit. Astron. Assn.*, 72 (1962), 30-34.

[15] Cf. W. Gleissberg, "80-year Cycle in Auroral Frequency Numbers," *Jour. Brit. Astron. Assn.*, 75 (1965), 230, and 76 (1966), 266.

112

16

Copyright © 1950 by the British Archaeological Association

Reprinted from pages 34–38, 43–46, and 48–49 of *British Archaeol. Assoc. Jour.*, ser. 3, **13**:34–49 (1950)

VISIONS IN NORTH-WEST EUROPE (A.D. 400–600) AND DATED AURORAL DISPLAYS

By D. JUSTIN SCHOVE, B.SC., F.R.MET.S.

INTRODUCTION

THE early records of northern lights are relevant to Chronology, to Science, and to History. Dates of the major displays, if determined for one chronicle, can, like comets and eclipses, be used to date others, and have thus significance as chronological criteria. The dates have scientific significance inasmuch as they indicate the variations of the sun and sunspot cycle; exact dates at which aurorae are likely to have occurred can be deduced from the known observations of sunspots and from the known way in which both phenomena recur at four-weekly intervals. Moreover, years in which these spectacles would be possible or impossible can often be distinguished scientifically by the known behaviour of the eleven-year cycle. Thirdly, the fact that the observations were made at all is proof that the annalist was truly recording events as they occurred, and the way in which they are described reflects the spirit of his time. In a religious epoch the chronicler is filled with religious fear; in a military age he sees terrestrial battles aided by celestial armies; whilst at certain naturalistic periods he describes simply and accurately what he really saw.

AURORAL CYCLES: THE DAY, THE MONTH, AND THE ELEVEN-YEAR PERIOD

Modern newspapers, referring to the appearance of sunspots, often predict the possibility of an aurora for the night after the sunspot crosses the central meridian of the sun. Such a display does not occur until the stream of particles, shot out from the sun, reaches the earth's outer atmosphere, twenty-four hours later. A similar delay presumably occurred in ancient times; in any case many coincidences between sunspots and aurorae can be noticed. As an instance of this, the Chinese in the third moon of A.D. 510, the 19th day, record two black 'clouds' [*sic*] seen in the sun, 'clouds in the sun' being one of their more scientific terms for sunspots; in the same moon of the same year Chinese records state that 'a light shone from the NE. on to the (palace?) courtyard', which is one of their more scientific accounts of an aurora.[1]

---

[1] These two items are from Chinese sources translated separately by Mr. D. Leslie, B.Sc., and Mr. Wang Lin, Graduate of the Central University, Nanking. The sunspot observations referred to here have been based on a paper sent to me by Mr. S. Kanda and translated by Mr. Leslie as 'Eastern Records of Sunspots and Sunspot Cycles', *Tokyo Astronomical Observatory Report*, October, 1932. The auroral observations were translated from various early Chinese encyclopaedias belonging to Dr. Joseph Needham. My thanks are due to him and to Mr. Wang Lin for the information. The Chinese data will be published separately when the sources have been fully located. I have since realized that my choice was unfortunate. The sunspot date is already in Julian form and should therefore read *March 19*. The aurora date is correct but the third *moon* is about a month later. There are nevertheless plenty of genuine instances. 'Clouds' is more strictly 'vapours', translated by Dubs as 'emanations'.

A great aurora is nearly always associated with a sunspot aiming a beam of rays and particles at the earth. On similar occasions today we know that short-wave radio communications are upset, not only on the night the rays are directed at us, but on the following night, as the particles strike our outer atmosphere and produce the northern lights. A sunspot has recently been defined as the most powerful short-wave transmitter known. Before the telescope a sunspot was regarded as a portent, and in China, where they are frequently visible during the hazy weather of the winter months, they were carefully noted in the annals, although the connexion with aurorae was never suspected. There are, however, many accounts of both phenomena in Chinese, Japanese, Korean, European, and even Islamic historical sources.

A typical sunspot travels round the sun as it rotates. If it is sufficiently persistent it may, some four weeks later, aim a second beam of particles at the earth. For instance, the Chinese observed a sunspot on October 27th, 1171, and on November 23rd, 1171, observed what they took to be another—but this was doubtless the same spot, probably the spot responsible for the aurorae seen at some unknown date in Europe in that year. In the same way in Russia an aurora was seen on January 13th, 1548, and another on February 9th in the same year.

Sunspots and aurorae vary in harmony not only with this monthly cycle, but also with a greater cycle which, for reasons not yet known, is usually about eleven years. In China, for example, sunspots were recorded near the years 300, 311, and 322, and there is evidence of aurorae about the same years. Nine such cycles make almost an exact century, and for several centuries the pattern is similar; as will be shown, northern lights are recorded about the years 500, 511, and 522. The last two digits follow the eleven-times table.

After A.D. 800 the eleven-year cycle gets out of step with the centuries, and the new pattern in a particular century is inverted so that years like 1855 and 1866 are actually years when sunspots and aurorae do *not* occur. We can therefore be fairly certain that 1955 will, like 1944 and 1933, be a year *without* great sunspots and aurorae, at least in temperate latitudes.[1]

Historians require more precise dates than can be given by the eleven-year rule. Many of the greater displays of aurorae were international phenomena. A spectacle seen in the Islamic world is almost certain to have been seen in northern Europe and eastern Asia on the same night. The exact date given in a Japanese chronicle can therefore be used for checking the chronology of Christian annals. The interpretation of aurorae by contemporaries is so varied that, but for the coincidence in time, it would be difficult to believe that all the phenomena were of the same nature. Fires in the sky, visions of angelic hosts, 'chasms' (Lat. *chasma*), flying dragons, showers of blood, are but a few of the European interpretations. Some of the 'flying saucers' recorded in the last few years undoubtedly belong to this same category. Readers can gain an insight into the vivid visions described by the ancients by studying the coloured

[1] See D. J. Schove, *Popular Astronomy*, lvi (May 1948), 133–8: 'Sunspot Epochs A.D. 188–1610'. Table I gives approximate dates of maxima since A.D. 300. For this I used what is the inaccurate but still standard chronology of continental aurorae, referred to in what follows as 'Fritz', i.e. H. Fritz, *Verzeichniss beobachteter Polarlichter*, Vienna, 1873.

paintings given by Gartlein, the old woodcuts in Hess, or observations of the aurora itself.[1]

The importance of the dates and the variability of interpretation make the identification of aurorae a subtle process. By itself the chronicler's account might refer equally to a comet, to shooting stars, to peculiar clouds, or even to a dream or delirium. However, a chronology of all natural phenomena, prepared from the combined records of the East and West, eliminates most of the ambiguities, as an important comet or aurora will be seen in various parts of the world.[2] By noting the coincidence of date, it has been possible to identify as aurorae many of the vague descriptions in the survey that follows.

PART I

The Roman Era   THE LEGENDARY PERIOD

No chronicles of Roman Britain remain. Annals were not easily made until the 'book' replaced the 'roll', and in Britain we learn more of the Roman period from archaeologists than historians.

It may be possible to determine the aurora cycle for this period by tree-ring analysis. Many timbers of Roman and Saxon date are being studied and dated in this country by Mr. A. W. G. Lowther; similar studies are being made in various parts of Europe and America. No detailed results are yet available.[3]

There are various references to northern lights in the classical writers, notably Aristotle, Livy, and Seneca, but neither Britain nor Gaul is mentioned specifically. The mention[4] of the bloody colour of the English Channel in A.D. 61 could refer to auroral reflection, but it was probably due to seaweed. It is interesting to note that the colour of lesser displays is usually described as bloody and, about 793, Alcuin's 'shower of blood' seen to the north from York on a fine night was a typical description of an aurora.

A.D. 400–520. *Holinshed's Coloured Rain*

The fifth and sixth centuries are known in England as the two lost centuries. In Ireland the situation is only slightly better. Most of the fragmentary records of natural phenomena are copied verbatim from continental (often Byzantine) annals.

The 'coloured rain', ascribed by Holinshed to A.D. 436, is interesting. Holinshed is, however, a late compilation and is particularly unreliable about dates. A 'Rain of Blood' at York was ascribed to A.D. 442 in the eighteenth century by Dr. Short, but it may be a duplication of the event of *c.* 793.[5] It is interesting, however, to note that a great aurora was

---

[1] C. W. Gartlein, *Nat. Geog. Mag.* xcii. 673–704; W. Hess, *Himmels- und Naturerscheinungen in Einblattdrucken des XV. bis XVIII. Jahrhunderts*, Leipzig, 1911.

[2] Part of this chronology, for the period 530–300 B.C., has already been published in the *Journ. Brit. Astr. Assn.* lviii (1948), 178–90, 202–4. See pp. 185–90. See also ibid. lxi (Dec. 1950), 21–25, 'The Earliest Dated Sunspot'.

[3] For the Swedish studies on this period see Ebba Hult de Geer, *Prehistoric Bulwark in Gotland Biochronologically Dated*, 2nd ed., Stockholm, 1935 (see Pl. VII), and *Den Planetariska Värvfysiken*, 1942 (see Pl. 4). These papers are numbers 22 and 69 of the *Stockholm Högskolas Geokronol. Inst.* Her tentative dates are much disputed but not disproved. The connexion with the sunspot cycle of Scandinavian trees seems slight: see D. J. Schove, 'Summer Temperatures and Tree Rings', *Scott. Geog. Mag.*, Edinburgh, lxvi (1950), pp. 37–42. See p. 40 and list of references.

[4] Cassius Dio lxii. 21, Loeb ed., viii, p. 83.

[5] Dr. Thomas Short, *A General Chronological History of the Air, Weather, Seasons, Meteors, &c.*, London (2 vols.), 1749. Most of his early dates are unreliable. See C. E. Britton, *A Meteorological*

recorded in different parts of China on the seventh moon of A.D. 441 and was presumably visible in Europe. In the annals of the Northern Wei dynasty it is recorded as a yellow light in the sky. Those of the contemporary Liu Sung dynasty likewise refer to a yellow light shining to the ground, and a medieval encyclopaedia states that heaven changed its colour. Each of the three references refers to the same year and the same month. This event occurred eleven years after another Chinese aurora (430 Moon 8) and is evidently reliable. In Europe its descriptions may have been confused with those of the comet of the end of A.D. 442.[1] The record of Holinshed and Short may therefore be based on a reliable source, but no such source is known to me.

Great aurorae definitely occurred in the 450s and, both in the East and the West, about the years 479 and 490, but they are not recorded in British chronicles.

About 501/2 sunspots were frequent in China and an international aurora was seen as far south as Edessa in Mesopotamia. From 510 to 514 sunspots and aurorae were seen several times in China, and a reference to the latter is given by Marcellinus Comes (quoted by Weiss) for 512 when he says that the sky was often seen to burn in the northern regions, presumably as seen from Byzantium. Possibly there was a detailed description in the works of Hesychios of Miletos, which, like many of the sixth-century sources, have been lost.

A.D. 521/2. *The Aurora of Columba's Birth*

The visions before the birth of St. Columba sound very much like descriptions of a typical auroral arc extending WNW.–ESE. According to Stokes's *Three Homilies* Columba's mother, Ethne, saw 'a great mantle . . . stretched from Insi-mod to Caer-Abrocc (Mayo in NW. Ireland to York) and every colour was present in it'.[2] According to the same authority the waiting-woman to Ethne saw this vision too—a sign that it was at least objective. It is asserted that 'She imagined that the birds of the air bore Ethne's bowels throughout the territories of Ireland and Scotland'. If this were indeed before the birth of Columba (which event is usually dated as 521, Thursday, December 7th) the aurora would probably have been seen in 521. According to the eleven-year cycle, northern lights would probably have taken place in 521 and 522.

In sixth-century China similar significance was attached to northern lights that were seen by a pregnant mother. The Chinese sources state (of a man who became a ruler of the period 532–50) that 'In the south of the dwelling-place of the emperor's father, there (appeared) on several occasions the strange scene of the red light and the purple clouds'. In this case it seems to be the period of the emperor's childhood that was meant. Probably this group of Chinese aurorae is that to which Kanda has given a mean date of autumn 522. One of these may have been identical with

Chronology to A.D. 1450, London (H.M.S.O.), 1937, pp. 4–5. Reliable for English but not for Irish dates. Except where otherwise stated, translations from Latin texts, if they refer to the British Isles, given below are derived from Britton.

[1] Cf. Hydatius, 126, *Chron. Min.* ii. 24, and Marcellinus Comes, 442. 1, *Chron. Min.* ii. 80, quoted by Dr. J. Weiss, *Elementarereignisse im Gebiete Deutschlands*, 2nd ed., Vienna, 1914.

[2] See A. O. Anderson, *Early Sources of Scottish History*, Edinburgh and London, 1922, p. 31 and note 2 (also p. lxxii), quoting 'Lebar Brecc' and referring also to the later text of the Lismore Lives.

the displays seen by Columba's mother. The aurorae of the next cycle are, however, specifically connected with the pregnancy of the mother of Ch'i Wen Hsuan Ti, and we are told that 'the light shone into the imperial bedroom as brightly as if it were daytime' (529/30) and in 530 Moon 7 when the prince was born 'there was a red light all over the room', several months before the appearance of Halley's comet. Later in the century the pregnancy of the mother of Liu Wu Chou is actually ascribed to the mysterious shafts of light directed towards her from the heavens.

Chinese biographers and historians at this time therefore took special note of aurorae, and it is possible therefore that Chinese sources may thus supply evidence for the date of Columba's birth! The stories of portents associated with the *births* of great men (as distinct from their deaths) are, however, sometimes misdated. This is often done by the next generation to add divine significance to the greatness which has meanwhile come upon them.

Sunspots are supposed by Humboldt to have been seen in Europe about 535/6, but no original source has been found and there is no Chinese confirmation of either sunspots or aurorae between 533 and 541, which period would seem to correspond with the usual lull. On August 28th, 532, in China 'Stars fell like rain', the typical description of shooting stars, but not of aurorae.[1] In Korea displays are also recorded about this time. This is presumably the same display dated 532 and recorded by Malalas of Antioch in his monastic chronicle written a generation later. This event was distinguished from the aurorae by the Chinese, but there is no information of either in western Europe. Even the eclipses in Bede (538 and 540) are part of the so-called 'international trade in marvels' and refer in reality to the Mediterranean, where they were certainly conspicuous.

[*Editor's Note:* In the original, the aurorae and sunspots of A.D. 540 to A.D. 581 are discussed. The Chinese aurorae were probably those now dated 1 December 430, 14 August 441, 13 December 478, 26 November 520, and 20 October 522. A radiocarbon increase in c. 540 (see Appendix C), when precisely dated, will help in dating weak maxima or prolonged minima.]

[1] Ed. Biot, *Catalogue des Étoiles Filantes . . . en Chine*, Paris, 1846, p. 18 (a translation from Ma-tuan-lin). Cf. also Zacharias of Mitylene, Book IX (Syriac chronicle), chaps. vii and xix.

Aurorae of 582–7

Three phenomena of 582 called by Gregory of Tours a 'comet', 'the heavens aflame', and 'real blood', respectively, are all in fact auroral.

The so-called 'comet' of January

'appeared in such a way that round about it there was a great blackness; it shone through the dark as if set in a cavity, glittering, and spreading abroad its hair. And there issued from it a ray of wondrous size which from afar appeared as the great smoke of a fire. It was seen in the western quarter of the heavens at the first hour of night.'

This has long been recognized as an excellent description of an aurora with the 'dark segment' (to be christened 'black cloud' by Stow) and characteristic rays.

'And on the holy day of Easter (i.e. March 29th, o.s., or April 2nd, n.s.), in the city of Soissons (north-east of Paris), men saw the heavens aflame, in such wise that there appeared two fires, the one greater, the other less. But after the space of two hours they were joined together, making a great beacon light before they vanished away.'

This again is an excellent description of a double auroral arc. It can hardly refer to the Lyrid shooting stars as has been supposed.

'In the territory of Paris, there rained real blood from the clouds, falling upon the garments of many men, who were so stained and spotted that they stripped themselves of their own clothing in horror. This portent was seen in three places within the territory of that city. In the territory of Senlis, a certain man, rising in the morning, found his house all spattered with blood within.'

Such showers of blood are frequently reported in the Middle Ages and frequently led to panic. Usually the blood is seen rather than felt, and, except for a few occasions when the effect is due to volcanic or other dust, it is seen to fall at night; it is in fact auroral. That this blood rain fell at night is implied by the story of the man at Senlis—it would seem that the men of Paris rose to find the bloodstains had vanished. The terror inspired can in fact be readily appreciated by anybody who has experienced rain falling on him when the night scene is lit by a reflected auroral glow.

Gregory's account of this is in the next sentence of the same paragraph describing the visions at Soissons. It may therefore be another interpretation of the same spectacle. Oddly enough, in the same year, 582, at the end of the summer, at the time of the great annual flood, the waters of the Yangtze had become as red as blood. Possibly this effect also is auroral in origin.

The phenomenon of 582 which led Gregory[2] to report 'A fiery light

---

[1] Gregory of Tours, v. 25 (33), pp. 203–4.   [2] Op. cit. v. 14 (21), p. 253.

was seen to traverse the sky' sounds like a meteor; so too, probably, was the phenomenon he describes for 31st January, 583 (o.s.), at Tours.[1] These meteors do not synchronize with 'the night the heavens opened from north-west to south-east and inside a blue and yellow colour rumbled like thunder' reported in China for the twelfth moon of the same year, which would seem to refer to a genuine aurora of the winter 583/4. This would seem to synchronize with the period referred to by Gregory of Tours, vi. 24 (33), about 583:

'In those days there appeared at midnight in the northern sky a multitude of rays which shone with an exceeding splendour; they came together, went apart again, and vanished in all directions. So brilliant was the heaven towards the north that it seemed the breaking of the dawn.'

About the month of December in 584 'a great light like a beacon traversed the heavens, illuminating the earth far and wide before the dawn. Rays also appeared in the sky; in the north a column of fire was seen for the space of two hours, as it were hanging from the heaven, with a great star above it. . . .'[2]

It is possible that the 'cloudless rain' [sic] reported in Japan for May 4th, 585, was a similar aurora.

In July 585 'at this time there appeared signs, fiery rays in the northern sky such as frequently appear. A brilliant light was seen to cross the heavens . . .' in France. In China on September 23rd 'Hundreds of shooting stars fell, dispersing in all directions', a phenomenon which certainly sounds meteoric. Nevertheless, the aurorae of this year must have been visible in China, notably those of October, and may be referred to in the ' "blood-red stuff" (which) fell in front of the palace' some time during this year, or the 'shooting stars' reported there for October 25th, 585.

In October of 585, at Trier, Gregory witnessed displays for three nights; from his observations of the last occasion about 7–8 p.m. he gave the first scientific account of the auroral corona:

'During our sojourn in this place, we beheld for two nights signs in the heaven, namely rays in the north so clear and splendid, that none such were ever seen before; on both sides, east and west, were blood-red clouds. On the third night, about the second hour, these rays appeared again; and while we gazed in wonder at them, lo! from the four quarters of the earth there rose others like them, and we saw them covering the whole sky. In the middle of the heavens was a gleaming cloud to which these rays gathered themselves as it were into a pavilion, the stripes of which, beginning broad at the bottom, narrow as they rise, and meet as it were in a hood at the top. In the midst of the rays were other clouds, flashing vividly as lightning. This was a great sign, and filled us with fear. For we looked that some disaster should be sent upon us out of heaven.'[3]

Fredegar (iv. 5) says of 586 in Gaul:

'In this year a sign appeared in the heavens. Fiery globe(s) falling (decedens) to the ground with sparks and a rumble (? rugeto).'

---

[1] Gregory of Tours, vi. 17 (25), p. 256.
[2] Op. cit. vii. 11, p. 293.
[3] Op. cit., viii. 17, pp. 342–3. Cf. also viii. 8 and 24.

Gregory of Tours (*Liber in gloria confessorum*, 102) says:

'A great ball of fire appeared, which rising from the east and rushing through the circle of the heavens stood above the diocese (? *ecclesia*) of Limoges.'[1]

These descriptions appear to refer to a meteorite, such as may have fallen in May and June, for the Korean sources state: 'In summer 5th moon, 8th year of King Shinpei, stars were observed to fall like rain.'[2]

In the winter of 586/7, according to Gregory: 'A brilliant light in the form of a serpent was seen to pass across the sky.' This could be a meteor, but in October of 587, after 'Rays were observed in the northern sky', he states: 'Certain persons declared they had seen snakes fall from the clouds.' There is one form of aurora which is shaped like a snake. On the other hand, the red meteor seen at Nanking in the fifth moon of 588 appears to have been a real meteor. In 590, probably in spring, 'so great a splendour shone upon the earth in the night (time) that you might deem it noonday; and in like manner fiery globes were seen often traversing the heavens and lighting up the earth. . . .'[3]

A.D. 594–600. *The Aurorae of Columba's Death*

The various visions described in Adamnan's *Life of Columba* can be, perhaps, identified with the foregoing phenomena, but so far I have refrained from making specific identifications, mainly because modern editors give insufficient clues for dating the events.

The aurorae associated with the death of St. Columba are in a different category. The event would seem to be datable, and aurorae were evidently seen in Ireland within a few months, as contemporaries firmly believed the rays seen to the north were the angels carrying the soul of St. Columba to Heaven.

Among the associated visions described by Adamnan (Book III, chapter 24) is that of an Irish saint, Lugud, son of Tailchan, at the monastery of Clonifinchoil (now Rosnarea, in the parish of Knockcommon, Meath, East Ireland). His vision, 'which at early dawn he told in great affliction to one called Fergnous', included the following statement, quoted by Adamnan:

'. . . I saw in the spirit the whole (Iona) island, where I never was in the body, resplendent with the brightness of angels; and the whole heavens above it, up to the very zenith, were illumined with the brilliant light of the same heavenly messengers, who descended in countless numbers to bear away his holy soul. . . .'

Adamnan himself adds:

'This vision above mentioned we have not only found in writing, but have heard related with the utmost freedom by several well-informed old men to whom Virgnous (i.e. Fergnous) himself had told it.'

He then gives another account of perhaps the same display seen in Donegal. Adamnan says:

'Another vision also given at the same hour under a different form was

---

[1] Translations from the original quotations given in Weiss, op. cit.
[2] From the Annals of the Shiragi dynasty quoted by Y. Iba in *Meteor Showers Chronicled in the Far East*, Kobe (Japan), 1933—the source for all Korean information used here.
[3] Op. cit. viii. 42, p. 363, ix. 5, p. 372, and x. 23, p. 459.

related to me—Adamnan—who was a young man at the time, by one of those who had seen it, and who solemnly assured me of its truth. . . .'

This statement is quoted as follows:

'On that night when St. Columba, by a happy and blessed death, passed from earth to heaven, while I and others with me were engaged in fishing in the valley of the river Fend (the Finn, in Donegal)—which abounds in fish— we saw the whole vault of heaven become suddenly illuminated. Struck by the suddenness of the miracle, we raised our eyes and looked towards the east, when, lo! there appeared something like an immense pillar of fire, which seemed to us, as it ascended upwards at that midnight, to illuminate the whole earth like the summer sun at noon; and after that column penetrated the heavens darkness followed, as if the sun had just set. And not only did we, who were together in the same place, observe with intense surprise the brightness of this remarkable luminous pillar, but many other fishermen also, who were engaged in fishing here and there in different deep pools along the same river, were greatly terri- fied, as they afterwards related to us, by an appearance of the same kind.'[1]

[*Editor's Note:* Further discussion of aurorae and visions of A.D. 594 to A.D. 660 has been omitted.]

CONCLUSIONS

The chronology of early medieval history can be clarified if the celestial visions associated with the saints can be identified with the spectacular comets and aurorae known to have been visible in particular years.

In the sixth century displays were particularly vivid and often extended into China and the Mediterranean.

The eleven-year sunspot cycle, familiar today, was equally predominant in the fifth and sixth centuries. Well-defined maxima are noted near the following years: . . . 430, 441, . . . 479, 490, 500, 511, 522, 532, . . . 555, 566, 577 . . . 643, 655, 664 . . . .

Exact observations of northern lights are available in this period from China (to *c.* 580), and for short periods in Gaul (570–90) and Ireland (660 +). Before A.D. 520 the accounts of displays in Britain and Gaul are probably borrowings from Italian and Byzantine chronicles, but there are semi-legendary accounts and accurate descriptions by Gregory of Tours

---

[1] Adamnan, edited by W. Reeves, *Life of Saint Columba*, Book III, chap. 24 (English transla- tion, Edinburgh, 1874, pp. 98–100).

between 541 and 600. The period 600–60 is a blank one in west European annals and the legends—if they really refer to aurorae—are wrongly dated. The period *c*. 660–*c*. 800 is notable for concise scientific accounts in the Irish and later the English annals, which confirm the impression that from this date the writing of chronicles was 'continuous with the flow of time'.

Many displays of northern lights were accepted as religious visions and were described in all sincerity as such by the saints, and are not the fabrications modern critics have supposed. Irish visions of Hell at this period, which influenced the thinkers and artists of the later Middle Ages (notably Dante), certainly used 'imagery' appropriate to auroral forms, and the reputation of the Celtic races for second sight may date from the great auroral spectacles witnessed by them in the sixth century.

My thanks are due to various friends of the Institute of Historical Research for items of useful information, to Mr. Eric Barker for suggestions, to the British Council, whose co-operation has led to my receiving weather chronologies from various countries, and to Mr. Leslie and Dr. Takeo Yamamoto for translations from Far Eastern sources.

An additional item has just come to my notice which shows that the connexion between aurorae and plague was familiar in the East as well as the West.

Celestial appearances in China in A.D. 599 Moon 6 were interpreted as people in the air clothed in robes of five colours, carrying such things as a sword, a fan, a club, and a jug of fire. The Grand Historiographer explained to the Emperor that these were signs of an epidemic. In 598 Moon 2 there had been a very severe plague. A quarter of the forces (Chinese!) attacking Korea by sea and land had died. I find no evidence of a great plague in 599, but the memory may have suggested plague as the meaning of the auroral portent.

[*Editor's Note:* We now know that Humboldt's sunspot of 535 to 536 was false—it was only volcanic activity (Schove, D. J., and R. W. Fairbridge, eds., 1983, *Ice-cores, Varves, and Tree-rings*, Balkema, Rotterdam, Netherlands)— and that data on the early Korean meteoric showers in 536 and 586 were borrowed from China and probably misdated. Further Chinese observations of meteoric and auroral phenomena are given by Keimatsu (1970–1976, *Kanazawa Univ. Ann. Sci.* **7**:1–10, **8**:1–16, **9**:1–36, **10**:1–32, **11**:1–36, **12**:1–40, **13**:1–32) and, together with Western visions, we can derive auroral clusters in c. 411, 418–419, 430–431, 441, 449–452, 459–464, 478, 486, and so forth. The dates of cycles given in Paper 20, although approximately correct from 300 to 650, have been tentatively revised from 650 to 770.]

ERRATUM

Page 45, line 3, the word "diocese" should be "church."

17

Reprinted from pages 295–299 and 302–304 of *British Astron. Assoc. Jour.*
69:295–304 (1959)

CHINESE AURORAE: I, A.D. 1048—1070

By Dr D. J. Schove* and Dr P. Y. Ho†

Received 1958 October 28

Chinese aurorae recorded in the Sung Dynastic History are presented in translation. Observation of the phenomena of the night sky was remarkably detailed in the later eleventh century, and the displays help us to date a weak sunspot maximum which has hitherto been dated only by interpolation. A weak maximum about 1050, or 1052 as previously interpolated, was followed by a long lull, and finally by a moderate maximum about 1067. The Chinese records confirm the activity recorded for this latter year by the Norman chronicler Gaimar. They also suggest that the magnetic meridian was NNW.–SSE. of N′W.–S′E. in the eleventh century, and that the last two digits of the maxima, at least in the later part of the century, tended to follow the 11-times table, as in 1088.

Introduction

Early sunspot maxima have been provisionally dated by one of us (Schove 1955, 1956) from the records of early aurorae but, in certain centuries such

* St David's College, Beckenham, Kent.
† Department of Physics, University of Malaya, Singapore.

as the eleventh, the lack of evidence made it difficult to be precise. A single maximum of 1052 was postulated between the maxima of 1038 and 1067, but there was no evidence available in 1954 to support it, and it still seemed possible (Cf. Schove, 1948, p. 249) to argue that two maxima could be inserted in the twenty-nine year interval.

In the eleventh century, the observations of meteors recorded in the Chinese histories are very detailed and indicate that a very careful watch was kept on the night sky. Decadal values from 1001–10 to 1091–1100 of meteors listed by Biot were thus 49, 33, 44, 33, 88, 105, 165, 247, 115 and 172; observations were especially detailed from 1058 to 1099. The hope was therefore expressed in 1955 that ' Some unpublished information (of aurorae or sunspots) in Chinese sources (especially the provincial histories) may later necessitate some revision ' of the dates. This hope has now materialized. A large number of aurorae have been discovered (by P.Y.H.) and the relevant translations (by P.Y.H.) will be published in a series of articles.

The source used is the Official History of the Sung Dynasty (A.D. 993–1279) known as the *Sung Shih*. The records of the Aurora Borealis were classified by the Chinese as two separate phenomena. Those described as ' red vapour', as in Nov./Dec. 1069 (the only one in the period of this paper) are included in the Wu Hsing Chih (Chapters on the Five Elements), Chapter 64, page 12 b ff. in the Chinese text. There are, however, numerous aurorae classified in the Section on ' Clouds and Vapours ' (yün-chhi in the Astronomical chapters), which is Chapter 60, page 13 of the Chinese text. Full translations of both sections have been made by one of us (P.Y.H.) and those relevant to the mid-century are now being presented here.

Distribution in Time

The clearest evidence that both phenomena relate to the aurora is the way in which they follow the 11-year cycle. Thus, in the period 1021—1099, remainders have been calculated by the method suggested previously (Schove 1955, p. 128), that is, by deducting multiples of 11 from the last two digits of the year. Remainders from 0 to 10 showed the following frequencies— 15, 12, 5, 5, 5, 1, 3, 5, 16, 8 and 14. In other words, the maxima tended to follow the eleven times table, such as 1088, and the minima tended to occur five years later, such as 1060; these particular examples correspond with years previously suggested (Schove 1955, p. 135).

Distribution and Direction of Displays

The seasonal distribution of Sung aurorae differs from that in Korea, where the maximum is March, and from that in Japan, where the maximum is September. The main maximum is November to December and there is a secondary maximum in July and August. The main minimum is January to April and there is a secondary minimum in September to October. The possibility that some of the December items refer to zodiacal light and that some of the summer instances are sunlight effects is being investigated, but in the period of the present paper the displays appear to be genuine. The absence of displays in January and February may be associated with the difficulties of observing in the cold winter months. Dates in the lunar month are indicated below in brackets beneath the name of the solar month. All dates are Old Style.

Colour and direction

The greater and red displays are collected mainly in the Five Element Chapters. In the other section the colours are often noted merely as ' dark sallow' or greyish, and it is probable that these descriptions refer to the faint yellow-green of aurorae of low intensity. Such displays would be apparent to experienced observers only, and some instances may not be truly aurorae.

No information as to the place of observation is included in the *Sung Shih*, but it seems probable that the displays were seen, not in the capital but nearer the northern boundaries of the Sung Empire, say near Peking, roughly 40°N 120°E. The geomagnetic latitude of this region is now only 30°N, and displays are very rare.

Directions in which the lights were seen in the eleventh century were most frequently north to north-west, but displays beyond the auroral zenith to the south were also frequent. There were slightly more displays on the WSW. half of the sky than the eastern half, and this is perhaps due to the greater cloudiness of the latter, although possible instances of zodiacal light may have been included. There seems to be a slight tendency for the displays about 1040 to be nearer north-west and those about 1090 to be nearer north, and palaeomagnetic studies (Cf. Cook and Belshe, 1958) may ultimately determine whether there was a real veer in the magnetic meridian between these two dates.

The weak Auroral maximum of c. 1052

There are no records of sunspots in the period of the present paper and no certain European, Japanese, or Korean records of the aurora between 1044 and 1064. The four displays given below therefore supply the only evidence for a sunspot cycle about 1050. This date differs only slightly from the interpolated value 1052 given previously (Schove 1955, p. 135), and it is consistent with the suggestion made in 1955 that there was only one maximum in the 29 years.

The aurorae from this series are as follows:—

1048 Feb. 14: On a 34th cyclical or *ting-yu* night in the first month of the
(28)   8th year of the Chhing-Li reign-period a dark vapour de-
         veloped a head and a tail reaching the horizon. It then
         gradually moved eastward and dispersed after a long time.

1048 Apr. 8: On a 28th cyclical or *hsin-mao* night in the second month of
(23)   the same year a dark vapour was developed at the west near
         the horizon. It measured 30 ft and dispersed after some time.

1052 Nov. 24: On a 39th cyclical or *jen-yin* night in the eleventh month of
(1)   the 4th year of the Huang-Yu reign-period a dark vapour

was developed in the east stretching from north to south.
It reached the horizon and penetrated the *Shen* lunar mansion
[in Orion] and *Hsien-Yuan* [in Leo].

1052 Dec. 13:   On a 58th cyclical or *hsin-yu* night in the same month of the
(20)   same year a white vapour was developed in the north.   It was
                 near the horizon and measured about 50 ft.   It passed *Pei-
                 Tou* [the Dipper] and dispersed after a long time.

It may seem surprising that four displays should be seen as far south as
China in what must have been a weak cycle.   However, displays in China
have been much more frequent throughout mediaeval history than has
hitherto been supposed, and paintings of the various forms exist from as early
as A.D. 1429.   It is possible that European sources in the primary sources of the
Mon. Germ. Hist. may include reference to the aurora of neighbouring dates;
the blood-rain noted in Armenia *c.* 1056 can hardly have been auroral in
origin, and, pending further discoveries, a date of 1050 rather than 1052 for
the maximum seems more plausible.

After 1052 there are no further aurorae until 1064, but from 1065 until
1069 displays were frequent in China.

Possible Aurorae in the British Isles 1048–54

Two possible instances of auroral activity in Europe in this cycle have been
noted since the rest of this paper was written.

The wild fire which in 1048, ' spread over Derbyshire and some other
places did much damage ' according to the Anglo-Saxon Chronicle, MS.D.
Simeon of Durham, who had access to other versions, in his Latin Chronicle
written between 1119 and 1130, amplifies this, saying ' fires in the air, commonly called woodland fires, destroyed towns and crops of standing corn . . .'
The fires would appear to have been real heath fires, but aurorae rather than
drought may have been considered as the cause.

If this hypothesis is correct, and it would seem also to explain the wild fire
of 1032, 1067 and 1077, the displays are not likely to have been those noted
in China, as in England the wild fire of 1048 came *after* the earthquake of
May 1.

A ' steeple of fire ' in 1054, according to the Chronicum Scotorum and the
Four Masters versions of the Irish Annals, would appear to have been inspired by an aurora.   This was later included among the wonders of Ireland
in the Book of Ballymote (fol. 140 b.   See Todd's note to p. 193 and p. 215
of his edition of the Irish version of Nennius).

' A steeple [the Irish word signifies a Church Tower] of fire was seen in
the air over Ros-Deala [i.e. Deala's Wood in S. of W. Meath] on the Sunday
[presumably the day after the Festival of the Saturday] of the festival of
St George (April 23) for the space of 5 hours;   innumerable black birds
passing into and out of it;   and one large bird in the middle of them;   and the
little birds went under his wings, when they went into the steeple.   They came
out and raised up a greyhound that was in the middle of the town, aloft in the

air, and let it drop down again, so that it died immediately, and they took up 2 cloaks and 3 shirts and let them drop down in the same manner. [the C.S. version says ' three garments ' and omits the cloaks]. The wood [the Irish word means wood in the sense of a plantation] on which these birds perished fell under them and the oak tree on which they perched shook with its roots in the earth.'

The year 1054 was also the year of the Supernova that became the Crab Nebula, observed and described in Japan and China, but not recorded in extant Arabic or European works.

[*Editor's Note:* Ibn Buṭlān, a Christian physician of Baghdad, described the Supernova and associated it with the Constantinople epidemic of 1054 and the Nile failure of 1055–1056 (see Brecher, K., and A. D. Lieber, 1978, *Nature* **273:**728–730). In the original, a discussion of the 27 displays of the next cycle (1064–1070) follows.]

Inexperienced observers of the aurora may often report a display when the luminosity is non-auroral. Confusion is less likely to arise on a cloudless, moonless night, but sources of confusion include reflections from the Earth on overcast nights with low clouds, the light of the Milky Way, or zodiacal light in the west near the point of sunset from November to February. (See Chapman *et al.*, 1956, in the I.G.Y. Instruction Manual, Appendix VII, pp. 100–101). Moreover, a report of any slight luminosity in a part of the sky outside the north-west to north-east sector, must be examined with care. Aurorae can and do appear in other regions of the sky, but in China they are rather rare. Peculiar and brilliant coloration of the sky at the time of sunset may also cause difficulty.

The records from the *Sung Shih* were examined for these various sources of confusion, and it was considered just possible that sunset effects, e.g. in July, or mid-winter zodiacal light might help to explain the bias towards displays in the western half of the sky. However, even the displays recorded as seen to the south would seem to be mostly true aurorae.

The Chinese observers of the eleventh century—or perhaps the authors and the block-makers of the *Sung Shih* in the eleventh century—may have slipped up occasionally. There is thus some inconsistency in the dates of the two 1070 displays, and possibly either the year or the diagnosis is wrong. Extensive displays—even faint ones visible to the skilled Chinese observers—are unusual within two years of sunspot minimum.

The records for 1072 June 27 and even that for 1073 June 6 may thus not be truly auroral, but they would in any case belong to the next cycle. This next cycle is represented by a boom of displays, dated 1074–77, when there were recorded in China alone no less than 24 displays, and these displays will be discussed in a later article. The aurorae from this cycle were sometimes observed in both Europe and China, and in China there are also records of sunspots.

(b) In Europe

The source of the display of about 1069 mentioned by Schoning (1760) has not been traced. However, Chinese records suggest that 1067 was the peak year for aurorae and indeed this is consistent with the only evidence in this cycle of aurorae from the British Isles. The source is the Norman-French History of Gaimar and is as follows:—

' But in coming from Normandy [the return of King William from Normandy is dated in the Anglo-Saxon Chronicle as 1067, December 6] some of his people perished in the sea. In this year, truly, several people saw a sign; in appearance it was fire: it flamed and burned fiercely in the air; it came near to the earth, and for a little time quite illuminated it; afterwards it revolved and ascended up on high, then descended into the bottom of the sea; in several places it burned woods and plains. There was no man who knew with certainty what this divined, nor what this sign signified. In the country of the Northumbrians this fire shoed itself; and in two seasons of one year were these demonstrations.'

This quotation has been seized upon by an author (H. T. Wilkins, 1955, p. 174) of a recent flying saucer book, but it was correctly diagnosed by Britton (1937).

In a previous discussion (Schove 1953, p. 66) the question arose as to whether any aurorae preceded the invasion of 1066. Halley's comet (Cf. Schove 1955, 1956) preceded the invasion, but the only aurorae recorded came afterwards.

The Four-Week cycle

The cycle of about four weeks in auroral activity is a reflection of the lunar cycle rather than the cycle of the Sun's rotation. Intervals of 27, 26 and 26 days occurred in 1067 and of 54 (27 × 2) days in 1048 or 58 (29 × 2) days in 1065 may nevertheless be mentioned. The day in the lunar month has been indicated in brackets.

In Sung times displays were seen more often at New Moon, but displays even at Full Moon were, in the second half of the year, surprisingly frequent. It is probable that some of the displays recorded between the 13th and 16th lunar days are not true aurorae. However, the seasonal distribution and the distribution of these displays in the (previously determined) 11-year cycle still show the same general features, so that most of them are real.

Acknowledgements

We should like to express our gratitude to Dr J. Needham, F.R.S., for giving us access to his collection of Chinese and western works. We should also like to thank Dr W. Bonser and Mr. T. O'Reifearteign for the information about the 1054 phenomenon and Miss C. M. Botley for her suggestions.

Bibliography

The *Sung Shih* (Official History of the Sung Dynasty—A.D. 960 to A.D. 1279) by Toktaga and Ouyang Hsüan in + 1345. The edition used is a lithographic copy of the original fourteenth century edition prioduced by the Commercial Press, Shanghai, in their *Po Na Pên* edition of the 24 Dynastic Histories.

S. Chapman, *et alii*, 1956, The I.G.Y. Instruction Manual Part II, Aurora and Airglow.

P. Y. Ho, 1957, ' The Astronomical Chapters of the *Chin Shu* ' inaug. diss. University of Malaya.

J. Needham, 1954, *Science and Civilization in China*, Vol. I., C.U.P.

J. Needham, in print, *Science and Civilization in China*, Vol. III, C.U.P.

G. Schöning, 1760, Om Nordlysets Aelda, *Selsk. Skr.* 8 (Brit. Mus. cat. No. 127, b. 16–27). New edition in preparation).

D. J. Schove, 1948, Sunspot Epochs, A.D. 188–1610, *Popular Astronomy*, **56**, May, pp. 247–252.

D. J. Schove, 1952, *Journ. Brit. Astr. Assn.*, **62**, pp. 38–42 and pp. 63–66.

D. J. Schove, 1955, The Sunspot Cycle, 649 B.C. to A.D. 2000 (with catalogue of sunspot maxima), *Journ. Geophys. Res.*, **60**, No. 2, pp. 127–146.

D. J. Schove, 1956, Sunspot Maxima since 649 B.C. (List), *Journ. Brit. Astr. Assn.*, **66**, No. 2, pp. 59–61.

D. J. Schove and W. A. G. Lowther, Tree-rings and Medieval Archaeology, *Medieval Archaeology*, 1.

D. J. Schove, 1959, *The Spectrum of Time* (In progress).

[*Editor's Note:* Professor Ho found many displays that are still unpublished. However, a display in 18 July 1052 and a list of uncertain phenomena in A.D. 1056–1066 have been included (in Chinese) by N. Dai and M. Chen (1980) in their "Chronology of Auroral Data in China, Korea, and Japan," *Kejishiwenji* **6**:87–146.]

18

Reprinted from *Isis* **70:**90–95 (1979)

ANCIENT AURORAE

R. Stothers

[*Editor's Note:* In the original, introductory material precedes this excerpt.]

What, then, are the phenomena that can safely be regarded as auroral (at least in most instances)? A systematic search of the classical literature reveals that most of the probable aurorae divide themselves neatly into just a few categories. These divisions are based on certain described forms, which reflect the ancients' view of aurorae and do not necessarily conform to the modern divisions. Therefore, it seems best to retain the original categories, since the ancient practice of using invariably the same descriptive formulae in reporting celestial prodigies suggests that the same physical phenomena (whatever they may be) are being reported over the centuries. In the order of their probable association with aurorae, the categories are:

X. Chasm (χάσμα, *hiatus* or *discessus*).

SF. Sky fire (οὐρανὸς φλεγυρός, *caelum ardens*).

NS. Night sun (ἥλιος or φῶς νυκτός, *sol* or *lux noctu*).

BR. Blood rain (ψεκὰς αἱματώδης, *pluvia sanguinea*).[29]

MR. Milk rain (ψεκὰς γαλάκτινη, *pluvia lactea*).[30]

B. Beam (δοκίς, *trabs*).

P. Pillar (κίων, *columna*).

T. Aurora-like torch (λαμπάς, *fax*).

K. Aurora-like comet (κομήτης, *stella crinita*).

---

[29] Some authors, including Cicero (*De divinatione* II 58), have ill-advisedly regarded "blood rain" as being in all cases simply contaminated water drops that have fallen to the ground; see, e.g., F. B. Krauss, *An Interpretation of the Omens, Portents, and Prodigies Recorded by Livy, Tacitus, and Suetonius* (Philadelphia: University of Pennsylvania Press, 1930), pp. 58–60. But, for one thing, the great frequency of reported "blood rains" argues against this particular interpretation. For another thing, the ancient scientific writers themselves have used the terms "red" and "bloody" in describing what we now know on other grounds to be auroral displays (Aristotle 342a35; Seneca I 14.2; Pliny II 97). Finally, the typical report that "blood rain" was seen in such and such a precinct can be read to mean simply that the viewer was stationed in that particular precinct. Except for the few cases where physical drops are specifically mentioned, I shall maintain an auroral interpretation of "blood rain."

[30] "Milk rain" is recorded only in Livy and his excerptors, and only for the period c. 265 to 92 B.C. Krauss, *An Interpretation*, pp. 65–66, interprets it as fallen raindrops, as he does "blood rain."

It should be emphasized that the ancient authors did not attribute to these categories a single underlying physical cause. This is a modern interpretation, based on a critical selection of sky phenomena that have been abstracted from a much larger number which the ancients described. In the previous auroral catalogues the last six categories have been ignored either partially or entirely.

It is impossible to be certain that every reported event belonging to each of these categories is an aurora. Sufficient detail in the ancient descriptions is nearly always lacking, and some of the events are probably due to other phenomena, such as noctilucent clouds, atmospheric dust, distant lightning, airglow, zodiacal light, meteor showers, and comets (in the modern sense). Of our three main ancient sources, Livy and his excerptor Obsequens only once report a "comet" and only once a "beam" (probably the terms were never used in the official Roman records and histories until Greek scientific knowledge became commonplace in Rome in the first century B.C.).[31] Thus, these two authors as well as our third main source, Cassius Dio, seem to use the word "torch" for all kinds of torchlike displays. Unless specific auroral properties are described, I have had to reject most of the reported "comets" and "torches."

The reports that are here accepted as being probably auroral are listed in Table 1.[32] A question mark placed after an assigned category indicates that the report as a whole, for one reason or another, cannot be regarded as auroral on its own merits. The collection of aurorae is quite homogeneous geographically: all the reported events occurred in Greece, Italy, or southern Gaul, with the exception of three questionable events, one in Egypt (30 B.C.), one in Judea (A.D. 30?), and one in Carthage (A.D. 212?).

The assigned dates depend in large part on the chronologies provided by the ancient authors reporting the events; for early Greece, either the annual Athenian archonship or the Olympiad and year number are usually reported, while for the Roman world the annual consulship, the emperorship and year number, or the year number since the founding of Rome is typically reported. Modern scholarship has been able to establish the necessary links between the ancient and modern systems of reckoning the years (by the help of ancient synchronisms and datable eclipses, for example). Since both the old Athenian year and the Roman year before 153 B.C. did not begin on January 1, and since there was a careless intercalation of months to fill the years before 45 B.C. as well as a poor tradition of the consular lists prior to circa 300 B.C. and even an occasional disagreement among the ancient authorities as to the date of an event, any modern attempt at exact dating is doomed to failure. Chronological accuracy is, in most cases, limited simply to the year of the event, the possible error of the date being ±1 year, especially for dates preceding the Julian calendar reform of 45 B.C.; dates in Table 1 that are more uncertain than this bear question marks. Of course, in any year more than one aurora may have been reported, although the general rarity of the reports makes this unlikely. Moreover, different

[31] The comet is Caesar's of 44 B.C. (Obsequens 68) and the "beam" occurred in 63 B.C. (Obsequens 61). A documented catalogue of "comets" and "torches" in classical literature has been assembled by W. Gundel, *s.v. Kometen,* in Pauly-Wissowa-Kroll, *Real-Encyclopädie der Classischen Altertumswissenschaft* (Stuttgart: Metzler, 1921), Vol. XI, Pt. 1, cols. 1143–1193.

[32] Lycosthenes lists a number of prodigies for the year 128 B.C. that I cannot locate in the ancient literature. He seems to imply that he drew these reports from Obsequens; but Scheffer thinks not, because they are not found in the Aldine edition of Obsequens. I have not included Far Eastern aurorae in Table 1; but some examples have been listed by S. Kanda, "Ancient Records of Sunspots and Auroras in the Far East and the Variation of the Period of Solar Activity," *Proceedings of the Imperial Academy of Japan,* 1933, *9*: 293–296, and by D. J. Schove, "Sunspots, Aurorae, and Blood Rain: The Spectrum of Time," *Isis,* 1951, *42*: 133–138.

Table 1. Documented catalogue of ancient auroral reports [a]

Year	Category [b]	References
B.C. 480	?	Pliny II 90 (K?); Lydus, *De mensibus* IV 73 (K?)
468/467	SF, B	Daimachus in Plutarch, *Lysander* XII 4 (SF); Charmander in Seneca VII 5.3 (B); Pliny II 149 (K?); Aristotle 344b34 (K?); Alexander of Aphrodisias, *ad loc.* (K?); Philoponus, *ad loc.* (K?); Olympiodorus, *ad loc.* (K?)
464/463?	SF	Livy III 5.14 (SF); Orosius II 12.2 (SF)
461?	SF	Livy III 10.6 (SF)
459?	SF	Dionysius of Halicarnassus X 2.3 (SF)
395/394?	B	Pliny II 96 (B)
373/371	B, T	Diodorus Siculus XV 50.2–3 (B, T); Callisthenes in Seneca VII 5.3 (B); *Parian Marble*, ep. 71 (T?); Pausanius VII 24.8 (T?); Aristotle 343b1, b18, 344b34 (K?); Alexander of Aphrodisias, *ad loc.* (K?); Philoponus, *ad loc.* (K?); Olympiodorus, *ad loc.* (K?); Aristotle in Seneca VII 5.4 (K?); Ephorus in Seneca VII 16.2–3 (K?)
350/349	X, BR, SF	Pliny II 97 (X, BR, SF); Lydus, *De ostentis* 10b (X, SF)
345/344	X, SF, T	Plutarch, *Timoleon* VIII 5–7 (X, SF, T); Diodorus Siculus XVI 66.3 (T); Pliny II 90 (K?)
265?	?	Orosius IV 5.1 (MR?); Paulus Diaconus II 16 (MR?)
223	SF, NS	Orosius IV 13.12 (SF, NS); Paulus Diaconus III 2 (SF,NS); Zonaras VIII 20 (SF, NS)
217	X, SF	Livy XXII 1.11–12 (X, SF); Orosius IV 15.1 (X); Paulus Diaconus III 9 (X); Plutarch, *Fabius Maximus* II 3 (X); Silius Italicus VIII 630–651 (X?, SF?, BR?, K?)
214	BR	Livy XXIV 10.7 (BR)
209	MR	Livy XXVII 11.5 (MR)
206	NS	Livy XXVIII 11.3 (NS)
204	NS	Livy XXIX 14.3 (NS)
200	SF	Livy XXXI 12.5 (SF)
198	SF	Livy XXXII 9.2 (SF)
197	NS	Livy XXXII 29.2 (NS)
183	BR	Livy XXXIX 46.5, 56.6 (BR); Obsequens IV (BR)
181	BR	Livy XL 19.2 (BR); Obsequens VI (BR)
172	BR	Livy XLII 20.5 (BR)
169	SF	Livy XLIII 13.3 (SF), 13.5 (BR?)
168	?	Seneca I 1.2 (K?)
166	NS, BR	Obsequens XII (NS, BR)
163	SF, NS, MR	Obsequens XIV (SF, NS, MR)
162	SF	Obsequens XV (SF)
147	SF	Obsequens XX (SF)
134	NS, BR	Obsequens XXVII (NS, BR)
130	MR	Obsequens XXVIII (MR)
128	BR	Obsequens (?) in Lycosthenes (BR, T?)
125	MR	Obsequens XXX (MR)
124	MR	Obsequens XXXI (MR)
118	MR	Obsequens XXXV (MR)
117	MR	Obsequens XXXVI (MR)
114	BR, MR	Pliny II 147 (BR,MR); Lydus, *De ostentis* 6 (BR, MR)

Year	Category[b]	References
113	SF, NS	Obsequens XXXVIII (SF); Pliny II 100 (NS)
111	MR	Obsequens XXXIX (MR)
108	MR	Obsequens XL (MR)
106	BR, MR	Obsequens XLI (BR, MR)
104	BR, MR	Obsequens XLIII (BR, MR, SF?); Plutarch, *Marius* XVII 4 (SF?); Pliny II 148 (SF?)
102	NS, BR	Obsequens XLIV (NS, BR)
95	MR	Obsequens L (MR)
94	SF	Obsequens LI (SF, T?)
93	X, SF	Obsequens LII (X, SF, BR?)
92	MR	Obsequens LIII (MR, BR?)
91	X, BR	Sisenna in Cicero, *De divinatione* I 99 (X, BR)
63	SF, B, K	Cicero, *In Catilinam* III 8 (SF); Cicero in Cicero, *De divinatione* I 18 (K, T?); Obsequens LXI (B); Dio XXXVII 25.2 (T?)
49	SF, BR	Lucan I 527–529 (SF, K?); Appian II 36 (BR); Dio XLI 14.3 (SF?); Pliny II 92 (K?)
48	B, P	Lucan VII 155–156 (B, P); Plutarch, *Caesar* XLIII 3 (T?); Appian II 68 (T?); Dio XLI 61.2 (T?); Zonaras X 9 (T?)
44	?	Ovid XV 788 (BR?)
42	SF, NS	Manilius I 907 (SF); Obsequens LXX (NS); Dio XLVII 40.2 (NS); Zonaras X 19 (NS); Vergil I 488 (K?)
32	T	Dio L 8.2 (T)
30	?	Dio LI 17.4–5 (BR?, K?)
B.C. 17	T	Obsequens LXXI (T); Dio LIV 19.7 (T)
A.D. 9	SF, P, K	Manilius I 901–902 (SF); Dio LVI 24.3–4 (SF, P, K)
14	SF, BR, K	Dio-Xiphilinus LVI 29.3 (SF, BR, K); Zonaras X 38 (SF, BR, K); Seneca VII 17.2 (K?)
30?	?	Pseudo-Pilate (NS?)
39?	BR	*Oracula Sibyllina* X 56–57 (BR)
50	SF	Dio-Xiphilinus LX 33.2 (SF); Zonaras XI 10 (SF)
54	BR	Dio-Xiphilinus LX 35.1 (BR)
76	K	Titus in Pliny II 89 (K)
185?	?	Lampridius, *Commodus* XVI 2 (SF?); Herodian I 14.1 (K?)
196	SF	Dio-Xiphilinus LXXV 4.6 (SF)
212?	?	Tertullian, *Ad Scapulam* III (SF?)
300?	?	*Oracula Sibyllina* XII 89–90 (BR?)
333	SF	Aurelius Victor XLI (SF)

[a]**Sources:** Alexander of Aphrodisias, *In Aristotelis Meteorologica*; Appian, *Civil Wars*; Aristotle, *Meteorologica*; Aurelius Victor, *Caesars*; Cicero, *De divinatione*, *In Catilinam*; Dio-Xiphilinus, *Roman History;* Diodorus Siculus, *Library of History*; Dionysius of Halicarnassus, *Roman Antiquities*; Herodian, *Ab excessu divi Marci*; Lampridius, *Vita Commodi (Historia Augusta)*; Livy, *Ab urbe condita*; Lucan, *Pharsalia*; Lydus, *De mensibus, De ostentis;* Manilius, *Astronomicon*; Obsequens, *Prodigiorum liber*; Olympiodorus, *In Aristotelis Meteorologica*; Anonymous, *Oracula Sibyllina*; Orosius, *Adversum paganos*; Ovid, *Metamorphoses*; Anonymous, *Parian Marble*; Paulus Diaconus, *Roman History*; Pausanias, *Description of Greece*; Philoponus, *In Aristotelis Meteorologica*; Pliny the Elder, *Naturalis historia*; Plutarch, *Parallel Lives*; pseudo-Pilate, *Report to Caesar*; Seneca, *Naturales quaestiones*; Silius Italicus, *Punica;* Tertullian, *Ad Scapulum;* Vergil, *Georgics;* Zonaras, *Annals.*

[b]**Categories:** B = beam; BR = blood rain; K = aurora-like comet; MR = milk rain; NS = night sun; P = pillar; SF = sky fire; T = aurora-like torch; X = chasm.

reports for the same year may, in some instances, refer to entirely different phenomena. On the other hand, not all the reports are of independent value, because later
authors have necessarily borrowed from their predecessors. Finally, it is typical that
more reports tended to be generated, or later remembered, during times of stress and
of other notable events.

A few comments about the three largest gaps in the auroral record seem to be
called for, since some authors, such as Schove, have identified these gaps with
aurorally quiet periods.[33] The first gap occurs in the Roman record between 459 and
223 B.C. Our main reference for this period, Livy's history, suffers both from a dearth
of reliable early records in his time (due in part to the burning of Rome by the Gauls
c. 390 and in part to the irregularity of the Roman pontifical annals before c. 300) and
also from the loss in postclassical times of those intermediate books of his history
that cover the years 292 to 220. However, aurorae did occur during at least the earlier
half of this long period, as is demonstrated by the four examples of Greek aurorae
preserved by later Roman writers. Yet not a single aurora from the fourth and fifth
centuries is reported in the great contemporary Greek histories that are still extant;
this silence is undoubtedly due to those historians' very sober attitudes toward
portents of all kinds. The second gap in the auroral record falls between 91 and 49
B.C. (with the exception of the year 63 B.C.). As may be judged by Obsequens' extracts
from Livy's history, the unprecedented civil wars of that period bred a growing public
disrespect for portents[34] and apparently interrupted the transmission to Rome of
reports of many aurorae that must nonetheless have been noted. Finally, the series of
gaps after A.D. 76 is at least partly due to the well-known paucity of historical records
for the late Roman Empire. In sum, I can find no good historical evidence either for
or against the supposition that the gaps in the record are associated with aurorally
quiet periods.

THE ANCIENT AURORAL CYCLE

Because of the fragmentary nature of the auroral record in Table 1, standard methods
of analyzing this record for possible periodicities fail. Thus, Nicolini, Schove, and
Link simply assumed in their work an eleven-year cycle of variability in analogy with
the modern auroral and sunspot cycles.[35] In part, their failure stemmed from not
having had an adequate catalogue. Schove listed, for the period before A.D. 300, only
thirteen auroral years that he regarded as suitable for mathematical analysis; of these,
a mere six lay in the well-documented interval 223–91 B.C. But I find thirty-six useful
auroral years in the latter time interval.

With the help of an appropriate method of time series analysis, I have recently
searched for possible cycles in the ancient auroral data. Since the results of this
analysis have already been presented,[36] it suffices here merely to summarize the main
points. First, there were sufficient data in the interval 223–91 B.C. to analyze separ-

[33]D. J. Schove, "The Sunspot Cycle, 649 B.C. to A.D. 2000," *Journal of Geophysical Research,* 1955,
60:127–146.
[34]For other causes see Krauss, *An Interpretation.*
[35]Nicolini, "Sull' andamento secolare dell' attività solare"; Schove, "The Sunspot Cycle"; F. Link,
"Manifestations de l'activité solaire dans le passé historique," *Planetary and Space Science,* 1964, *12*:
333–348. Pliny (II 97) originally suggested that "chasms," "beams," and the like were periodic phenomena,
but in this opinion he was simply following Pythagorean and Chaldean cometary tradition.
[36]R. Stothers, "Solar Activity Cycle during Classical Antiquity," *Astronomy and Astrophysics* (in
press). This research has depended heavily on the classics collections of the Columbia University Libraries
and the New York Public Library.

ately the categories of "sky fire," "night suns," "blood rain," and "milk rain." These categories showed virtually the same period of cyclical variation, suggesting that they were merely different manifestations of the same phenomenon. The mean period was 11.5 years, with a scatter of less likely periods ranging from 8 to 13 years. A longer period of 80 to 100 years was also present in the data. Second, it was found that the average frequency of visible aurorae near Rome was approximately three per decade. Since these results resemble so closely the characteristics of modern aurorae, it would seem that the second century B.C. was very similar to our own century as far as aurorae are concerned.

[*Editor's Note:* The Conclusion has been omitted. In the paper cited as footnote 36 *(Astronomy and Astrophysics* **77:**121-127) Stothers provides spectral analyses showing that *Sky Fire* and *Night Suns* definitely responded to cycles of the order of 11 years, and *Blood Rain* and *Milk Rain* marginally repsonded to cycles of the order of 11 years.]

19

AURORAL NUMBERS SINCE 500 B.C.

BY D. J. SCHOVE*

Received 1961 March

Variations of sunspot activity are reflected in variations in auroral frequency in middle latitudes of Europe and Asia. Estimated values of auroral frequency reflecting the 'maximum' of successive 11-year cycles were published in the *Journal of Geophysical Research* (Schove 1955) and the estimated dates of the maxima have been briefly tabulated separately (Schove 1956a) in our *Journal*. The information about historical aurorae was collected as part of the so-called Spectrum of Time project (cf. Schove 1961a). Several members of the B.A.A. (notably Miss C. M. Botley and Dr Ho) are kindly contributing to this project by collecting the various dated or datable references to astronomical and other natural phenomena in the historical sources from different parts of the world.

The curve presented in 1955 has since been found[†] to show similarities with curves of radiocarbon content of tree-rings, curves which reflect the varying initial activity in different centuries. This is not perhaps surprising, for radiocarbon in the atmosphere is produced by cosmic rays, and cosmic rays in our upper atmosphere are already known to vary in quantity and quality with the ordinary eleven-year sunspot cycle. It now seems as if the major variations in auroral- and sunspot-activity are reflected in major variations in the radiocarbon produced in our upper atmosphere, directly or indirectly, by the Sun. Irregular long waves in solar activity may, therefore,

* Dr D. J. Schove, St. David's College, Beckenham, Kent.

† Willis 1961, Stuiver 1961. This was first pointed out to me by Professor Fairbridge.

be reflected in corresponding waves in the curve of the differences between the true age and the radiocarbon age of individually dated tree-rings.

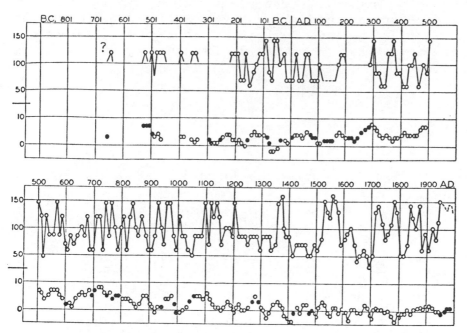

FIG. 1—Curves showing intensity and phase, B.C. 650 to A.D. 2000
Upper curve: Auroral intensity (sunspot numbers from A.D. 1750) of maxima
Lower curve: Phase (remainder) of minima; o = reliable, ● = uncertain

[Reproduced from Schove, 1955b, by kind permission of *J. Geophys. Res.*]

The mean intensity for each decade was also estimated in 1955—when any hope of climatological and tree-ring correlations seemed unlikely. These decadal values, not published at the time, are now presented (with a few minor revisions in the light of information received from Dr Ho and from Miss C. M. Botley) in Table 1. The auroral numbers are again (cf. Schove 1955, p. 137) intended to be comparable with sunspot numbers, but this time with mean sunspot numbers for the decade (or, for most practical purposes, with mean numbers over an eleven-year period).

In the calculation of these auroral numbers, it was first assumed that auroral frequencies in different parts of the world have varied with sunspot numbers before 1749 in much the same way as they have since, and that in Europe the quantitative relationship has been constant; it was then found necessary to suppose that aurorae must have been several times as frequent in medieval China, Korea and Japan (from which countries many of the records derive) as would have been expected from the modern map of isochasms (cf. Newton 1958 Figure 29) originally prepared by Fritz.

It is also necessary to make assumptions about the quality of the record. The numbers of aurorae recorded represent only a fraction of those which must have occurred and it is necessary to have a fairly clear idea as to the value of this fraction. Here the Spectrum of Time information helps in our assessment. In the third century A.D., for instance, the coverage is very poor and even our knowledge of comets is inadequate. On the other hand, as is evident from our previous paper (Schove and Ho 1958), in the eleventh century the Chinese astronomical observers were so well trained and kept such a close watch on the night sky that they frequently observed aurorae, even in weak cycles.

TABLE I

Century	0	1	2	3	4	5	6	7	8	9	Probable error	
				Decades								
V BC	40	55	50	70	65	less	than	60			15	
IV BC	65	60	—	—	—	—	—	—	—	—	15	
III BC	—	—	—	—	—	70	—	—	65	65	10	from 220 BC
II BC	60	40	40	60	30	30	40	50	60	70		
I BC	60	50	30	60	70	50	30	60	40	40	15	from 30 BC
I AD	30	50	40	30	40	60	50	30	40	30	20	to AD 80
II AD	50	—	—	—	—	—	—	50	50	50	15	to AD 180
III AD	—	—	—	—	—	—	—	—	—	50		to AD 300
IV AD	70	60	50	30	40	60	60	70	50	60	10	
V AD	50	40	50	55	55	60	40	50	50	50	15	
VI AD	70	60	45	50	50	50	70	60	70	70	10	AD 561/600
VII AD	40	60	60	30	50	40	40	60	50	30	15	AD 601/660
VIII AD	30	50	50	40	60	50	70	60	50	60	10	AD 661 to
IX AD	60	40	50	75	60	60	50	70	50	40		AD 1070
X AD	40	40	60	50	45	40	70	70	60	50		
XI AD	70	60	50	50	25	25	50	60	50	75	5	AD 1071 to 1240
XII AD	50	70	80	80	40	50	45	65	60	50		
XIII AD	70	50	60	30	50	50	45	45	40	30	10	AD 1241 to 1520
XIV AD	60	40	45	45	30	50	60	80	65	40		
XV AD	40	25	25	35	40	40	35	25	30	30		
XVI AD	35	40	60	65	60	65	70	75	60	45	5	from 1521 to
XVII AD	50	50	60	50	25	30	25	30	25	20		1750
XVIII AD	25	35	60	60	40	36	57	70	72	28		
XIX AD	26	22	33	67	57	46	53	40	35	45		
XX AD	37	41	42	54	74	94						

All these assumptions mean that, as sunspot numbers, our figures are somewhat aribtrary. A general unknown error evidently affects all data before about 1850 and this error is likely to become greater before 1700 and greater still before A.D. 1000. Thus, the mean value for 1951–60, 94 (kindly supplied by Mr Laurie), although the highest value in our table, may well have been exceeded in various decades prior to 1740.

On the other hand, estimates of the auroral numbers made in accordance with those assumptions—from the sunspot and auroral records of Europe and China separately were often found to be remarkably consistent and this consistency is reflected in the smallness of the 'probable error' specified in

138

the right-hand column of the table. As the previously mentioned general error is presumably constant within a particular century, this new error can be regarded as the error affecting the (exact) departure of the decade-mean from the (inexact) mean for the century concerned.

It will be apparent from my earlier papers (Schove 1948b, 1951c, 1950d, 1951e) that information is often very scanty up to A.D. 550. The probable errors are thus large, and indeed little significance can be attached to smoothed values before A.D. 650. On the other hand, decades after A.D. 650 are well-documented throughout medieval times in at least one of the three Far Eastern countries: Japan, Korea or China. They are also well-documented, except for certain periods prior to 1090, in the West.

The nature of the auroral evidence in a weak period can be gauged from the joint paper by Schove and Ho (1958) in this journal. In a strong decade, such as the 1120s, the international displays (cf. Botley 1958) are much more numerous, and, occasionally, important aurorae occur as late as five years after the maximum (cf. Botley 1956, Newton 1958 p. 163, cf. Visser 1959). It is hoped to publish soon, as joint papers with my collaborators, some of this detailed evidence. In the meantime, further extracts, in the original language and in English, relating to the primary sources of observed displays would be appreciated for the collections of the Spectrum of Time.

The variations of solar activity produce characteristic variations in the circulation pattern of the Earth's atmosphere through what I term the pressure parameter (Schove 1961g, 1961b), although for some reason the parameter changes are often out of phase with the 11-year sunspot cycle. It is hoped to calculate agreement coefficients between the auroral numbers, the pressure parameter and indices of rainfall, temperature and tree-ring growth (cf. Schove 1961a, Table 1). Relations between auroral numbers and weather had seemed unlikely in 1955, when the data were originally prepared, so that the inevitable subjective aspect of the figures could hardly influence any results so determined.

Other results of geophysical interest can be determined from the documentary information. Naked-eye sunspots and other aurorally-productive regions travelled round the Sun in the twelfth century in $27\frac{2}{3}$ days just as they do (cf. Newton 1958, p. 40 and 162) today. The direction of magnetic North was different, but the changes in the meridian can be gauged (cf. for modern Britain, McInnes and Robertson 1958, p. 171–3) from the recorded changes in direction of auroral displays. Thus in England in the early twelfth century 'holy men' walked due North towards the celestial light, whereas in China the contemporary observed change of auroral direction from NW towards N noted by Schove and Ho in 1958 is not inconsistent with the new palaeomagnetic evidence (Watanabe 1959) from pottery in nearby Japan; Watanabe (his Plate III Figure 15) found that magnetic declination there changed gradually from NNW about 1075 to N about 1225.

The decadal data may prove useful in testing the significance of the cycles briefly noted in 1955 (p. 141–2). Vague, interrelated cycles were noticed in both period and activity. The 80-year cycle (80 to 90, according to Willett, but 89, according to Visser 1959) in sunspot length noted by Gleissberg (cf. 1958) seemed so real since 1610 that it was taken into account in estimating the dates of some earlier minima, but my adoption in 1955 of Gleissberg's value of 78-years was perhaps arbitrary. It is possible that, if maxima with probable errors of 3 or more are excluded from the analysis, as originally suggested (Schove 1955, p. 137), the true length and significance of some cycle between 65 and 90 years can be determined. The cycle of about 165-years noted since 1510 might seem to be double the Gleissberg cycle, but no cycle of this length was noticed in the medieval period. Moreover, the Gleissberg cycle did not seem to be clearly marked in the intensity pattern, whereas, on the other hand, a 200-year cycle seemed significant (cf. also Dewey 1960). Solar activity was apparently greater in even centuries (such as the present twentieth century) although this rule has been clear only since A.D. 300.

Further information of historical aurorae is still welcome, especially for the third, thirteenth and fifteenth centuries A.D. and for periods 590–640, 770–805, 870–900, 1090–1120. Tree-ring and varve studies may eventually throw light on the sunspot cycle (cf. Schove 1961, Stuiver 1961) but at present these historial records are essential scientific evidence needed to piece together an important aspect of the Spectrum of Time. My thanks are due again to all those who have contributed.

The 'auroral numbers' given previously (Schove 1955, p. 136–137) are approximately equivalent to the sunspot numbers in the maximum (calendar) year. The (twelve-month) smoothed Zurich sunspot numbers are slightly greater.

Bibliography

BOTLEY, C. M. 1957 'Some great Tropical Aurorae', *J. Brit. astron. Ass.* **67**, 188–191.

DEWEY, E. R. 1958 'The Length of the Sunspot Cycle'. *J. Cycle Res.* **7**, No. 3, 70–91.

1960 'The 200-year Cycle in the Length of the Sunspot Cycle'. *J. Cycle Res.* **9**, No. 2, 67–82.

GLEISSBERG, R. A. 1958 'The eight-year Sunspot Cycle'. *J. Brit. astron. Ass.* **68**, No. 4, pp. 148–152.

McINNES, B. and 1959 'The Auroral display of 1958 September 4–5'.
K. A. ROBERTSON *J. Brit. astron. Ass.* **69**, 167–176.

NEWTON, H. W. 1958 *The Face of the Sun* (Pelican Books), London.

SCHOVE, D. J. 1948b 'Sunspots and Aurorae, 500–250 B.C.' *J. Brit. astron. Ass.* **58**, pp. 178–190 and pp. 202–204.

1950d 'Visions in North-West Europe and dated Auroral Displays (A.D. 400–600)'. *J. Brit. Archaeol. Ass.* 3rd ser. 13, pp. 34–49.

1951c 'The earliest dated Sunspot', *J. Brit. astron. Ass.* **61**, (Dec. 1950) pp. 22–24 and pp. 126–128.

1951e 'Sunspots, Aurorae and Blood Rain', *Isis*, **42**, pp. 133–138.

1955b 'The Sunspot Cycle, 649 B.C. to A.D. 2000', *J. Geophys. Res.*, **60**, No. 2, pp. 127–146.

	1961a	'Solar cycles and the Spectrum of Time since 200 B.C.'.
	1961b	Both papers have appeared in the Monograph, 'Solar Variations, Climatic Change, and Related Geophysical Problems' now published by the New York Academy of Sciences. *Annals NY Acad. Sci.*, New York, 1961, **95**, 605–622.
	1961e	The Spectrum of Time. *J. Brit. astron. Ass.*, **71**, 320–322.
	1961g	The Major Pressure Oscillation, *c.* 1875–*c.* 1960. *Geofisica pura e applicata*, Milano, **49**, 255–263.
SCHOVE, D. J. with Ho, Dr P. Y.	1958	'Chinese Aurorae, 1. A.D. 1048–1070', *J. Brit. astron. Ass.* **69**, pp. 295–304.
STUIVER, M.	1961	Monograph of the New York Academy of Science (cf. Schove 1961b above).
VISSER, S. W.	1959	*Medelingen en Verhandelingen*, **75**, The Hague.
WATANABE, N.	1959	The Direction of Remanent Magnetism of Baked Earth . . . in Japan. *J. Fac. Sci., Univ. Tokyo*, **5**, (Vol. 11), 1–188.
WILLIS, E. H.	1961	Monograph of the New York Academy of Science (cf. Schove 1961b above).

[*Editor's Note:* Since this paper was written, charts by Siscoe (Paper 36; see also Figs. 4 and 5 in our Introduction) and Dai and Chen (1980, Chronology of Historical Auroral Data in China, Korea, and Japan, *Kejishiwenji* **6**:87–146) confirm the main results (in Chinese) of Table I in this paper. Slight revisions have, however, been incorporated in our Appendix C, column *h*, but only where new auroral evidence has been found.

Comparison with ^{14}C data reveals that there is an average delay of about 17 years in the maximum effect of outstanding peaks, for example, between the A-max (Paper 39) and ^{14}C-min. In Table I no allowance has been made for the changing position of the auroral oval (see Feynman and Silverman in *Exploration of the Polar Upper Atmosphere,* Deehr and Holtet, eds., Reidel, Dordrecht, Holland, 1980), which may have been further south in Eurasia in the early sixteenth century.

Palaeomagnetic results are summarized by Creer in *Ice-cores, Varves and Tree-rings,* Schove and Fairbridge, eds., Balkema, London, 1983.

Decadal values of sunspots for 1961–1970 and 1971–1980 are 60 and 67 (see Table 1 in our Introduction).]

Part IV

SYNTHESIS

Editor's Comments
on Papers 20 Through 23

20 **SCHOVE**
The Sunspot Cycle, 649 B.C. to A.D. 2000

21 **MAUNDER**
The Prolonged Sunspot Minimum, 1645–1715

22 **EDDY**
The Maunder Minimum

23 **ANONYMOUS**
Excerpts from Junk in Space: What Goes Up

> Scissors-and-paste historians study periods; they collect all the
> extant testimony about a certain limited group of events, and
> hope in vain that something will come of it. Scientific historians
> study problems: they ask questions, and if they are good histori-
> ans they ask questions which they see their way to answering.
> (Collingwood, 1946, p. 281)

We have considered the catalogs of sunspots and aurorae separately. We now consider what problems can be solved by considering
them in combination. The first problem is the detection, numbering,
and magnitude of successive solar cycles. The author's attempt in
1955 to go back to 649 B.C. is presented as Paper 20. The date 649 B.C.
is based on the statement "It rained gold in China" in what is thought
to be the genuine Bamboo Annals (Professor D. M. Keightley, personal communication). This could have been either an aurora or a
meteor shower.

Revisions have been made since Paper 20 was written but
changes are small except for data for about A.D. 700. Those for the
period since A.D. 1500 are given in Tables 1, 2, and 3 in our
Introduction. Minima for the period before 1500 are revised in our
Appendix A, which replaces Table 3 in Paper 20; some strong maxima
have been redated in our Appendix B one year earlier (cf. Schove,
1980) than in Paper 20. The reliability of the revisions is assessed
as follows.

pre-250 B.C.: speculative. For the corrections to Livy's dates see Ogilvie (1965) and Paper 2.

250 B.C.–A.D. 45: reliably counted and errors in dates generally less than 2 years.

A.D. 70–A.D. 285: still uncertain.

A.D. 285–A.D. 600: reliably counted and dated ($\pm$ 2 years).

A.D. 600–A.D. 750: A tentative revision has been made.

A.D. 770–A.D. 960: generally reliable but evidence weak c. 900 and c. 950.

A.D. 960–A.D. 1500: reliably counted but dates still uncertain c. 1200–1230, c. 1300–1350, c. 1400–1430 and c. 1470–1500.

The doubtful mean given in the table on page 140 of Paper 20 now seems to be A.D. 691. This table (with the third column corrected) implies a mean 14-cycle length of about 155.2 years, a better hypothesis than the original assumption of 9 cycles to a century (cf. Paper 9).

The short research note by Henkel (1972) is mentioned because his diagram, neatly but unintentionally, highlights the erroneous variability of the cycle lengths in Table 2 of Paper 20. In revising both Wolf's seventeenth century dates and my own medieval dates, I have assumed that the cycle length from minimum to minimum lies between 9 and 13.

The Maunder Minimum of sunspots and aurorae is the topic of Papers 21 and 22. This minimum is clearly illustrated in curves and auroral numbers (see Figs. 5, 8, and 10 in our Introduction). The lull in sunspot activity in the mid-1650s was noted by Huygens and Hevelius (see Schove, 1983). Huygens wrote to Hevelius on 16 September 1658 that "for the space of three years now (the sun) has remained without spots, which at other times, were so frequently to be seen" (Huygens, 1889). Hevelius in his reply of 26 October 1658 gave details of spots occurring in 1654 (25 August and 18–19 September) and in 1657 (22–26 December). This Latin correspondence (see Huygens, 1889) is not noticed in recent literature and implies that spots were seen more often *before* 1655. The lull in the mid-1660s was mentioned by Marvell in his 1667 "Poem: The Last Instructions to a Painter" (in Margoliouth, 1971) and again by Hevelius in 1668 (see Weiss and Weiss, 1979), but the general lull from 1645 to 1699 was first emphasized by Spoerer (1887, 1889).

Little notice was taken of Spoerer's papers until Maunder's second paper of 1922 (Paper 21; cf. Maunder, 1890). The story is well told by Eddy (Paper 22, 1976) who has "mildly questioned" whether the 11-year cycle was in existence at all during this period. Eddy is a senior scientist at the High Altitude Observatory, National Center for Atmospheric Research, Boulder, Colorado, and his work has been the subject of television and radio programs. Maunder's extreme view has not been supported by other investigations (cf. Schroeder, 1979; Botley, 1979; Landsberg, 1979; Xu and Jiang, 1979; Vitinsky, 1978). Thus Gleissberg and Damboldt refer to the work of von Guericke (1672) and the dissertation of Ettmueller (1693), who had observed sunspots in 1689 and 1690. Nevertheless, except for a display in January 1693 there was indeed a lull in auroral displays in New England in 1645 to 1699 (see Rizzo and Schove, 1961; Eather, 1980) and in the light of Eddy's evidence (Paper 22), the writer has now modified the sunspot and auroral magnitudes given in 1955, but possible double maxima with a prolonged Gnevyshev gap have *not* been introduced in Tables 1 and 2 (in our Introduction).

Prediction of a sunspot maximum, *once the cycle has started,* is usually successful as Waldmeier (1935, 1939, 1968) has neatly demonstrated (Fig. 2 in our Introduction). The difficulty of prediction before the *preceding* minimum led Waldmeier to postulate his *eruption hypothesis,* according to which each cycle is independent and unpredictable. Some longer predictions have been made, partly inspired by spectral analyses and by trends believed to belong to the 80-year cycle. Gleissberg had thus predicted cycles 19 and 20 successfully although he, and most other experts listed, were unsuccessful in 1971 (see data on our p. 15), and remained unsuccessful until 1976 (cf. Kane, 1978), in predicting the strength of the present cycle 21. The success of the writer's predictions in 1955 of a strong maximum based on the 200-year cycle was accidental: my timing was more than 3 years off since the intervening cycles were stronger and shorter than expected. Many forecasting rules are now available (Vitinsky, 1965; Rubashev, 1964) and Bumba is the leader of a working group on long-term solar activity predictions (see Donnelly, ed., 1980). Failure to allow for the effects of a sunspot maximum (see King-Hele, 1963) once resulted in the fall to earth of the heavy satellite laboratory *Skylab* in July 1979 (Paper 23).

The effects of the sunspot maximum in reducing the pressure level of the upper troposphere (200 mb) in the 1950s was shown in a table (Schove, 1967) this sunspot maximum even affected the Western tropics. (In column c of the table in Schove, 1967 *west* should be *east.*)

The mean length of the sunspot cycle has been variously estimated in Table 1 of this part.

Nicolini and Wittmann in effect omit one (not specified) of the cycles listed in Paper 20 but the auroral evidence seems to indicate that since A.D. 300 there have never been *less* than nine cycles in a century.

Table 1. Mean Length of Sunspot Cycles According to Different Authors

Solar Activity	Years	Source
sunspots since 1610	11.1	Wolf, 1852 and Paper 9
Far Eastern spots and aurorae	10.83	Kanda, 1933
sunspots since A.D. 300	11.086 ± .005	Schove, Paper 20
sunspots since A.D. 300	11.13	Nicolini, 1977
sunspots only	11.135	Wittmann, 1978
Early Chinese spots	10.5	Yunnan Observatory, 1977
classical aurorae	11.5	Stothers, 1979

REFERENCES

Botley, C. M., 1979, The "Maunder Minimum" in Perspective, in *Yearbook of Astronomy*, P. Moore, ed., Faber & Faber, London, pp. 187–191.

Collingwood, R. G., 1946, *The Idea of History*, Clarendon Press, Oxford, England, 339p.

Donnelly, R. F., ed., 1980, *Solar-Terrestrial Predictions Proceedings*, vol. 3, Part A, United States Department of Commerce National Oceanic and Atmospheric Administration, Boulder, Colo., 128p.

Eather, R. H., 1980, *Majestic Lights: The Aurora in Science, History, and the Arts*, American Geophysical Union, Washington, D.C., 321p.

Eddy, J. A., 1976, The Sun Since the Bronze Age, in *Physics of Solar Planetary Environments*, D. J. Williams, ed., American Geophysical Union, Washington, D.C., pp. 958–972.

Ettmueller, M. E., 1693, *De maculis in sole visis*, Wittenberg. (2nd ed., 1697, published at Giessen.)

Henkel, R., 1972, Evidence for an Ultra-long Cycle of Solar Activity, *Solar Physics* **25:**498–499.

Huygens, C., 1889, *Oeuvres (Leyden)* **2:**241, 262–263.

Kanda, S., 1933, Ancient Records of Sunspots and Aurorae in the Far East and the Variation of the Period of Solar Activity, *Imperial Academy Japan Proc.* **9:**293–296. (Sine curve solution.)

Kane, R. P., 1978, Predicted Intensity of the Solar Maximum, *Nature* **274:**139–140.

King-Hele, D. G., 1963, Decrease in Upper Atmosphere Density Since the Sunspot Maximum of 1957–1958, *Nature* **198:**832–834.

Landsberg, H. E., 1979, *Some Previously Unrecorded Observations of Sunspots During the Maunder Minimum and Auroras Not Listed in Standard Catalogs*, Institute for Physical Science and Technology Technical Note BN-911, University of Maryland, College Park, Md.

Margoliouth, H. M., ed., 1971, *The Poems and Letters of Andrew Marvell*, 3rd. ed., Oxford.

Maunder, E. W., 1890, Professor Spoerer's Researches on Sunspots, *Royal Astron. Soc. Monthly Notices* **50:**251–252. (Short note only.)

Nicolini, T., 1977, Sull' andamento secolare dell' attivita solare, *Accad. Sci. Fis. e Mat. (Soc. Naz. Sci. Lettere ed Arti Napoli) Rend.*, ser. 4, **43:**1–11.

Ogilvie, R. M., 1965, *A Commentary on Livy,* books 1–5, Clarendon Press, Oxford, England, 788p.

Rizzo, P. V., and D. J. Schove, 1961, Early New World Aurorae, 1644, 1700, 1719, *British Astron. Assoc. Jour.* **72:**396–397.

Rubashev, B. M., 1964, *Problemy solnechnoy aktivnosti,* Nauka, Moscow & Leningrad, 362p. (*Problems of Solar Activity,* translated in NASA TTF 244.)

Schove, D. J., 1967, Anomalies at the 200 mb Level in the Tropics, 1950–1962, *Royal. Meteorol. Soc. Quart. Jour.* **93:**137–138.

Schove, D. J., 1980, The 200-, 22- and 11-year Cycles and Long Series of Climatic Data, Mainly Since A.D. 200, in *Sun and Climate,* Centre National d'Etudes Spatiales, pp. 87–100.

Schove, D. J., 1983, *The Sunspot Minimum of the Mid-1650s, Annales Geophysicae.*

Schroeder, W., 1979, Auroral Frequency in the Seventeenth and Eighteenth Centuries and the Maunder Minimum, *Jour. Atmos. and Terrest. Physics* **41:**445–446.

Spoerer, F. W. G., 1887, Ueber die Periodicitat der Sonnenflecken seit dem Jahre 1618, *Astronomische Gesell. Vierteljahrsschrift* **22:**323–329.

Spoerer, F. W. G., 1889, Ueber die Periodicitat des Sonnenflecken seit dem Jahre 1618, *Kgl. Leopoldinisch-Carolinische Deutsch. Akad. Naturf. Nova Acta* **53:**280–324.

Stothers, R., 1979, Solar Activity Cycle During Classical Antiquity, *Astronomy and Astrophysics* **77**(1–2):121–127.

Vitinsky, Y. I., 1965, *Solar Activity Forecasting,* Israel Program for Scientific Translations, Jerusalem, 129p. (*Prognozy solnechnoi aktivnosti,* Akad. Nauk Izd. SSSR, Leningrad, 151p.) (Standard work.)

Vitinsky, Y. I., 1978, Comments on the So-called Maunder Minimum, *Solar Physics* **57:**475–478.

von Guericke, O., 1672, *Experimenta nova* (*ut vocantur*) *magdeburgica de vacuo spatio,* Amstelodami, Amsterdam.

Waldmeier, M., 1935, Neue eigenschaften der sonnenflecken kurve, *Astronomische Mitt.* **133:**105–130.

Waldmeier, M., 1939, Sunspot Activity, *Astronomische Mitt.* **138:**470.

Waldmeier, M., 1968, Sonnenfleckenkurven und die methode des sonnenaktivitatsprognose, *Astronomische Mitt.* **286:**1. (See Fig. 2 in our Introduction.)

Weiss, J. E., and N. O. Weiss, 1979, Andrew Marvell and the Maunder Minimum, *Royal Astron. Soc. Quart. Jour.* **20:**115–118.

Wittmann, A., 1978, The Sunspot Cycle Before the Maunder Minimum, *Astronomy and Astrophysics* **66:**93–97.

Wolf, R., 1852, Bestimmung der Epochen für das Minimum und Maximum der Sonnenflecken-Bildung, *Naturf. Gesell. Bern Mitt.*

Xu, Z., and Y. Jiang, 1979, the Solar Activity of the Seventeenth Century Viewed in the Light of the Sunspot Records in the Local Topographies of China, *Nanking Univ. Annals. Phys. Sci.* **2:**33–38. (In Chinese.)

Yunnan Observatory, 1977, A Recompilation of our Country's Records of Sunspots Through the Ages and an Inquiry into Possible Periodicities in Their Activity, *Chinese Astronomy* **1:**347–359. (First published in Chinese in *Acta Astron. Sinica* **17:**217–227, 1976.)

20

Reprinted from *Jour. Geophys. Research* **60**:127–146 (1955)

THE SUNSPOT CYCLE, 649 B.C. TO A.D. 2000

By D. Justin Schove

[*Editor's Note:* Since this paper was written a revision of the 1701–1749 part of Table 1 on page 128 has been prepared (see Table 1 in our Introduction). In Table 2 on page 132 the maxima dated -501 and -272 should be deleted, and the dates A.D. 80–250, A.D. 650–750, and A.D. 1210–1235 should be revised (see our Appendix B).

On page 136 maximum 1937.4 is 3.9 years rather than 2.9 years after 1933.5, and it would have been better in the next column to use R_M rather than annual values. In Figure 1 I now suggest that the curve of phase (remainder) often responds inversely to the solar wind (see our Appendix C, column *j*). Errors of 11 as indicated in section 11 are possible in the third century A.D. and in the early eighth century A.D. Radiocarbon evidence suggests that the aurorally weak period of 440–400 B.C. lasted until c. 350 B.C. (see T. Wigley in *Sun and Climate*, 1980, Centre National d'Etudes Spatiales, p. 14).

Sunspot cycles in 1955–1980 proved to be even stronger and earlier than expected.]

ABSTRACT

Annual sunspot numbers since 1700 and the known maxima and minima since 1610 show a similarity of pattern from century to century. This suggests that the mean cycle is approximately 11-1/9 years.

The records of sunspots and aurorae enable magnitudes and dates of sunspot maxima since at least A.D. 300 to be estimated. The constancy of the mean cycle over long periods enables the number of missing maxima to be calculated, and, using certain general principles, a table of minima complete since at least 200 B.C. can be established. A 78-year cycle appears to exist in the length of the sunspot cycle and an irregular cycle of about 200 to 205 years may exist in auroral intensity. A characteristic pattern in even centuries enables some predictions to be made for the next 50 years. Intervals between intense maxima in the range 200 to 1,000 years apart show clusters at certain values; these values are close to multiples of 11.11. Intervals between well-dated maxima since A.D. 300 are often slightly less than such multiples, for example, 554 instead of 555; from B.C. 200 to A.D. 300, intervals are slightly greater. In classical and early medieval times the cycle was thus slightly less. The variability of the sunspot cycle is only apparent. The fundamental rhythm of 11.1 years (together with the 78-year cycle) is constant through the centuries; temporary aberrations are partly due to variations in sunspot intensity, inasmuch as active cycles tend to become "early" and weak cycles "late."

1. *Introduction.*

Sunspot numbers form a long and reliable series since A.D. 1749, and using the calculations of Wolf [see 1 of "References" at end of paper] can be extended (Table 1), on an annual basis, back to A.D. 1700. The 11-year solar cycle has been

TABLE 1—*Annual sunspot numbers, A.D. 1700–1953 (information partly supplied by H. W. Newton)*
—Part A

Note: The value for 1700 was (6); values in parentheses are less certain than the others.

Period	Schove remainder										
	1	2	3	4	5	6	7	8	9	10	11/0
1701–1711	(12)	(19)	26	39	60	32	23	12	9	(3)	0
1712–1722	0	2	12	31	50	65	(62)	42	31	29	(25)
1723–1733	(12)	24	43	80	112	(100)	(75)	(50)	(31)	(12)	(6)
1734–1744	(20)	(37)	(72)	82	(106)	98	(75)	(44)	23	18	(6)
1745–1755	12	25	44	62	81	83	48	48	31	12	10
1756–1766	10	32	48	54	63	86	61	45	36	21	11
1767–1777	38	70	106	101	82	67	35	31	7	20	93
1778–1788	154	126	85	68	39	23	10	24	83	132	131
1789–1799	118	90	67	60	47	41	21	16	6	4	7
1800	15										
Mean.........	39	48	56	64	71	66	46	35	29	28	32
Smoothed mean	41	48	54	61	61	56	49	41	34	33	35

TABLE 1 *(Continued)*—Part B

Period	Schove remainder										
	1	2	3	4	5	6	7	8	9	10	11/0
1801–1811	34	45	43	47	42	28	10	8	3	0	1
1812–1822	5	12	14	35	46	41	30	24	16	7	4
1823–1833	2	9	17	36	50	63	67	71	48	27	9
1834–1844	13	57	121	138	103	86	63	37	24	11	15
1845–1855	40	61	99	124	96	67	65	54	39	21	7
1856–1866	4	23	55	94	96	77	59	44	47	31	16
1867–1877	7	37	74	139	111	102	66	45	17	11	12
1878–1888	3	6	32	54	60	64	63	52	25	13	7
1889–1899	6	7	36	73	85	78	64	42	26	27	12
1900	9										
Mean.........	12	29	55	82	77	67	54	42	27	16	9
Smoothed mean	24	37	51	62	67	64	53	41	30	21	19

TABLE 1 (*Concluded*)—Part *C*

Period	Schove remainder										
	1	2	3	4	5	6	7	8	9	10	11/0
1901–1911	3	5	24	42	63	54	62	49	44	19	6
1912–1922	4	1	10	47	57	104	81	64	38	26	14
1923–1933	6	17	44	64	69	78	65	36	21	11	6
1934–1944	9	36	80	114	110	78	65	36	21	11	6
1945–1955	33.2	92.6	151.6	136.3	134.7	83.9	69.4	31.5	13.9	( )	( )
Sum.........	55	152	310	403	434	409	345	228	148	......	......
Mean.........	11	30	62	81	87	82	69	46	30	......	......

reliably traced back to about 1610, when sunspots were discovered through the telescope. No explanation of the cycle has proved satisfactory, and current estimates of its mean length vary according to the period selected for investigation. Prediction of future sunspot numbers and cycles is important but harmonic analysis has proved unsuccessful and there has been little agreement among scientists as to a suitable basis for forecasting.

Records of aurorae and sunspots extend back to at least the fifth century B.C., and an attempt to trace the solar maxima of intervening centuries appears to offer a basis for theoretical and mathematical investigations.

2. *Sunspot Numbers since A.D. 1700.*

From A.D. 1749, the annual sunspot numbers in Table 1 are the unaltered Zurich figures as calculated by Wolf and his successors. It is not generally known that Wolf also extended his calculations back to 1700, and indeed his estimates [1] have suffered neglect, because they were never revised to the same basis as his later values. Thus, the values for 1749 and 1750 were given by Wolf in 1868 as 63.8 and 68.2. As these two years are now credited with sunspot numbers of 80.9 and 83.4, Wolf's earlier figures have likewise been increased by 24.4 per cent for tabulation here. Many values—here indicated by parentheses—were obtained by interpolation or data that Wolf considered inadequate; on the other hand, he indicated that his figures for certain years (1705, 1709, 1716, 1717, 1719, 1749, and 1750) were especially reliable.

The evidence of auroral observations suggests that these revised values are still too low in the 1720's and 1730's. At the same time, for many purposes the figures are useful, and in particular they show clearly the phase of the 11-year sunspot cycle.

3. *The Sunspot Cycle since A.D. 1600.*

Annual figures are not available for the seventeenth century, but the Zurich dates of maxima and minima from 1610 onwards have been calculated with fair reliability, and these dates have been included in my Tables 2 and 3.

151

4. *The Length of the Sunspot Cycle.*

The well-known 11-year cycle has varied in practice between 8 and 16 years; so far in the present century, it has been a cycle of 10 years. The length of the *mean* cycle is variously regarded by different authors as 10.83, 11.1 and 11.4 years. It certainly differs little from 11.11 or 100 divided by 9, because maxima and minima tend to occur within corresponding groups of years in each century. Thus, maxima occur in 1639, 1738, 1837, and 1937, whilst minima occur in 1610, 1712, 1810, and 1913.

5. *The Phase of the Sunspot Cycle.*

In accordance with the rule just indicated, the years of sunspot minima tend to be approximately divisible by 11. Examples are 1655, 1666, 1755, 1833, and 1933.

The phase of a particular cycle can be loosely defined by the *remainder* when multiples of 11 have been deducted from the last two digits of the years of minima. Thus, in the past three centuries, a remainder of 5 has been typical. The phase of a maximum near the turn of a century is ambiguous. Thus, 700 may be regarded as a seventh-century maximum with remainder 1 or an eighth-century maximum with remainder 0.

6. *Prediction of Future Cycles.*

Prediction of sunspot cycles is hampered by the variable length of the period between successive maxima. The last great maximum was expected by some scientists to fall in 1949, a year with a typical remainder of 5, but for some reason it arrived early, in 1947, a year with a remainder of 3 [2]. Some success is being achieved in the prediction of sunspot numbers from month to month and year to year [3]; Gleissberg has predicted that the next minimum will lie between March and July 1955 and will be followed by another intense maximum between August 1958 and August 1959.

The ultimate test of all theories and formulae must lie in prediction. A successful prediction of a single maximum would not be decisive; a formula relating to the maxima before A.D. 1600 would, however, be subject to proof or disproof, provided that the dates of early maxima can be determined from historical evidence.

7. *Sunspot Cycles before A.D. 1600.*

A preliminary attempt to determine the years of sunspot maxima was made in the nineteenth century by Fritz, whose results [4] are those generally quoted. His analysis was however somewhat uncritical, and the evidence was fitted rather arbitrarily into an 11-year scheme. (*a*) *Sunspots* and (*b*) *aurorae* are, of course, indicators of the cycle, but (*c*) *hail* and (*d*) *wine harvests* in Central Europe were also used in the analysis. It is now clear that no valid conclusions about sunspot cycles can be gleaned from references to weather and harvests. The results of (*e*) *tree-ring analysis* and (*f*) *earthquakes* may possibly prove useful eventually. The data from tropical trees used by de Boer [5] and from earthquakes in different parts of the world used by Davison [6] lead to maxima in the sixteenth century that are not inconsistent with those determined from auroral records and presented here. At the same time, my own investigations of tree-ring data do not give grounds for optimism [7].

The rather shaky foundations of the final table of Fritz must now be disregarded. On the other hand, his tables for sunspots and aurorae separately were recently combined and retabulated [8], because many of his results have been confirmed by the present extended investigation.

8. *Ancient Aurorae and Sunspots.*

There are numerous records of sunpots seen with the naked eye and of ancient displays of northern lights [9] in different parts of the world. These were collected initially as by-products of a collection of meteorological chronologies, chronologies which have been sent to me in connection with the so-called "Spectrum of Time" project [10].

9. *The Estimation of Years of Maxima.*

The evidence of aurorae and sunspots can be used to prepare a table of years of maxima subject to certain assumptions. Thus, in the preparation of Table 2, it has been assumed that

(a) The time between successive maxima was never less than 8 and never more than 16 years.

(b) There are 9 maxima in 100 years.

These two assumptions have proved valid since A.D. 1515 [8, 11]; indeed, despite the irregularity of individual cycles, large departures from eleven do not persist for many cycles. There are 30-year groups when the cycle is longer (for example, centred c. 1650/55, 1720, 1805/10, 1885/90) and intervening groups when it is shorter (for example, centred c. 1685, 1760/65, c. 1845, c. 1935). The mean length of seven cycles has lain between 10 and 12 years for at least two and one-half centuries.

These two generalizations, together with the principles outlined below, enabled the maxima of Table 2 to be determined. Some doubt still exists about the dates in parentheses; dates before 502 B.C. depend only on somewhat dubious evidence for aurorae in China, and dates from 502 to 460 on the B.C. dating implied by Livy. The Greek evidence for the fifth century B.C. has not been used, as it has not been possible to differentiate between comets and aurorae [12].

10. *The Numbering of Minima and Maxima.*

The system of numbering adopted is a decimal one, based on the phase-rule already indicated. Thus, in recent centuries, the year endings of sunspot minima tend to follow the pattern

$$\text{'00} \quad \text{'11} \quad \text{'22} \quad \text{'33} \quad \ldots \quad \text{'88} \quad \text{'99/00}$$

and they are indicated by

$$.0 \quad .1 \quad .2 \quad .3 \quad \ldots \quad .8 \quad .0$$

An additional minimum involves the inclusion of a term ending in .9. The intervening maxima are thus denoted by decimals .05, .15, .25 ... .85, .05. In early centuries, it is the maxima that follow the eleven-times table rule.

TABLE 2—*Sunspot cycles since B.C. 649* (November 1954)

Number of max.	Year of max.	Probable error	Remainder	Intensity of maximum	Years since p. max.	Years since p. min.	Min.	Rem.
−6.45	−648	3	8	S(?)	..	..	(−653)	3
−5.25	−522	3	12	S(?)	..	5	(−527)	7
−5.15	(−512)	4	11	..	(10)	4	(−516)	7
−5.05	−501	2	11	S(?)	(11)	4	(−505)	7
−4.85	(−491)	4	9	W–M	(10)	5	(−496)	4
−4.75	−181	2	8	S	10	5	−486	3
−4.65	−471	2	7	S	10	3	−474	4
−4.55	−461	2	6	S	10	4	−465	2
−3.85	−393	3	7	S	..	6	−397	3
−3.75	.....	..	..	..	..	..	−386	3
−3.65	.....	..	..	..	..	..	(−375)	..
−3.55	.....	..	..	..	..	..	(−365)	..
−3.45	−349	3	7	S	..	5	−354	2
−3.35	−340	3	5	S	9	4	−344	1
−3.25	.....	..	..	..	..	..	−332	2
−2.85	(−293)	4	9	..	..	..	(−298)	2
−2.75	(−283)	4	8	..	..	..	(−288)	1
−2.65	−272	3	6	X	..	(5)	−277	1
−2.55	(−261)	4	6	..	..	..	−266	1
−2.45	(−249)	4	7	..	..	..	(−254)	2
−2.35	−236	3	9	X	..	6	−243	3
−2.25	−223	3	11	X	13	7	−230	4
−2.15	−214	2	9	S	9	5	−219	4
−2.05	−205	2	7	S	9	5	−210	2
−1.85	−192	2	8	S	13	7	−199	2/1
−1.75	−182	3	7	WM	10	5	−187	2
−1.65	−172	3	6	WM	10	5	−177	1
−1.55	−163	3	4	S	9	4	−167	0
−1.45	(−149)	4	(7)	W	(14)	(5)	(−154)	2
−1.35	−135	3	10	M	(14)	6	−141	4
−1.25	−125	3	9	MS	10	4	−129	5
−1.15	−113	2	10	S	12	6	−119	4
−1.05	−104	2	8	S	9	4	−108	4
−0.85	−91	2	9	SS	13	5	−96	4
−0.75	(−82)	4	7	M	(9)	4	(−86)	3
−0.65	(−72)	3	6	WM	(10)	5	(−77)	1
−0.55	−62	2	5	SS	10	7	−69	−2
−0.45	−53	2	3	SS	9	5	−58	−2
−0.35	−42	3	3	MS	11	4	−46	−1
−0.25	−27	2	7	S	15	5	(−32)	2
−0.15	−16	3	7	WM	11	5	−21	2
−0.05	(− 5)	4	7	WM	(11)	(6)	−11	1

TABLE 2—*Sunspot cycles since B.C. 649*—Continued

Number of max.	Year of max.	Prob-able error	Re-main-der	Intensity of maximum	Years since p. max.	Years since p. min.	Min.	Rem.
+0.05	(A.D.8)	4	(8)	W or M	13	5	(3)	3
0.15	20	3	9	S	12	5	15	4
0.25	(31)	4	(9)	W or M	(11)	(5)	26	4
0.35	42	3	9	W or M	(11)	5	37	4
0.45	53	3	9	S	11	6	47	3
0.55	65	3	10	S	12	5	60	5
0.65	(76)	4	(10)	W or M	(11)	6	(70)	4
0.75	(86)	4	(9)	W or M	(10)	6	(80)	3
0.85	(96)	4	(8)	W or M	(10)	5	(91)	3
1.05	105	3	5	M or S	9	4	101	1
1.15	(118)	4	7	W or M	(13)	6	112	1
1.25	(130)	4	8	W or M	(12)	6	(124)	2
1.35	(141)	4	8	W or M	(11)	6	(135)	2
1.45	(152)	4	8	W or M	(11)	6	(146)	2
1.55	(163)	4	8	W or M	(11)	6	(157)	2
1.65	175	3	9	M or S	(12)	5	170	4
1.75	186	3	9	S	11	4	182	5
1.85	196	3	8	S	10	4	192	4
2.05	(208)	4	8	X	(12)	5	203	3
2.15	(219)	4	8	X	(11)	5	(214)	3
2.25	(230)	4	8	X	(11)	5	(225)	3
2.35	(240)	4	7	X	(10)	5	(235)	2
2.45	(252)	4	8	X	(12)	5	(247)	3
2.55	(265)	4	10	X	(13)	5	(260)	5
2.65	(277)	4	11	X	(12)	5	(272)	6
2.75	290	3	13	M or S	13	6	(284)	7
2.85	302	1	14/13	SS	12	6	296	8
2.85	302	1	13	See previous row				
3.05	311	1	11	M	9	4	307	7
3.15	321	1	10	M	10	4	317	6
3.25	330	3	8	W	9	4	326	4
3.35	342	2	9	W	12	6	336	3
3.45	354	2	10	S	12	6	348	4
3.55	362	3	7	S	8	4	358	3
3.65	372	1	6	SS	10	4	368	2
3.75	387	3	10	M	15	7	380	3
3.85	396	3	8	M	9	5	391	3
4.05	410	4	10	W	14	6	404	4
4.15	(421)	4	(10)	W	11	5	416	5
4.25	430	1	8	M or S	9	4	426	4
4.35	441	1	8	M or S	11	4	437	4
4.45	452	2	8	S	11	4	448	4
4.55	(465)	4	(10)	W	(13)	6	459	4
4.65	479	1	13	M or S	(14)	7	472	6
4.75	490	1	13	M	11	6	484	7
4.85	501	1	13/12	SS	11	6	495	7

TABLE 2—*Sunspot cycle since B.C. 649*—Continued

Number of max.	Year of max.	Prob- able error	Re- main- der	Intensity of maximum	Years since p. max.	Years since p. min.	Min.	Rem.
4.85	501	1	12	See previous row				
5.05	511	1	11	S	10	4	507	7
5.15	522	2	11	W or M	11	5	517	6
5.25	531	1	9	S	9	5	526	4
5.35	542	3	9	M	11	4	538	5
5.45	557	1	13	M	15	6	551	7
5.55	567	1	12	SS	10	5	562	7
5.65	578	2	12	M	11	5	573	7
5.75	585	3	8	S	7	3	582	5
5.85	597	2	9	W or M	12	5	592	4
6.05	(607)	4	(7)	W	10	5	(602)	2
6.15	618	2	7	M	11	5	613	2
6.25	628	3	6	W–M	10	5	623	1
6.35	642	2	9	M	14	5	637	4
6.45	654	2	10	M–S	12	5	649	5
6.55	665	2	10	M	11	5	660	6
6.65	677	2	10	S	12	6	671	5
6.75	(689)	4	13	W	12	5	684	7
6.85	(699)	4	12	W	10	6	(693)	5
7.05	714	1	14	S	15	7	(707)	7
7.15	724	1	13	S	10	5	719	8
7.25	735	3	13	W	11	5	730	8
7.35	745	1	12	SS	10	6	739	6
7.45	754	2	10	M	9	5	(749)	5
7.55	765	1	10	SS	11	4	761	6
7.65	(776)	4	10	MS	(11)	6	(770)	4
7.75	(787)	4	10	W	(11)	5	(782)	5
7.85	(798)	4	10	MS	(11)	5	(793)	5
8.05	809	1	9	S	11	5	804	4
8.15	821	3	10	W	12	6	815	4
8.25	829	2	8	S	8	4	825	4
8.35	840	1	7	SS	11	4	836	3
8.45	850	1	6	MS	10	4	846	2
8.55	862	1	7	M	12	6	856	1
8.65	872	3	6	S	10	4	868	2
8.75	887	3	10	M	15	5	882	5
8.85	898	3	10	W	11	5	893	5
9.05	907	2	7	W	8	5	902	2
9.15	917	2	6	M	11	5	912	1
9.25	926	1	4	SS	9	5	921	−1
9.35	938	2	5	MS	12	4	934	1
9.45	(950)	3	6	WM	12	5	(945)	1
9.55	963	2	8	SS	13	4	959	4
9.65	974	3	8	SS	11	4	970	4
9.75	986	2	9	M	12	4	982	5
9.85	(994)	2	6	W	8	(4)	(990)	2

TABLE 2—*Sunspot cycles since B.C. 649*—Continued

Number of max.	Year of max.	Probable error	Remainder	Intensity of maximum	Years since p. max.	Years since p. min.	Min.	Rem.
10.05	1003	1	3	S	9	(5)	(998)	−1/−2
10.15	1016	1	5	M	13	5	1010	−1
10.25	1027	1	5	M	11	5	1022	0
10.35	1038	3	5	W	11	4	1034	+1
10.45	(1052)	4	8	WW	14	5	(1047)	3
10.55	1067	4	12	M	15	7	(1060)	5
10.65	1078	2	12	M	11	7	1071	5
10.75	1088	1	11	M	10	6	1082	5
10.85	1098	1	10	SS	10	6	1092	4
11.05	(1110)	3	10	WM	12	4	1106	6
11.15	1118	1	7	SS	8	3	1115	4
11.25	1129	1	7	S	11	5	1124	2
11.35	1138	1	5	SS	9	4	1134	1
11.45	1151	1	7	S	13	6	1145	1
11.55	1160	2	4	WM	9	5	1155	0
11.65	1173	1	7	MS	13	6	1167	1
11.75	1185	1	8	MS	12	5	1180	3
11.85	1193	1	5	M	8	3	1190	2
12.05	1202	1	2	SS	9	3	1199	0/−1
12.15	1219	3	8	M	17	7	1212	1
12.25	1228	1	6	M	9	4	1224	2
12.35	1239	3	6	M	11	6	1233	0
12.45	1249	3	5	WM	10	5	1244	0
12.55	1259	2	4	M	10	3	1256	1
12.65	1276	3	10	M	17	7	(1269)	3
12.75	1288	1	11	M	12	6	1282	5
12.85	1296	3	8	W	8	5	(1291)	3
13.05	1308	2	8	M	12	7	1301	1
13.15	1316	2	5	M	8	5	1311	0
13.25	1324	2	2	M	8	5	1319	−3
13.35	1337	3	4	W	13	5	1332	−1
13.45	1353	2	9	WM	16	7	1346	+2
13.55	1362	2	7	SS	9	4	1358	3
13.65	1372	1	6	SSS	10	4	1368	2
13.75	1382	2	5	MS	10	4	1378	1
13.85	1391	3	3	M	9	5	1386	−2
14.05	1402	1	2	M	11	6	1396	−3/−4
14.15	(1413)	3	2	WW	11	6	1407	−4
14.25	(1429)	4	7	WM	16	8	(1421)	−1
14.35	(1439)	3	6	WM	10	5	1434	+1
14.45	1449	4	5	WM	10	6	1443	−1
14.55	1461	3	6	WM	12	4	1457	2
14.65	(1472)	4	6	WW	11	4	1468	2
14.75	(1480)	4	3	WW	8	4	(1476)	−1
14.85	1492	4	4	WM	12	4	(1488)	0

TABLE 2—*Sunspot cycles since B.C. 649*—Concluded

Number of max.	Year of max.	Probable error	Remainder	Estimated sunspot number	Intensity of max.	Years since p. max.	Years since p. min.	Min.	Rem.
15.05	1505	3	5	(60)	W	13	7	1498	−1/−2
15.15	1519	1	8	(80)	M	14	7	1512	+1
15.25	1528	1	6	(150)	SS	9	3	1525	+3
15.35	1539	1	6	(130)	S	11	4	1535	+2
15.45	1548	1	4	(120)	S	9	5	1543	−1
15.55	1558	1	3	(160)	SS	10	5	1553	−2
15.65	1572	1	6	(150)	SS	14	5	1567	+1
15.75	1581	1	4	(130)	S	9	3	1578	+1
15.85	1591	2	3	(70)	WM	10	4	1587	−1
16.05	1604.5	2	4	(80)	WM	13	5	1599.5	0/−1
16.15	1615.5	1	4	(90)	M	11	4.7	1610.8	−0.7
16.25	1626.0	1	3.5	(100)	MS	10.5	7	1619.0	−3.5
16.35	1639.5	1	6	(70)	WM	13.5	5.5	1634.0	+0.5
16.45	1649.0	1	4.5	(40)	WW	9.5	4	1645.0	+0.5
16.55	1660.0	1	4.5	(50)	WW	11.0	5	1655.0	−0.5
16.65	1675.0	1	8.5	(60)	W	15.0	9	1666.0	−0.5
16.75	1685.0	1	7.5	(50)	WW	10.0	5.5	1679.5	+2
16.85	1693.0	1	4.5	(30)	WWW	8.0	3.5	1689.5	+1
17.05	1705.5	1	5.0	(50)	WW	12.5	7.5	1698.0	−1.5/−2.5
17.15	1718.2	1	6.7	(130)	S	12.7	6.2	1712.0	+0.5
17.25	1727.5	1	5.0	(140)	SS	9.3	4.0	1723.5	+1.0
17.35	1738.7	1	5.2	(110)	S	11.2	4.7	1734.0	+0.5
17.45	1750.3	..	5.8	80	M	11.6	5.3	1745.0	+0.5
17.55	1761.5	..	6.0	90	M	11.2	6.3	1755.2	−0.3
17.65	1769.7	..	3.2	110	S	8.2	3.2	1766.5	0
17.75	1778.4	..	+0.9	150	SS	8.7	2.9	1775.5	−2.0
17.85	1788.1	..	−0.4	130	S	9.7	3.4	1784.7	−3.8
18.05	1805.2	..	+4.7	50	WW	17.1	6.9	1798.3	−1.2/−2.2
18.15	1816.4	..	4.9	50	WW	11.2	5.8	1810.6	−0.9
18.25	1829.9	..	7.4	70	WM	13.5	6.6	1823.3	−0.8
18.35	1837.2	..	3.7	140	SS	7.3	3.3	1833.9	+0.4
18.45	1848.1	..	3.6	120	S	10.9	4.6	1843.5	−1.0
18.55	1860.1	..	4.6	100	MS	12.0	4.1	1856.0	+0.5
18.65	1870.6	..	4.1	140	SS	10.5	3.4	1867.2	+0.7
18.75	1883.9	..	6.4	60	W	13.3	5.0	1878.9	+1.4
18.85	1894.1	..	5.6	90	M	10.2	4.5	1889.6	+1.1
19.05	1907.0	..	6.5	60	W	12.9	5.3	1901.7	+1.2
19.15	1917.6	..	6.1	100	MS	10.6	4.0	1913.6	+2.1
19.25	1928.4	..	5.9	80	WM	10.8	4.8	1923.6	+1.1
19.35	1937.4	..	3.9	110	S	9.0	3.6	1933.8	+0.3
19.45	1947.5	..	3.0	150	SS	10.1	3.3	1944.2	−0.2
19.55	(1958.5)	2	(3)	...	SS	....	(4)	(1954.5)	(−1)
19.65	(1972.5)	2	(6)	...	M/S	....	(6)	(1966.5)	(0)
19.75	(1984.5)	2	(7)	...	SS	....	(6)	(1978.5)	(+1)
19.85	(1994.5)	2	(6)	...	S	....	(5)	(1989.5)	(+1)
20.05	(2004.5)	2	(5/4)	...	M/S	....	(4)	(2000.5)	(1/0)
20.15	(2014.5)	2	(3)	...	M	....	(5)	(2009.5)	(−2)
20.25	(2025.5)	2	(4)	...	..	....	...	......	.....

11. *Probable Error.*

Wherever a sunspot cycle has been interpolated without definite evidence, the year has been indicated in parentheses and the probable error has been regarded as 4. A probable error of 3 indicates ambiguity in the interpretation. For purposes of mathematical analysis, the dates of cycles of both these kinds should be excluded.

Modern investigations suggest that the phase of the auroral cycle is slightly later than the sunspot cycle [12], and the great aurorae associated with intense maxima (S) occur between $S - 2$ and $S + 3$ years. On the other hand, sunspot maxima that are weak or "late" (that is, with high remainders), or those near the onset of an aurorally weak period, often appear to be associated with double or diffuse auroral maxima, and in these cycles the minima are more readily dated than the maxima. There may be persistent bias in our dating of ± 0.3 year either for these reasons or for the inevitable personal equation.

12. *Remainders.*

In order to avoid minus signs, the "remainders" in Table 2 have been expressed as follows:

    (a) Up to A.D. 800 on a scale from 5 to 15 (instead of −6 to +4)
    (b) After A.D. 800 on a scale from 1 to 13

13. *Strength.*

The strength of each cycle can be readily gauged from the evidence for aurorae considered in relation to other observations of natural phenomena in the historical sources.

		Annual mean sunspot number
SSS	= Extremely strong	> 160
SS	= Very strong (150, 140, 140)	145
S	= Strong (110, 130, 120, 110)	120
MS	= Moderately strong (100, 100)	100
M	= Moderate (90, 90, 80)	85
WM	= Moderately weak (70)	70
W	= Weak (60, 60)	60
WW	= Very weak (50, 50)	50
WWW	= Extremely weak	< 45
X	= Unknown	

The figures in parentheses are corresponding sunspot numbers of maxima since 1750, and the figures at the right-hand side are the equivalent auroral numbers used in the curve (Fig. 1) for the earlier period.

14. *The Calculation of Years of Minima.*

In aurorally rich periods, such as the sixteenth century, the years of sunspot minima are defined by groups of years without notable displays, as extensive aurorae do not occur within two years of sunspot minimum [14].

In aurorally weaker periods, the position of a minimum in relation to the dates of neighbouring maxima can be estimated from certain rules, based on the behaviour of the sunspot cycle since 1610. Thus,

159

(a) The *minimum preceding* a given *maximum* usually occurs

4 years before, if the maximum is to be strong or very strong,
5 years before, if it is to be moderate or moderately strong, and
6 years before other (that is, weaker) maxima.

TABLE 3—*Sunspot minima arranged by centuries* (November 1954)

Period	.1	.2	.3	.4	.5	.6	.7	.8	.0
−700+	..	..	..	(47)	(57)	..	..	..	...
−600+	..	..	..	..	..	(73)	(84)	(95)	(104) i.e., 497 B.C. (Livy)
−500+	14	26	35	46	..	..	..	..	103 i.e., 398 B.C.
−400+	14	(25)	(35)	46	56	68	(..)	(..)	(102)
−300+	(12)	23	34	(46)	57	70	81	90	101 i.e., 200 B.C.
−200+	13	23	33	(46)	59	71	81	92	104 i.e., 97 B.C.
−100+	14	23	31	42	54	68	79	(91)	(103) i.e., A.D.3
0+	15	26	37	47	60	(70)	(80)	91	101
100+	12	(24)	(35)	(46)	(57)	70	82	92	203
200+	(14)	(25)	(35)	(47)	(60)	(72)	(84)	96	307
300+	17	26	36	48	58	68	80	91	404
400+	16	26	37	48	59	72	84	95	507
500+	17	26	38	51	62	73	82	92	(602)
600+	13	23	37	49	60	71	84	(93)	(707)
700+	19	30	39	(49)	61	(70)	(82)	(93)	804
800+	15	25	36	46	56	68	82	93	902
900+	12	21	34	(45)	59	70	82	(90)	(998)
1000+	10	22	34	(47)	(60)	71	82	(92)	1106
1100+	15	24	34	45	55	67	80	90	1199
1200+	12	24	33	44	56	(69)	82	(91)	1301
1300+	11	19	32	46	58	68	78	86	1396
1400+	7	(21)	34	43	57	68	(76)	(88)	1498
1500+	12	25	35	43	53	67	78	87	1599
1600+	10.8	19.0	34.0	45.0	55.0	66.5	79.5	89.5	1698.0
1700+	12.0	23.5	34.0	45.0	55.2	66.5	75.5	84.7	1798.3
1800+	10.6	23.3	33.9	43.5	56.0	67.2	78.9	89.6	1901.7
1900+	13.6	23.6	33.8	44.2	(54)	(66)	(78)	(89)	(2000)

160

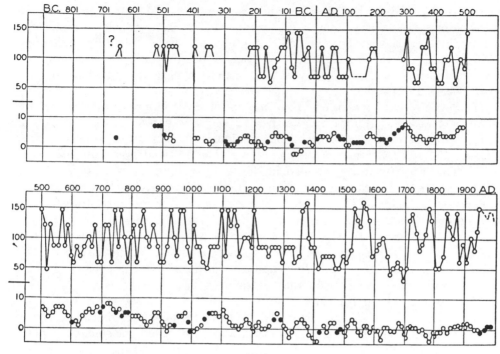

FIG. 1—Curves showing intensity and phase, B.C. 650 to A.D. 2000
Upper curve: Auroral intensity (sunspot numbers from A.D. 1750) of maxima
Lower curve: Phase (remainder) of minima; o = reliable, ● = uncertain

The formula $7 - (0.03)n$, where n is the sunspot number at maximum, holds for the strictly-dated intervals from minima to maximum since 1850.

(*b*) The intervals indicated by the preceding rule should nevertheless be increased by 1 to 1-1/2 years near the transition from an aurorally weak to an aurorally rich series of cycles.

(*c*) The minimum *following* a *weak* maximum often occurs about six years afterwards. The minimum following a *strong* maximum often occurs about seven years afterwards.

(*d*) The period from maximum to the following minimum is nevertheless only about five years near the transition from an aurorally weak to an aurorally rich group of cycles.

(*e*) The period of the sunspot cycle measured from one minimum to the next behaves with much more regularity than the period between consecutive maxima. Since 1715, it has invariably lain between 13.6 and 9.0 and, if—as seems possible—the Zurich minima at present dated 1634 and 1698 could be redated 1633 and 1699, these limits would have been valid since 1610. It is reasonable in any case to assume as a first approximation that the "remainders" of the consecutive minima do not differ by more than 3.

The minima can by these rules be dated with some precision back to 1512; intermittently they can be traced back to c. 500 B.C.

The presence of a great cycle of 78 years or seven sunspot cycles has been postulated by Gleissberg and is certainly apparent in the data since 1610. Its presence in a preliminary table of the medieval data came to me as a surprise, but its phase in classical times helped to confirm the rule—implied by the table— of 90 to 91 cycles to a millenium. Nevertheless, before this cycle was taken for granted, the time occupied by 7 or 14 consecutive cycles was investigated, with results (since 1512) as follows:

(*f*) Seven cycles, measured from minimum to minimum, usually occupy between 77 and 79, but always more than 72 and less than 83 years.

(*g*) Fourteen cycles, measured from minimum to minimum, usually occupy between 154 and 158 years, but always more than 150 and less than 162 years.

(*h*) The longer values mentioned in the two preceding rules, that is, 83 and 162 years, are approached in periods measured forwards from the conclusion of an aurorally rich period. The length of 14 cycles in the Zurich figures thus reached its maximum (159.5 years) when measured from 1785 to 1944.

(*i*) The shorter values, 72 and 150 years, are conversely approached in periods which end near the conclusion of an aurorally rich period. Minima such as 1586 and 1785 are thus "early" and 1679 and 1712 are "late." The sunspot cycle is longer in aurorally weak periods and shorter in very active periods.

(*j*) The mean sunspot cycle for periods of the order of 500 years has varied between 11.03 and 11.14, and, since 200 B.C., there have been between 90 and 91 cycles to a thousand years. The justification for these statements is given below (§20).

The following series of dates, based on means of three successive minima in Table 3, is thus considered in relation to an arithmetical progression of the form

$$1932 - (155.2)n$$

Mean minimum	Arithmetical progression	Observed minus formula
(−)	−551.2	( )
(−397)	−396.0	+1
−241	−240.8	+1
− 86	− 85.6	0
+ 70	+ 69.6	0
(225)	224.8	(0)
380	380.0	0
538	535.2	+3
(695)?	690.4	(−5)
846	845.6	0
999	1000.8	+2
1156	1156.0	0
1310	1311.2	+1
(1467)	1466.4	(−1)
1621.3	1621.6	+0.3
1775.6	1776.8	+0.8
1933.9	1932.0	−1.9

The differences are remarkably small, and the positive signs are usually associated with active "early" maxima.

These rules made possible the calculation of the *number* of minima missing at the various breaks. Table 3 seems to offer a unique and correct solution of the number of cycles since at least 250 B.C. It may be noted that identical totals are obtained from applying the same rules to the various maxima listed by Fritz [8], Kanda [15], and Nicolini [16]. It must also be added that, although the total *number* of cycles can be ascertained in this way, some of the more anomalous cycles—as in the mid-first century B.C. or the seventh to eighth centuries A.D.— may have been "smoothed" or wrongly placed in an attempt to keep within the limits obtaining since 1610. Errors of about 11 in certain remainders may thus occur, and the dates of maxima tabulated with probable errors 3 or 4 in Table 2 should be omitted in rigorous analysis.

15. *The Mean Cycle.*

The last two digits of years of *minima* have roughly followed the eleven-times table since cycle A.D. 800 [8], and before A.D. 800 the *maxima* behaved in the same way. Thus, except at the transition, the mean cycle has been approximately 11.11. We have seen that the length of 14 cycles appears to be 155.2, corresponding to a mean cycle (since B.C. 398) of 11.09, so that it may be supposed that near the transition at A.D. 800 the sunspot cycle was shorter than usual.

The mean values of the "remainders," as divided into centuries in Table 2, are set out in Table 4. The impression is given of a cycle slightly greater than 11.11 up to A.D. 800, followed by a cycle somewhat less. The mean length is calculated from the final maxima of each century (605 B.C. ... A.D. 1894); a deviation from the nine-cycle pattern in a particular century would introduce a ten per cent error in the mean.

16. *The 78-year Cycle.*

The 78-year cycle [17] is clearly shown since 1610 by an alternation of periods of shortening (c. 1650-1700, c. 1725-65, c. 1890-1930) and lengthening cycles (c. 1700-25, c. 1765-1810, c. 1845-90). The half-amplitude of this "Gleissberg" cycle is about ±1 year measured by "remainders," and the extreme values of the length of the ordinary sunspot cycle measured from minimum to minimum are close to the minimum remainders of this 78-year cycle. Preliminary investigations revealed that this greater period of seven ordinary cycles was in existence before 1610, and these principles were therefore utilized in the final Tables 2 and 3.

In addition to the 78-year cycle, a more important longer cycle of about 160 to 170 years could fit well the known changes in cycle length *since* c. 1510. Thus, the mean cycle has been

> 10 years from 1560 to 1590
> 1750 to 1790
> 1900 to 1950
> but 12 years from 1600 to 1670
> and 1780 to 1820

This alternation appears, however, to arise indirectly from the two-century cycle in auroral *activity* mentioned below, and late maxima (high remainders) at about A.D. 300, 495, 690, 890, 1080, 1290, 1520, 1680, 1830 (and 1900) follow the longer cycles of the weaker periods. No cycle of 160 to 170 years is evident before c. 1510,

D. JUSTIN SCHOVE

TABLE 4—*The sunspot phase through the centuries*

Century	Remainder		Difference	Cycle length implied in Table 2··
	Maximum	Minimum		
7th B.C.	(8)	(3)	(5)	. . . .
6th B.C.	(11)	(7)	(4)	(11.4)
5th B.C.	(7)	(3)	(4)	(10.8)
4th B.C.	(6)	(2)	(4)	(11.1)
3rd B.C.	(7)	(3)	(4)	(11.0)
2nd B.C.	7.7	(2.5)	5.1	11.2
1st B.C.	(6)	(0.9)	5.1	(11.0)
1st A.D.	9.0	(3.7)	5.3	(11.2)
2nd A.D.	(7.8)	(2.5)	(5.2)	(11.1)
3rd A.D.	(9.7)	(4.4)	(5.2)	11.8
4th A.D.	8.8	3.9	4.9	10.4
5th A.D.	10.3	(5.0)	5.2	11.7
6th A.D.	10.4	5.8	4.7	10.7
7th A.D.	9.3	(4.3)	5.2	(11.3)
8th A.D.	(11.3)	(6.0)	5.3	(11.0)
9th A.D.	8.1	3.3	4.8	(11.1)
10th A.D.	6.5	2.1	4.4	(10.7)
11th A.D.	(7.9)	(2.3)	(5.5)	(11.5)
12th A.D.	6.8	2.2	4.5	10.5
13th A.D.	6.7	1.5	5.1	11.4
14th A.D.	5.4	+0.3	5.1	10.5
15th A.D.	(4.5)	(−0.6)	(5.2)	11.2
16th A.D.	5.0	+0.2	4.8	11.0
17th A.D.	5.2	−0.2	5.4	11.3
18th A.D.	4.1	−0.6	4.8	10.5
19th A.D.	5.0	−0.1	4.9	11.8
20th A.D.	(5.2)	(+0.7)	(4.5)	(11.1)

but, if our placing of the cycles is correct, the sunspot period has been shorter in the even centuries.

17. *Fluctuations in Auroral Activity.*

The fluctuations between aurorally rich and aurorally weak periods follow sunspot fluctuations closely, and in Figure 1 and the last part of Table 2 have been interpreted numerically. No cycle of the length of 78 years could be traced in auroral *intensity*, although periodicities about 85 and perhaps 65 years were vaguely indicated.

A more definite cycle of two centuries may be inferred from the rule that *activity has been greater in even than in odd centuries.* This rule is indicated in the number of dates of maxima of either sunspots or aurorae which Fritz was able to determine before 1610. Seventy-eight belonged to even and only 41 to odd centuries [8]. Since the beginning of the Christian era, the ninth century forms the only odd century in which activity was as intense as in the centuries immediately preceding and following.

18. *The Active Periods of Even Centuries.*

The aurorally rich periods of the even centuries characteristically have two maxima with a short intervening lull. Central dates of such maxima—which correspond to the short lull—can be identified as follows: A.D. 1755, 1555, 1350, 1160, 955, 755, 540, 340 . . . . The cycle of auroral activity thus appears to be 200 to 205 years in length. The dates of the two major peaks are often about a half-century apart, and, although irregularities exist, they can be traced as follows:

Earlier peak	Later peak	Difference	
(?) 520 B.C.	(?) 460 B.C.	Unknown	
(310 B.C.)	(260 B.C.)	No evidence	
92 B.C.	54 B.C.	(4 × 9.5)	38 years
2nd century A.D.	. . . .	Unknown	
A.D. 302	372	(7 × 10)	70 years
A.D. 511	567	(6 × 9)	56 years
A.D. 724?	765	(5 × 10)	51 years
A.D. 926	974	(4 × 12)	48 years
A.D. 1118	1173	(5 × 11)	55 years
A.D. 1324	1372	(4 × 12)	48 years
A.D. 1528	1572	(4 × 11)	44 years
A.D. 1727	1778	(5 × 10)	51 years
A.D. 1947	(1995?)	(5 × 10)	c.50 years

These double maxima appear to be the only basis for the cycle of 55-56 years conceived by Fritz [18].

The mean behaviour of the sunspot cycle in even centuries provides a clue to prediction of future cycles. The pattern since A.D. 900 has been

Year	(Rem.)	Typical intensity	Years since previous maximum
'07	(7)	Weak	. .
'17	(6)	Strong	10
'27	(5)	Strong	10
'38	(5)	Strong	11
'50	(6)	Moderate	12
'61	(6)	Strong	11
'72	(6)	Very strong	11
'82	(5)	Strong	11
'91	(3)	Weak Moderate	8
'03	(4/3)	Weak Moderate	12

The probable error in the dating is less than two years; the intensity pattern progresses slowly forward and is subject to much irregularity. Table 1 has been continued on Gleissberg's assumption that a third intense and early maximum will shortly occur. The weak "central" maximum has thus been dated c. 1970 instead of c. 1960. The effect of the 78-year cycle has also been included.

The dates of aurorally rich periods—in both odd and even centuries—may be summarized as follows:

(?) c. 649 B.C.	651– 682
(?) c. 523 B.C.	743– 772
(?) 502–455 B.C.	826– 880
(?) 395–340 B.C.	920– 930
217–162 B.C.	961– 979
113– 89 B.C.	1000–1018
66– 43 B.C.	1095–1152 V. strong
c. A.D. 20 (?)	1200–1205
A.D. 43–68	1245–1264
Early 2nd cent. A.D.(?)	1307–1325
A.D. 174–196	1352–1393 V. strong
Third century A.D.(?)	1520–1582 V. strong
299–322	1616–1640 Relatively strong
351–375	1716–1746 V. strong
430–458	1770–1789 V. strong
499–514	1830–1880
565–586	1930–1960(?) V. strong
c. 620(?)	

19. *Quiescent Periods of the Odd Centuries.*

Since A.D. 200, long periods with very little auroral activity have occurred in the odd centuries only, notably in the third, eleventh, fifteenth, and seventeenth. The high remainders at the conclusion of some of these cycles have already (§16) been noted in relation to the two-century cycle. No regular pattern can be discerned as typical of odd centuries. Aurorally weak periods can be dated as follows:

c.440–400 B.C.		1037–1066	Very weak
160–135 B.C.		1279–1306	
1st, 2nd, 3rd cents. A.D.	Exact limits unknown	1326–1351	
401–428		1404–1431	Very weak
458–477		1468–1516	Very weak
601–615		1641–1670	Very weak
683–713	Very weak	1681–1715	Very weak
813–825		1791–1816	
891–916		1883–1908	

20. *Intervals between Intense Maxima.*

The dating of the major maxima is clearly established, and, pending harmonic analysis of the intensity curve, the intervals between the great peaks have been studied. No obvious periodicities are apparent, but clusters occur at particular values such as 244 and 353. As an example, we may consider 554.

$$
\begin{array}{ll}
\text{A.D.} \quad 501- 54 \text{ B.C.} & = 554 \text{ years} \\
\text{A.D.} \quad 926-372 & = 554 \text{ ``} \\
\text{A.D.} \ 1117-566 & = 551 \text{ ``} \\
\text{A.D.} \ 1527-974 & = 553 \text{ ``} \\
[\text{Cf. also A.D. } 1938-1382 & = 556 \text{ ``} \\
\text{and A.D.} \quad 54-(?)502 \text{ B.C.} & = 555 \text{ ``} \ ]
\end{array}
$$

That this was significant and not accidental was shown by the calculation of "remainders," defined in the sense of §5, but applied to the intervals. Their distribution was not accidental. In the range 200 to 300 years, there were actually ten instances like 244 with zero remainder but only one (247) leaving a remainder 3.

The explanation can only lie in the constancy of the mean cycle over long periods. The intervals noted must correspond to integral numbers of "mean cycles" and may be split up as follows:

$$
\begin{aligned}
244 &= 22 \times 11.09 & \text{Rem.} & \quad 0 \\
353 &= 32 \times 11.03 & \text{Rem.} & \quad -2 \\
375 &= 34 \times 11.03 & \text{Rem.} & \quad -2 \\
480 &= 43 \times 11.16 & \text{Rem.} & \quad +3 \\
554 &= 50 \times 11.08 & \text{Rem.} & \quad -1 \\
632 &= 57 \times 11.09 & \text{Rem.} & \quad -1 \\
743 &= 67 \times 11.09 & \text{Rem.} & \quad -1 \\
763 &= 69 \times 11.06 & \text{Rem.} & \quad -3 \\
1005 &= 91 \times 11.04 & \text{Rem.} & \quad -6
\end{aligned}
$$

Similar clusters were noted at 93 (94?), 109 (110?), 115 (116?), and 117 (118?) cycles. Considering all intervals from 200 to 1,000, those yielding remainders 8, 9, 10, 0, and 2 were most frequent and those yielding remainder 3 least frequent. In this way, the value of the mean cycle as 11.1 and the assumption of between 90 and 91 cycles per millenium were confirmed.

An independent check on this was provided by intervals (200 to 1,000 years) of the main maxima selected by Fritz. The dates are less reliable, but have the merit of being more evenly distributed. Moreover, Fritz himself believed that the mean cycle was greater than 11.1 and that there were 120 (not 122) cycles between A.D. 503 and 1848. However, even in the dates selected by Fritz, remainders of 9, 10, 0, and 1 were the most frequent and clusters could be identified as follows:

Years	Remainder	Number of cycles	Mean period
309	9	28	11.04
410	10	37	11.08
475	9	43	11.05
542	9	49	11.06
720	9	65	11.08
854	10	77	11.09
898	10	81	11.09
911	0	82	11.11

A separate examination was made of the probable and established maxima between B.C. 649 and A.D. 302. The remainder 0 was then the most frequent and the tendency for cycles slightly greater than 11.11 was noted. The following results seemed most probable.

$$
\begin{aligned}
\text{649 B.C. to 215 B.C.} &= 434 \text{ years} = (?) \ 39 \times 11.13 \\
\text{215 B.C. to A.D. 196} &= 410 \text{ years} = \quad 37 \times 11.08 \\
\text{A.D. 196 to 302} &= 106 \text{ years} = \quad \ 9 \times 11.8 \ (sic)
\end{aligned}
$$
$$(\text{N.B. Not } 10 \times 10.6)$$

We thus have

$$
\begin{aligned}
\text{649 B.C. to A.D. 1947} &= 2595 \text{ years} = 234 \times 11.09 \\
\text{215 B.C. to A.D. 1947} &= 2161 \text{ years} = 195 \times 11.08
\end{aligned}
$$

The integral numbers (that is, of cycles 11.1 years) which had appeared in these three investigations themselves showed a tendency to clusters—in this case, near multiples of seven. This was independent evidence of the 78-year cycle, but no evidence of longer cycles was apparent.

21. *Conclusion.*

There is thus a fundamental rhythm involved in the production of sunspot cycles. The mean period of this rhythm is 11.1, and the rhythm itself is subject to a major cycle of seven sunspot cycles (77-1/2 or 78 years). Individual cycles in active periods tend to appear earlier in relation to the fundamental period. This is especially true of maxima, but is also true of minima. Conversely, the length of the cycle is longer in aurorally weak periods.

22. *Acknowledgments.*

I should like to thank all those who have kindly sent or translated the historical records of aurorae that form the basis of this paper. Many of these collaborators—in the so-called "Spectrum of Time" project—are named in previous articles. Prof. S. Chapman kindly read the typescript and made some helpful suggestions.

References

[1] R. Wolf, Astron. Mitt., Zurich, **24**, 111 (1868).

[2] D. J. Schove, The sunspot-cycle before 1750, Terr. Mag., **52**, 233-238 (1947).

[3] A. F. Cook, On the mathematical characteristics of sunspot-variations, J. Geophys. Res., **54**, 347-353 (1949); W. Gleissberg, The predictability of the probable features of the sunspot cycle, Astroph. J., **109**, 321-326 (1949); W. Gleissberg, Zs. Astroph., Berlin, **29**, 1-S (1951); A. G. McNish and J. V. Lincoln, Prediction of annual sunspot numbers, Washington, D.C., U.S. Dept. of Comm., Nat. Bur. Stan., Central Radio Prop. Lab., CRPL Rep. No. 1-1 (1947).

[4] H. Fritz, The periods of solar and terrestrial phenomena, translation by W. W. Reed, Mon. Weath. Rev., **56**, 401-407 (1928). [Published originally in Vierteljahrsch. Natf. Ges., Zürich, Heft 1, 1893.]

[5] H. J. de Boer, Tree-ring measurements and weather fluctuations in Java from A.D. 1512, Proc. Kon. Nederland. Akad. Wetensch., Amsterdam, Ser. B, Vol. LIV, No. 3, 194-209 (1951) and Vol. LV, 386-394 (1952).

[6] C. Davison, Studies in the periodicity of earthquakes, London (1938).

[7] D. J. Schove, Geog. Ann., Stockholm, **36**, 40-80 (1954).

[8] D. J. Schove, Sunspot epochs, A.D. 188-1610, Pop. Astr., **56**, 247-252 (1948); see Table 1.

[9] D. J. Schove, J. Brit. Astr. Assn., **61**, 22-25, 126-128 (1951); **62**, 38-43, 63-66 (1952); and **63**, 266-270, 321-325 (1953); also J. Brit. Archaeol. Assn., 3rd Ser., No. 13, 34-49 (1950).

[10] D. J. Schove, Sunspots, aurorae and blood rain, Isis, **42**, 133-138 (1951).

[11] D. J. Schove, The auroral cycle in the 16th century (not yet published).

[12] D. J. Schove, Sunspots and aurorae (500-250 B.C.), J. Brit. Astr. Assn., **58**, 178-190 (1947).

[13] L. Harang, The aurora, New York, John Wiley and Sons, Inc. (1951). [Vol. I, International Astrophysical Series.]

[14] H. W. Newton and A. S. Milsom, The distribution of great and small geomagnetic storms in the sunspot cycle, J. Geophys. Res., **59**, 203-214 (1954).

[15] S. Kanda, Ancient records and sunspots and auroras in the Far East and the variation of the period of solar activity, Proc. Imp. Acad. Tokyo, **9**, No. 7, 293-296 (1933).

[16] T. Nicolini, Il periodo medio dell'attività solare in relazione alle osservazioni antiche e moderne, R. Oss. Astrofs. di Capodimonte, Napoli, Contributi Astronomici, Ser. 11, Vol. 11, No. 7, 79-88 (1942).

[17] W. Gleissberg, Terr. Mag., **49**, 243-244 (1944).

[18] H. Fritz, Das Polarlicht, Leipzig, Brockhaus, p. 161 (1881).

21

The Prolonged Sunspot Minimum, 1645–1715.

By E. WALTER MAUNDER, F.R.A.S.

It is just one-third of a century since Prof. F. W. G. Spoerer, the veteran observer of sunspots, published two important papers* with regard to " the sunspot cycle," in which he drew attention to evidence pointing to a long-continued " damping-down " of the solar activity which began in the middle of the seventeenth century. It fell to my duty to supply a short note on these two papers to the Council of the Royal Astronomical Society for their Annual Report for the year 1890,† and, four years later, I gave a fuller account of the first paper, in *Knowledge* for August 1, 1894 (p. 173).

It appears to me that this discovery of Spoerer's has not received as much attention as it deserves, and I would ask permission of the Association to summarise the main facts.

---

* " Ueber die Periodicität der Sonnenflecken seit dem Jahre 1618 " (*Royal Leopold–Caroline Academy*, 1889), *and* " Sur les différences que présentent l'Hémisphère nord et l'Hémisphère sud du Soleil " (*Bulletin Astronomique*, 1889, February).

† *Monthly Notices Roy. Astr. Soc.*, Vol. I., p. 251. *Also* Vol. LVI., p. 213.

[*Editor's Note:* In Spoerer's 1894 paper he gives 1609.3 and 1612.5 as tentative dates for minimum and maximum, and a detailed analysis of the heliographic latitudes of the 1621–1627 sunspots and 1642–1644 sunspots.

Additional European sunspots of 1678–1688 and 1700–1710 are noted by H. E. Landsberg, 1980, Variable Solar Emissions, the Maunder Minimum, and Climatic Temperature Fluctuations, *Arch. Meteorol. Geophys. Bioklimatol.,* ser. B, **28:**181–191.]

The invention of the telescope at the opening of the seventeenth century led, in 1610 and 1611, to the discovery of sunspots, and to their systematic study by Galileo, Scheiner and others. The discovery of the rotation of the Sun, the determination of its period, and of the position of the solar axis, soon followed; but there does not seem to have been, at first, any suspicion that the changes of the solar spots, as to their numbers, areas, positions, and movements, proceeded in a series of undulations. That was not fully apprehended until the Sun had been under telescopic observation for more than two centuries. The reason for the delay in the discovery of what we now know as " the sunspot cycle " was simply that at first it does not seem to have been working in what we should now consider to be its normal manner; and before the middle of the seventeenth century the supply of sunspots would seem to have run out.

The history of sunspot observations from 1610 onwards may be briefly summarised as follows :—The first years after the discovery showed some considerable spot-activity, which had quieted down to a minimum by 1619, but recovered again later up to a well-marked maximum in 1625. The next minimum fell in 1634, and was followed by a maximum in 1639. If the solar activity had then been conforming to its recent type, the next maximum would have been due to commence about 1650, but very few spots appeared; indeed, from November 14, 1652, to August 12, 1654, no spots were reported at all. Then—after a single group, which ran its course from February 9 to February 21, 1655—the next observation appears to be that of a large spot observed by the Hon. Robert Boyle from April 27 to May 9, 1660.

Except for some small spots noted in 1661, by Hevelius in February and by Fogelius in October, no similar observations appear to have been made until 1671, when Cassini reported a large one as seen on August 9. This seemed to have caused a widespread sensation, Cassini reporting that " it is now about twenty years since that Astronomers have not seen any considerable *Spots* in the Sun, though before that time, since the Invention of Telescopes, they have from time to time observed them."

Oldenburg, the Secretary of the Royal Society, and Editor of the *Philosophical Transactions*, was so impressed with the rarity of such opportunities, that on learning of Cassini's observation he thought it desirable to publish in the *Phil. Trans.* Boyle's account of the spot of April 27, 1660, together with a small illustration,* and the remark that at Paris " the excellent Signior *Cassini* hath lately detected again *Spots* in the Sun, of which none have been seen these many years that we know of." There are also not a few notices during the years from 1648 to 1660 that the Sun was then spot-free, and Picard, who

---

* *Phil. Trans.*, No. 74, p. 2216, for April 27, 1671. (Reproduced also in the *Observatory*, Vol. VI., p. 274.)

independently discovered Cassini's spot of 1671, declares that
he " was so much the better pleased at discovering it since it
was ten whole years since he had seen one, no matter how great
the care which he had taken from time to time to watch for
them." In 1665 Weigel, at Jena, confirms the blank condition
of the Sun in this period with much particularity. From 1671
to the end of the century, spots were very occasionally observed.
Spoerer gives a list of these, which I have somewhat condensed
as under :—

1672.—Nov. 12–22, South Latitude, 13°.

1674.—Aug. 29–31.

1676.—June 26–July 1, S. Lat., 13°; Aug. 6–14, S. Lat., 6°;·
Oct. 30–Nov. 1; · Nov. 19–30; and Dec. 16–18.
Three returns of the same, S. Lat., 5°.

1677.—Same spot observed in fourth rotation; another,
April 10–12.

1678.—Feb. 25–March 4, S. Lat., 7°; May 24–30, S. Lat., 12°.

1680.—Spots observed in May and June and August.

1681.—Spots observed in May and June.

1684.—Kirch and Cassini observed a spot through four rotations.
April–July, S. Lat., 10°.

1686.—April 23–May 1, S. Lat., 15°; Sept. 22–26.

1687.—Cassini could find no spots, though observing carefully.

1688.—May 12, S. Lat., 13°.

1689.—July 19–22; Oct. 27–29. These spots are reported as
ephemeral.

1689.—March to 1695 May.—De la Hire reports that he found no
spots.

1695.—May 27. De la Hire says : " It is a long time since anything so great as these have appeared." No spots
until 1700 November.

1699.—Last year wholly without spots.

1700.—Nov. 7–13, S. Lat. 10°; Dec. 30–1701, Jan. 2, S. Lat. 3°.

1701.—March 29–31, S. Lat. 12°; Oot. 31–Nov. 10, S. Lat. 12°.

1702.—Two groups in May and one in Dec. at S. Lat. 10°, 12°
and 11° respectively.

1703.—Three groups in S. Lat. 2° (seen in two rotations), 19°
and 10° respectively.

1704.—Several spots in February in S. Lat. 8° and 13°; March
19–21, S. Lat. 10°; Nov. 25–29 in S. Lat. 9°.

1705.—Several spots were seen; one in April was in the northern
hemisphere, the first reported in this hemisphere
in the whole period since 1645.

1706.—Five or six spots, all in S. Lat.

1707.—Nine separate groups, all in south latitude, except one,
which appeared on Nov. 27 in N. Lat. 16°. Cassini and
Maraldi remark that they do not remember having
seen any in this hemisphere before, except that of
April, 1705.

1708.—Six spots in rather low southern Latitudes.

1709.—Three spots in S. lat., 16°, 11°, and 6° in Jan. and Feb. From Feb. 9–Aug. 18 no spots. Two spots in S. Lat. 6° or 7° in Aug. and Nov.

1710.—Cassini, De la Hire and Maraldi saw only one spot in this year, that in Oct. in S. Lat. $12\frac{1}{2}$°.

1710, Oct. 29—1713, May 18, Wurzelbauer saw no spots, although he observed " daily."

1713.—One spot, May 18–26, S. Lat. 16°.

1714–1715.—Many spots appeared in high latitudes in the northern hemisphere.

But in 1716 the " Histoire de l'Académie " states : " This year has had still more spots than the preceding, and perhaps no other year has had so many. There have been twenty-one different appearances of them, counting only as a single appearance that of several different spots at a time. Only the months of February, March, October and December have been free from them . . . . The phenomenon of two different spots at the same time has entirely ceased to be rare."

With this great increase in activity, which gave rise to a yet more decided maximum in 1718, the long dearth came to an end. It would seem to have commenced when the maximum of 1639— a low maximum itself—had fairly died down—that is to say, somewhere about 1645—and it was broken only by the very rare apparition of a single spot now and then, and by a feeble revival in 1703–7. It ended, as we have seen, with the setting in of the maximum 1715–20.

Thus, for close upon 70 years, the ordinary progress of the solar cycle, as we have been accustomed to it, was in abeyance— in abeyance to such a degree that the entire records of those 70 years combined together would scarcely supply sufficient observations of sunspots to equal one average year of an ordinary minimum such as we have been accustomed to during the past century.

It may be objected that, bearing in mind the feeble instruments of the seventeenth century and the paucity of observers, it may well have happened that many spots passed unnoticed. But no great instrumental power is needed to detect the presence of a sunspot, and, indeed, during the year 1919 four observers of the Solar Section, using only their unassisted sight, together obtained observations of spots on 119 days out of 259. In other words, spots were seen, or believed to have been seen, on 46 per cent. of the days of observation, and that not in a year of full maximum. Besides, telescopes were certainly more powerful during " the seventy years' dearth " than in the days of Galileo and Scheiner.

Nor were the observers few. In England there were Boyle, Derham, Flamsteed, Gray, Halley, and Hooke ; in Holland, Huyghens ; in France, De la Hire, Maraldi, Picard, and, above all, Cassini ; in Germany, Siverus, Vagetius, Vogelius, Weigel, and Wurzelbauer. Indeed, Derham, writing in 1711, well anticipates this objection : " There are," he says, " doubtless great intervals sometimes when the Sun is free, as between the years 1660 and

1671, 1676 and 1684, in which time, Spots could hardly escape the sight of so many' Observers of the Sun, as were then perpetually peeping upon him with their Telescopes in *England, France, Germany, Italy,* and all the World over, whatever might be before from Scheiner's time." And the fact that a new spot was apt to be independently discovered by more than one observer shows that the watch was carefully kept.

The unanimity with which, when the earlier days of revival had come, all observers speak of the greater frequency of spots as quite a new experience, · is irrefragible evidence that the change was a very real one. The " Histoire de l'Académie," calling attention to this very change, states : " Since 1695, for example, up to 1700, none were seen. Since 1700 our histories have been ' full of them, up to 1710, when only one was seen."

It ought not to be overlooked that, prolonged as this inactivity of the Sun certainly was, yet the few stray spots noted during " the seventy years' dearth,"—1660, 1671, 1684, 1695, 1707, 1718,—correspond, as nearly as we can expect, to the theoretical dates of maximum. If I may repeat the simile which I used in my paper for *Knowledge* in 1894, " just as in a deeply inundated country, the loftiest objects will still raise their heads above the flood, and a spire here, a hill, a tower, a tree there, enable one to trace out the configuration of the submerged champaign," so the above-mentioned years seem to be marked out as the crests of a sunken spot-curve.

The late Miss Agnes Clarke contributed a note to *Knowledge* in September, 1894, p. 206, which may well serve as an appendix to the above. She writes :—" There is strong, although indirect, evidence that the ' prolonged sun-spot minimum ' was attended by a profound magnetic calm. This evidence is to be found in the auroral records of the time. For the connection between the occurrence of auroræ and the magnetic condition of the earth is so close, that the absence of one kind of disturbance may safely be held to betoken the absence of the other.

" Now in England, during the whole of the seventeenth century, not an auroral glimmer was chronicled. Stowe recounts that on the 14th and 15th of November, 1574, ' the heavens from all parts did seem to burn marvellous ragingly,'* and the next similar occurrence took place on March 17, 1716. Upon his observations of this fine display, Halley founded his magnetic theory of auroræ.† . . . The event created an extra-ordinary sensation throughout the country, some slight and partial sky-illuminations in 1706 and 1709 having escaped general notice.

" On the Continent, the auroral blank was much less complete than in this country. Gassendi's aurora of September 12, 1621, was seen as far south as Aleppo ; Cromerus registered the passage of ' whole armies ' across the sky in 1629 ; in 1640,

---

* "Annales of England," p. 678.

† *Philosophical Transactions,* Vol. XXIX., p. 407. *See also* E. J. Lowe's " Chronology of the Seasons," quoted by Dr. Garnett in *Nature* Vol. III., p. 46. For a fuller account of auroral history during this period we may refer to the *Edinburgh Review,* No. 336 (October, 1886), pp. 418–421.

south polar lights were visible in Chile every night of February and March, and some corresponding appearances were noted in northern latitudes. By the middle of the century, however, polar lights had virtually died out everywhere, except, perhaps, in northern Scotland, where the ' merry dancers ' were seen without surprise in 1691. But even in Iceland and Norway they became so rare as to be considered portentous, and their reappearance at Copenhagen in 1709 was greeted with consternation and amazement. De Mairan, in his ' Traité de l'Aurore boréale,' makes the curious remark that a great extension of the zodiacal light attended the auroral outburst of 1716.

" As regards the solar corona during the ' prolonged minimum,' it appears probable that . . . its radiated structure was in abeyance, but there is positive proof that the inner corona maintained at least its average brightness in 1666. The partial solar eclipse of June 22 in that year, being viewed through Boyle's 60-ft. telescope by Hooke, Pope (Professor of Astronomy in Gresham College), and others, ' there was perceived a little of the limb of the moon without the disc of the sun, which seemed to some of the observers to come from some shining atmosphere about the body either of the sun or moon.' " *

* *Philosophical Transactions*, Vol. I., p. 295.

Reprinted from *Science* **192:**1189–1202 (1976)

The Maunder Minimum

*The reign of Louis XIV appears to have been a time of
real anomaly in the behavior of the sun.*

John A. Eddy

It has long been thought that the sun is a constant star of regular and repeatable behavior. Measurements of the radiative output, or solar constant, seem to justify the first assumption, and the record of periodicity in sunspot numbers is taken as evidence for the second. Both records, however, sample only the most recent history of the sun.

When we look at the longer record—of the last 1000 years or so—we find indications that the sun may have undergone significant changes in behavior, with possible terrestrial effects. Evidence for past solar change is largely of an indirect nature and should be subject to the most critical scrutiny. Most accessible, and crucial to the basic issue of past constancy or inconstancy, is a long period in the late 17th and early 18th centuries when, some have claimed, almost no sunspots were seen. The period, from about 1645 until 1715, was pointed out in the 1890's by G. Spörer and E. W. Maunder. I have reexamined the contemporary reports and new evidence which has come to light since Maunder's time and conclude that this 70-year period was indeed a time when solar activity all but stopped. This behavior is wholly unlike the modern behavior of the sun which we have come to accept as normal, and the consequences for solar and terrestrial physics seem to me profound.

The author is an astronomer on the Special Projects staff of the High Altitude Observatory, National Center for Atmospheric Research, Boulder, Colorado 80303.

The Sunspot Cycle

Surely the best-known features of the sun are sunspots and the regular cycle of solar activity, which waxes and wanes with a period of about 11 years. This cycle is most often shown as a plot of sunspot number (Fig. 1)—a measure of the number of spots seen at one time on the visible half of the sun (*1*). Sunspot numbers are recorded daily, but to illustrate long-term effects astronomers more often use the annual means, which smooth out the short-term variations and average out the marked imprint of solar rotation.

There is as yet no complete physical explanation for the observed solar cycle. Modern theory attributes the periodic features of sunspots to the action of a solar dynamo in which convection and surface rotation interact to amplify and maintain an assumed initial magnetic field (*2*). Dynamo models are successful in reproducing certain features of the 11-year cycle, but with these models it is not as yet possible to explain the varying amplitudes of maxima and other long-term changes.

The annual mean sunspot number at a typical minimum in the 11-year cycle is about six. During these minimum years there are stretches of days and weeks when no spots can be seen, but a monthly mean of zero is uncommon and there has been only 1 year (1810) in which the annual mean, to two-digit accuracy, was

zero. In contrast, in the years around a sunspot maximum there is seldom a day when a number of spots cannot be seen, and often hundreds are present.

Past counts of sunspot number are readily available from the year 1700 (*3*), and workers in solar and terrestrial studies often use the record as though it were of uniform quality. In fact, it is not. Thus it is advisable, from time to time, to review the origin and pedigree of past sunspot numbers, and to recognize the uncertainty in much of the early record.

A Brief History

Dark spots were seen on the face of the sun at least as early as the 4th century B.C. (*4*), but it was not until after the invention of the telescope, about 1610, that they were seen well enough to be associated with the sun itself. It would seem no credit to early astronomers that over 230 years elapsed between the telescopic "discovery" of sunspots and the revelation of their now obvious cyclic behavior. In 1843, Heinrich Schwabe, an amateur, published a brief paper reporting his own observations of spots on the sun for the period 1826 to 1843 and pointing out an apparent period of about 10 years between maxima in their number (*5*).

Rudolf Wolf, director of the Observatory at Bern and later at Zurich, noticed Schwabe's paper and shortly after set out to test the result by extending the limited observations on which the 10-year cycle was based. In 1848 he organized a number of European observatories to record spots on a regular basis and by a standard scheme, thus inaugurating an international effort which continues today. Wolf also undertook a historical search and reanalysis of old data on the sun in the literature and in observatory archives. More than half of the record of sunspot numbers in Fig. 1, and all of it before 1848, is the result of Wolf's historical reconstruction. The most reliable part of the curve thus comes after 1848, when it is based on controlled observations. Wolf found descriptions and drawings of the sun which allowed him to reconstruct daily sunspot numbers 30 years into the past—to 1818—although, unlike the real-time data, they came from a thinner sample and with less certain corrections for observers and conditions. He was able to locate sufficient information on the more distant past to allow reconstructed "monthly averages" of the sunspot number (that is, a minimum of one observation per month) to 1749, and approximate "annual averages" from more scattered data to 1700 (3). The reliability of the curve, and especially of its absolute scale, may be graded into four epochs: reliable from 1848 on, good from 1818 through 1847, questionable from 1749 through 1817, and poor from 1700 through 1748.

Wolf collected data to extend the historical curve the final 90 years to the telescopic discovery of sunspots in 1610 (6). He published estimated dates of maxima and minima for 1610 through 1699 but not sunspot numbers. That he elected to discontinue sunspot numbers at 1700 may be significant: perhaps he felt he had reached the elastic limit of the sparse historical record at the even century mark; it could also be that at 1700 he ran into queer results. In this article I shall point out that the latter probably applies. It seems fair to assume that, once he had confirmed and refined Schwabe's cycle, Wolf was biased toward demonstrating that the sunspot cycle persisted backward in time (7); thus, when the cycle appeared to fade, especially in dim, historical data, he would have been inclined to quit the case and to call it proven. In any event we should be especially skeptical of the curve in its thinnest and oldest parts (1700 through 1748), and to question anew what happened before 1700.

Even though we are aware of the varying quality of the Wolf sunspot record, most of us probably take it as evidence of a truly continuous curve, much like the sample of a continuous wave form that we see on the screen of an oscilloscope. We assume that, just as Schwabe's 17-year sample was enough to reveal the cycle's existence, so the 260-year record in Fig. 1 is adequate to establish its likely perpetuation to the future and extension through the past. Reconstructions of the solar cycle have been estimated from indirect data to the 7th century B.C. in the Spectrum of Time Project (STP) of D. J. Schove, but these heroic efforts are of necessity based on far from continuous information and are built on the explicit assump-

tion of a continued 11-year cycle (8–11). Recent insights into the physical basis for the sunspot cycle and its origin in the fluid, outer layers of the sun give us new cause to suspect that at least some of the features of the present sunspot cycle may be transitory. If we accept the solar dynamo, we must allow that any of its coupled forces could have changed enough in the past to alter or suspend the "normal" solar cycle. Indeed, there is now evidence that solar rotation has varied significantly in historic time (12).

The "Prolonged Sunspot Minimum"

The possibility that sunspots sharply dropped in number before 1700 was pointed out rather clearly by two well-known solar astronomers in the late 19th century. In papers published in 1887 and 1889 the German astronomer Gustav Spörer called attention to a 70-year period, ending about 1716, when there was a remarkable interruption in the ordinary course of the sunspot cycle and an almost total absence of spots (13). Spörer was studying the distribution of sunspots with latitude and had found evidence that the numbers of spots in the northern and southern hemispheres of the sun were not always balanced. To check this observation he had consulted historical records, including Wolf's, and was surprised at what he found in the data of the 17th and early 18th centuries. Not long after, Spörer died. Meanwhile, E. W. Maunder, superintendent of the Solar Department, Greenwich Observatory, took up the case. In 1890 Maunder summarized Spörer's two papers for the Royal Astronomical Society and in 1894 gave a fuller account in an article entitled "A Prolonged Sunspot Minimum" (14, 15). In his second paper Maunder provided more details and pointed out that to acknowledge this unusual occurrence was to admit that the solar cycle and the sun itself had changed in historic time, and could again. He stressed that the reality of a "prolonged sunspot minimum" had important implications not only for our understanding of the sun but also for studies of solar-terrestrial relations.

It is not obvious that anyone in solar physics listened. In any case, nearly 30 years later, at 71, Maunder tried again with another paper of the same title on the same subject (16). Included were quotations from a paper by Agnes Clerke who had claimed that during the "prolonged sunspot minimum" there was also a marked dearth of aurorae (17). Maunder offered as well the interesting

conjecture that the long delay between the telescopic discovery of sunspots and Schwabe's discovery of the solar cycle may have been due in part to this temporary cessation of the solar cycle during a part of the interim.

In their five papers Spörer and Maunder made the following striking assertions: (i) that for a 70-year period, from approximately 1645 to 1715, practically no sunspots were seen; (ii) that for nearly half of this time (1672 through 1704) not a single spot was observed on the northern hemisphere of the sun; (iii) that for 60 years, until 1705, no more than one sunspot *group* was seen on the sun at a time; and (iv) that during the entire 70-year period no more than "a handful" of spots were observed and that these were mostly single spots and at low solar latitudes, lasting for a single rotation or less; moreover, the total number of spots observed from 1645 to 1715 was less than what we see in a single active year under normal conditions.

Maunder supported these claims with quotations from the scientific literature of the period in question. The editor of the *Philosophical Transactions of the Royal Society*, in reporting the discovery of a sunspot in 1671 (in the middle of the "prolonged sunspot minimum"), had written that (15, p. 173) ". . . at Paris the Excellent Signior Cassini hath lately detected again Spots in the Sun, of which none have been seen these many years that we know of." (Following this, the editor went on to describe the last sunspot seen, 11 years before, for those who might have forgotten what one looked like.)

Cassini's own description of his 1671 sighting reads as follows (15, p. 174): ". . . it is now about 20 years since astronomers have seen any considerable spots on the sun, though before that time, since the invention of the telescopes they have from time to time observed them." Cassini also reported that another French astronomer, Picard, ". . . was pleased at the discovery of a sunspot since it was ten whole years since he had seen one, no matter how great the care which he had taken from time to time to watch for them" (16, pp. 141–142). And when the Astronomer Royal, Flamsteed, sighted a spot on the sun at Greenwich in 1684, he reported that "[t]hese appearances, however frequent in the days of *Scheiner* and *Galileo*, have been so rare of late that this is the only one I have seen in his face since *December* 1676" (15, p. 174).

Maunder did not have to look hard to find support for the strange case, for an absence of sunspots in the latter part of the 17th century had been matter-of-factly reported in astronomy books written before Schwabe's discovery of the cycle (*18*). William Herschel had mentioned it in 1801 (*19*). Herschel's source of information was LaLande's three-volume opus, *Astronomie*, of 1792, in which dates and details are given of the anomalous absence of sunspots, including some of the quotations that Maunder later used (*20*). Thus, neither Maunder nor Spörer had "discovered" the "prolonged sunspot minimum." These authors, like myself, were simply pointing back to an overlooked and possibly important phenomenon which in its time had not seemed unusual but which looms large in retrospect.

Questions

Maunder's assessment of the significance of the "prolonged sunspot minimum" was probably not an exaggeration. If solar activity really ceased or sank to near-zero level, it places a restrictive boundary condition on physical explanations of the solar cycle and suggests that a workable mechanism for solar activity must be capable of starting, and maybe stopping, in periods of tens of years. It labels sunspots as possibly transitory characteristics of the sun, and by association also flares, active prominences, and perhaps the structured corona. One of the enigmas in historical studies of the sun is the long delay in the naked-eye discovery of the chromosphere (*21*) and the lack of any ancient descriptions of coronal streamers at eclipse (*22, 23*). It may be more than curious coincidence that the discovery of the chromosphere (1706), the first description of the structured corona (1715), and a lasting, tenfold jump in the number of recorded aurorae (1716) all came at the end of the Maunder Minimum, when, it seems, the solar cycle resumed, or possibly began, its modern course. If Maunder's "prolonged sunspot minimum" really happened, it provides damning evidence (*24*) in the protracted debate over the production of sunspots by planetary gravitational tides, for through the years between 1645 and 1715 the nine planets were, as always, in their orbits. Finally, as Maunder stressed, this apparent anomaly in the sun's history, if real, offers a singularly valuable test period for studies of the connection between solar activity and terrestrial weather. If the Maunder Minimum really occurred, it may define a minimum of a long-term

envelope of solar activity which could be more important for terrestrial implications than the 11-year modulation that has for so long occupied attention in solar-terrestrial studies (*25*).

It seems worthwhile to open, once again, the case of the missing sunspots, for it was never really solved. All the early work was based almost entirely on the same piece of evidence: the paucity of sunspot reports in the limited literature of the day. Spörer's original papers and Maunder's expansions of them leaned heavily on a lack of evidence in archival records and journals, and on contemporary statements that it had been a long time between sunspot reports. But in the words of a modern astronomer, absence of evidence is not evidence of absence (*26*). How good were the observers in the 17th century, and how good the observing techniques? How constant a watch was kept? How many spots were missing, and when? New evidence has come to light in the 50 years since Maunder's time: we now have better catalogs of historical aurorae, compilations of sunspot observations made in the Orient, a fuller understanding of tree-ring records, and a new tool in atmospheric isotopes as tracers of past solar activity. New understanding of the sun since Maunder's day can sharpen our assessment of the facts in the case: we now know the relationship of sunspots to solar magnetic fields and something of the relation of magnetic fields to the corona, and can thus examine more critically the evidence from total solar eclipses during the time.

Solar Observations in the 17th Century

History has left an uncanny mnemonic for the dates of the Maunder Minimum: the reign of Louis XIV, *le Roi Soleil*, 1643 through 1715. This was also the time of Milton and Newton; by 1642 Brahe, Kepler, and Galileo were gone. Astronomical telescopes were in common use and were produced commercially; they featured innovations and important improvements over the original miniature models which in 1612 had sufficed to distinguish umbrae and penumbrae in sunspots and by 1625 had been used to find the solar faculae. During the Maunder Minimum the Greenwich and Paris observatories were founded, and Newton produced the reflecting telescope; it was also the age of the long, suspended, and aerial telescopes with focal lengths that stretched to 60 meters and apertures of 20 centimeters and more (*27*). The more usual telescopes turned on the sun had focal lengths of 2 to 4 meters and apertures of 5 to 10 centimeters, which would describe most solar telescopes used in the 18th and 19th centuries as well. To observe sunspots then, as today, one projected the solar image on a white screen placed at a proper distance behind the eyepiece (Fig. 2). The image scale was adequate to permit one to see and to sketch not only spots of all sizes but their features and their differences; observers recorded details of white-light faculae, penumbral filaments, satellite sunspots, and most of the observational detail known of sunspots today (Fig. 3).

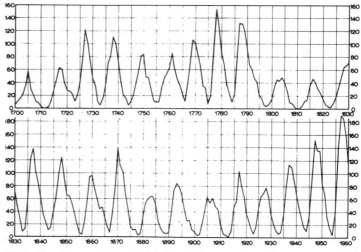

Fig. 1. Annual mean sunspot number, *R*, from 1700 to 1960. [From (*3*); courtesy of M. Waldmeier]

Fig. 2. Illustration of the technique used in the early 17th century for the observation of sunspots in which the solar image is projected on a screen [from a contemporary account by Scheiner (*31*)]. [By permission of the Houghton Library, Harvard University]

During the Maunder Minimum the same astronomers who observed the sun discovered the first division in Saturn's ring (in 1675) and found five of Saturn's satellites (1655 through 1684); the former discovery attests to an effective resolution of almost 1 arc second and the latter to an acuity to distinguish an 11th-magnitude object less than 40 arc seconds from the bright limb of the planet. During the 17th century astronomers observed seven transits of Venus and Mercury, which implies a certain thoroughness and a knowledge of other spots on the sun at the time. Römer determined the velocity of light (1675) from precise observations of the orbits of Jupiter's satellites. During the same century at least 53 eclipses of the sun—partial, annular, or total—were observed, including some in Asia and the Americas. It is significant that not one solar eclipse that passed through Europe was missed (*28, 29*).

Active astronomers of the time included Flamsteed, Derham, Hooke, and Halley in England, both of the Huyghens in Holland, Hevelius in Poland, Römer in Denmark, the Cassinis, Gassendi, de la Hire, and Boulliau in France, Grimaldi and Riccioli in Italy, and Weigel and von Wurzelbau in Germany, to name but a few. And astronomers of that era were generous in their definition of astronomy and still included the sun among objects of respectable interest. During the years when the Cassinis were pursuing their investigations of Saturn in Paris, they also wrote scientific articles on their observations of the sun and sunspots (*30*). In 1630 Christopher Scheiner published a massive book, the *Rosa Ursina*, on sunspots and faculae and methods of observing them (*31*), and Hevelius produced in 1647 a detailed appendix on sunspots and

a chapter on solar observation in his *Selenographia* (*32*).

In 1801 William Herschel commented that instrumental and observational shortcomings could explain most of the sunspot dearth between 1650 and 1713, and that, had more modern equipment been turned on the sun, many more spots would have been found (*19*); but we have little cause to think that he had looked very far into the matter, which then seemed of minor import, long before the discovery of the sunspot cycle. Maunder did not cite Herschel's dissenting view, but trumped it anyway, with a quotation from the more contemporary English astronomer William Derham, who in 1711 had given his view on whether observers of the time could have missed the spots (*16*, pp. 143–144):

There are doubtless great intervals sometimes when the Sun is free, as between the years 1660 and 1671, 1676 and 1684, in which time, Spots could hardly escape the sight of so many Observers of the Sun, as were then perpetually peeping upon him with their Telescopes in *England, France, Germany, Italy,* and all the World over.

It seems clear that on this question Derham was right and Herschel wrong and that during the period of the Maunder Minimum astronomers had the instruments, the knowledge, and the ability to recognize the presence or absence of even small spots on the sun. And I might add that it does not take much of a telescope to see a sunspot.

Was a continuous watch kept on the sun? This is quite another question, and one for which direct evidence is lacking. Scheiner (1575–1650) and Hevelius (1611–1687) for at least a number of years made daily drawings of the sun and sunspots, but we cannot assume that this

dutiful practice was continued by successors without interruption for 70 years. There were no organized or cooperative efforts, so far as we know, to keep a continuous diary of the sun, as is done today. But the motives of astronomers, then and now, are much the same: when a surprising dearth of sunspots was reported, as it was on repeated occasions during the span, we can expect that it would have inspired a renewed search to find some. In this respect it is significant that new sunspots were reported in the scientific literature as "discoveries," and that the sighting of a new spot or spot group was cause for the writing of a paper (*30*). This practice, were it followed today by even a few owners of 5-centimeter refractors, would produce an intolerable glut of manuscripts in the minimum years of the sunspot cycle and an avalanche in the years of maximum.

Comparisons with the present time are dangerous: toward the end of the 17th century the first learned societies were founded and the first journals came into existence. These journals were limited in number and scope and restricted in authorship and in that time bore little resemblance to the scientific periodicals we read and rely on for thorough coverage today. Absence of evidence may be a limited clue in such circumstances, as may uncontested and possibly unrefereed reports. Moreover, prevailing ideas of what something is influence how it is observed and reported. Sunspots were not thought to be what we know they are today. The original theological opposition to spots on the sun had been assuaged long before 1645, but, throughout the period of the Maunder Minimum and until Wilson's observations in 1774 (*33*), a prevalent concept of sunspots was that they were clouds on the sun, and who keeps a diary of clouds? Finally, we can suspect that sunspots, like all else in science, went in and out of vogue as objects of intense interest. After the initial surge of telescopic investigation, sunspots may have drifted into the doldrums of current science. If this is so, Scheiner's massive tome may have been in part to blame: the *Rosa Ursina* must have been considered a bore by even the verbose standards of its day, and it may have smothered initiative for a time (*34, 35*).

Aurorae

Records of occurrence of the aurora borealis and aurora australis offer an independent check on past solar activity since there is a well-established correla-

178

tion between sunspot number and the number of nights when aurorae are seen. The physical connection is indirect: auroral displays are produced when charged particles from the sun interact with the earth's magnetic field, resulting in particle accelerations and collisions with air molecules in our upper atmosphere. Aurorae register, therefore, those particle-producing events on the sun (such as flares and prominence eruptions) which happen to direct their streams toward the earth. Since these events arise in active regions on the sun, where there are also sunspots, we find a strong positive correlation between reported numbers of the two phenomena.

Aurorae are especially valuable as historical indicators of solar activity since they are spectacular and easily seen, require no telescopic apparatus, and are visible for hours over wide geographic areas. They have been recorded far back in history as objects of awe and wonder.

An increase in the number of reported aurorae inevitably follows a major increase in solar activity, and a drop in their number can generally be associated with the persistence of low numbers of sunspots, with certain reservations. As with sunspots, aurorae will not be seen unless the sky is reasonably clear, and an absence of either on any date in historical records could be due simply to foul weather. For the period of our interest we can exclude the possibility of years or decades of persistent continental overcast, since this would constitute a significant meteorological anomaly which would certainly have been noted in weather lore or cited by astronomers of the day (*36*).

In fact, the period between 1645 and 1715 was characterized by a marked absence of aurorae, as was first pointed out by Clerke. "There is," she wrote, "... strong, although indirect evidence that the 'prolonged sunspot minimum' was attended by a profound magnetic calm" (*17*, p. 206). Historical aurora catalogs (*37*, *38*) confirm her assessment that there were extremely few aurorae reported during the years of the Maunder Minimum. Far fewer were recorded than in either the 70 years preceding or following.

Auroral occurrence is a strong function of latitude, or more specifically of distance from the geomagnetic poles. Analyses of auroral counts in the modern era (*39*) lead us to expect a display almost every night in the northern "auroral zone"—a band of geomagnetic latitude which includes northern Siberia, far-northern Scandinavia, Iceland, Greenland, and the northern halves of

Canada and Alaska. But this region is also an area of sparse historical record for the 17th century, and it should probably be excluded from consideration for the present purpose. In a more populous band just south of this zone—which includes Sweden, Norway, and Scotland— we expect aurorae on 25 to about 200 nights per average year, the higher number at higher latitude. Progressively fewer are expected as we move south. For most of England, including the London area, we expect to see an average of 5 to 10 aurorae per year, or roughly 500 in 70 "normal" years. In Paris we can expect about 350 in the same period, and in Italy perhaps 50. From England, France, Germany, Denmark, and Poland, where astronomers were active during the Maunder Minimum, we might have expected reports of 300 to 1000 auroral nights, by the statistics of today. Fritz's historical catalog (*37*) lists only 77 aurorae for the entire world during the years from 1645 to 1715, and 20 of these were reported in a brief active interval, from 1707 to 1708, when sunspots were also seen. In 37 of the years of the Maunder Minimum not a single aurora was reported anywhere. Practically all reported aurorae were from the northern part of Europe: Norway, Sweden, Germany, and Poland.

For 63 years of the Maunder Minimum, from 1645 until 1708, not one was reported in London. The next, on 15 March 1716, moved the astronomer Edmund Halley to describe and explain it in a paper that is now classic (*40*). He was then 60 years old and had never seen an aurora before, although he was an assiduous observer of the sky and had long wanted to observe one.

The auroral picture, which seems clear at first glance, is muddied by subjectivity and by the obscurity of indirect facts from long ago. Historical catalogs cannot record aurorae but only reports of aurorae. Clerke did not mention that auroral counts from all centuries before the 18th are very low by modern standards. The 77 events noted during the Maunder Minimum actually exceed the number recorded in any preceding century except the 16th, for which there are 161 in Fritz's catalog. By contrast, 6126 were reported in the 18th century and about as many in the 19th century (*41*).

The really striking feature of the historical record of aurorae (Fig. 4) is not so much the drop during the Maunder Minimum but an apparent "auroral turn-on" which commenced in the middle 16th century and surged upward dramatically after 1716. Were the historical record of

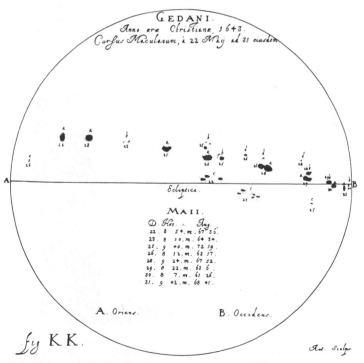

Fig. 3. A 17th-century drawing of the sun and sunspots by Hevelius (*32*). [By permission of the Houghton Library, Harvard University]

179

uniform quality (and it is not), this apparent "switching on" of the northern (and southern) lights would loom as the most significant fact of recent solar-terrestrial history. In truth, it must in part at least reflect the general curve of learning which probably holds for all of life in northern Europe at the time. The Renaissance came to auroral latitudes later than to the Mediterranean, and the envelope we see in Fig. 4 may be but its shadow. The effect is large, however, and a part of it could well represent a real change in the occurrence of aurorae on the earth, and, by implication, a change in the behavior of the sun. It is important that

auroral reports do not increase monotonically with time as a learning curve might imply; the number reported rose in the 9th through 12th centuries and then fell off.

The separation of the physical from the sociological in Fig. 4 is a question of major importance in studies of the sun and earth. An acceptable solution would involve starting with a new and careful search for auroral data, particularly from northern latitudes, in the New World, Old World, and Orient. It must include careful allowance for superstition and vogues and restrictions in observing aurorae, shifts of population, and the possi-

bly important effects of single events, such as the development of the printing press (about 1450), or Gassendi's decription of the French aurora of 1621 (38, p. 15) and Halley's paper in 1716 (40). One suspects that the dramatic jump in the number of reported aurorae after 1716 was a direct result of this important paper of Halley, which put the auroral phenomenon on firm scientific footing so that more aurorae were looked for and more regular records were kept.

As for the Maunder Minimum, its presence in the auroral record is surely real, appearing in Fig. 4 as a pronounced pause in the already upward-sweeping curve. Had Maunder looked first at Fritz's auroral atlas, he could have hypothesized a "prolonged sunspot minimum" from auroral evidence alone.

Sunspots Seen with the Naked Eye

Spots on the sun were seen with the naked eye long before the invention of the telescope (42) and were particularly noted in the Far East, where a more continuous record survives. They offer another check on the reality of an extended sunspot minimum, since naked-eye reports of sunspots might be expected were there any strong solar activity at the time. Large spots and large spot groups can be seen with little difficulty when the sun is partially obscured and reddened by smoke or haze, or at sunset or sunrise; small groups or small spots are beyond the effective resolution of the eye and cannot be seen. Thus reports of naked-eye sightings are biased toward times of enhanced solar activity, and attempts have been made to establish the epochs of past maxima in the solar cycle from naked-eye sunspot dates (43, 44).

Pretelescopic sunspot observations probably come almost wholly from accidental observation. In Europe reports are rare and fragmentary (4). It is from the Orient, where sunspots were deemed important in legend and possibly in augury, that we find more extensive and useful records. But here, too, the numbers are small and can only be used as a very coarse indicator of past solar activity.

In 1933 (5 years after Maunder's death), Sigeru Kanda of the Tokyo Astronomical Observatory compiled a comprehensive list of 143 sunspot sightings from ancient records of Japan, Korea, and China, covering the period from 28 B.C. through A.D. 1743 (43). Most came after the 3rd century, so that the long-term average was about one sighting per decade. Were they distributed regularly (or just at solar maxima), we would thus

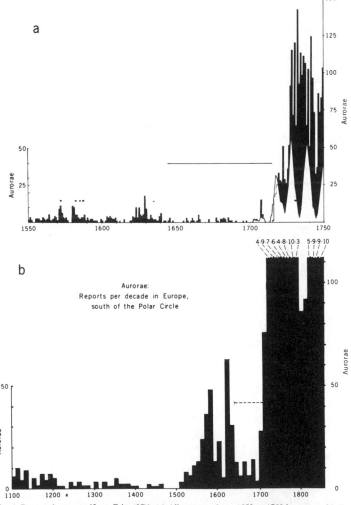

Fig. 4. Reported aurorae [from Fritz (37)]. (a) All reports, from 1550 to 1750 by year, with the annual mean sunspot number superposed as white curves at the right and Far East aurorae (43, 44, 46) shown as solid squares. (b) Reports per decade in latitudes 0° to 66°N; counts after 1715 must be multiplied by the numbers shown at the top right of the plot. The period of the Maunder Minimum is shown in each diagram as a horizontal line.

expect six or seven events during the Maunder Minimum. It is significant that none was recorded between 1639 and 1720—a Far East gap that matches Western Hemisphere data very well.

As with aurorae, the evidence is necessary but not sufficient. Social practices or pressures could have suppressed observation or recording of spots during the time (45), leading to an apparent but unreal dearth. Moreover, the sunspot gap from 1639 to 1720 is neither the only nor the longest in Kanda's span of reports: there were 84 years without any reports of sunspot sightings ending in 1604, 117 years ending in 1520, and 229 years ending in 808 (Fig. 5a).

We may extend the naked-eye data in a sense by adding dates of reported aurorae in Japan, Korea, and China. All of these lands lie at low auroral latitudes, where displays are expected no more than once in 10 years. As in the case of sunspots seen with the naked eye, aurorae reported in the Orient are presumed to sample only intense solar activity. And, as with the sunspot sightings, no Far East aurorae were reported during the Maunder Minimum, and more specifically between 1584 and 1770 (43, 44, 46). The oriental data (sunspots and aurorae) confirm that there were no intense periods of solar activity during the Maunder Minimum and probably no "normal" maxima in the solar cycle.

We may use the long span of oriental sunspot data as a coarse check on possible earlier occurrences of prolonged sunspot minima, or other gross, long-term modulations of sunspot activity. Of particular note is an intensification of

sunspot and aurora reports in the 200-year period centered at around 1180, which is about halfway between the Maunder Minimum and a more extended period of absence of Far East sunspots and aurorae in the 7th and early 8th centuries. As I will show below, the naked-eye maximum coincides with a similar maximum of solar activity in the ^{14}C record. If this is a real long-term envelope of solar activity, its period is roughly 1000 years. We may be measuring only social effects, but, as with historical European aurorae, the subject is one of potential importance which deserves more specific attention by historians.

Carbon-14 and the History of the Sun

Modern confirmation for Maunder's "prolonged sunspot minimum" may be found in recent determinations of the past abundance of terrestrial [14]C. Carbon and its radioactive isotopes are abundant constituents of the earth's atmosphere, chiefly as carbon dioxide (CO_2). When CO_2 is assimilated into trees, for example, the carbon isotopes undergo spontaneous disintegration at well-known rates. Thus, by a technique now well established, it is possible to determine the date of life of a carbon-bearing sample, such as wood, by chemical measurement of its present ^{11}C content and comparison with a presumed original amount. The method requires a knowledge of the past abundance of ^{14}C in the atmosphere, and this value is found by analyzing, ring by ring, the ^{14}C content of trees of known chronology. The history

of relative ^{14}C abundance deviations is now fairly well established and serves as the basis for accurate isotopic dating in archeology (47–50).

The ^{14}C history is useful in its own right as a measure of past solar activity, as has been demonstrated by a number of investigators (51, 52). The isotope is continuously formed in the atmosphere through the action of cosmic rays, which, in turn, are modulated by solar activity. When the sun is active, some of the incoming galactic cosmic rays are prevented from reaching the earth. At these times, corresponding to maxima in the sunspot cycle, less than the normal amount of ^{14}C is produced in the atmosphere and less is found in tree rings formed then. When the sun is quiet, terrestrial bombardment by galactic cosmic rays increases and the ^{14}C proportion in the atmosphere rises. There are other terms in the ^{14}C equilibrium process, as well as significant lags, but, were there a prolonged period of quiet on the sun, we would expect to find evidence of it in tree rings of that era as an abnormally high abundance of [14]C.

Such is the case. The first major anomaly found in the early studies of ^{14}C history was a marked and prolonged increase which reached its maximum between about 1650 and 1700 (53), in remarkable agreement in sense and date with the Maunder Minimum. The phenomenon, known in carbon-dating as the DeVries Fluctuation, peaked at about 1690 and is the greatest positive excursion found in the ^{14}C record—corresponding to a deviation of about 20 parts per mil from the norm. Subsequent stud-

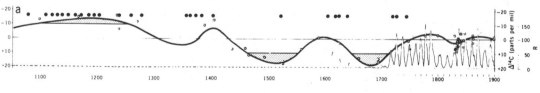

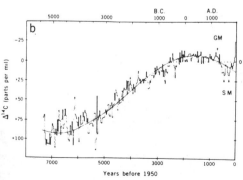

Fig. 5. (a) History of deviations in the relative atmospheric ^{14}C concentration from tree-ring analyses for the period 1050 to 1900 (54): single open circles, Northern Hemisphere data; double open circles, Southern Hemisphere data (a heavy line has been drawn through the Southern Hemisphere data); closed circles, dates of reported sunspots seen with the naked eye from Kanda (43). The annual mean sunspot number, R, is shown as a light solid line where known for the period after 1610, from Waldmeier (3) and this study. Periods when the relative ^{14}C deviation exceeds 10 parts per mil are shaded. They define probable anomalies in the behavior of the solar cycle: 1100 to 1250, Grand Maximum; 1460 to 1550, Spörer Minimum; 1645 to 1715, Maunder Minimum. (b) Measured ^{14}C deviation (in parts per mil) since about 5000 B.C., with observed (smoothed) curve of sinusoidal variation in the earth's magnetic moment [from (61), figure 2]. At about A.D. 100 the magnetic moment reached a maximum of about 10^{24} gauss per cubic centimeter. Shorter-term ^{14}C excursions attributed in this article to solar cause are marked with arrows: *M*, Maunder Minimum; *S*, Spörer Minimum; *GM*, Grand Maximum in the 12th to 13th centuries. The sharp negative ^{14}C deviation at the modern end of the curve is the Suess effect, due to fossil fuel combustion.

ies have established the DeVries Fluctuation as a worldwide effect.

Figure 5a shows a curve (open circles and heavy line) of the relative deviation in the ^{14}C concentration based on recent measurements of tree rings (54), plotted with increasing concentration downward for direct comparison with solar activity; also shown are annual numbers of sunspots, from (3) and the present work (light line), and the years of early naked-eye sunspot sightings from Kanda (closed circles) (43). The three quantities give a wholly consistent representation of the Maunder Minimum. We also note a clustering of naked-eye sunspot sightings at times when the ^{14}C record indicates greater than normal activity, and a general absence of them when the ^{14}C record indicates less than normal activity. Where annual sunspot numbers are plotted, the ^{14}C curve seems a fair representation of the overall envelope of the sunspot curve. It thus seems valid to interpret the ^{14}C record as an indicator of the long-term trend of solar activity and of real changes in solar behavior in the distant past, before the time of telescopic examination of the sun (55–57).

We may calibrate the ^{14}C curve for this purpose by noting that the years of the Maunder Minimum define a time when the relative deviation of ^{14}C exceeded 10 parts per mil. If we can make allowance for other effects on ^{14}C production and equilibrium, we may infer that, whenever the ^{14}C deviation exceeded ± 10 parts per mil, solar activity was anomalously high or low, with the Maunder Minimum corresponding to a definition of "anomalous." We must remember that the ^{14}C indications will tend to lag behind real solar changes by periods of 10 to 50 years, because of the finite time of exchange between the atmosphere and trees. By this criterion there have been three possible periods of marked solar anomaly during the last 1000 years: the Maunder Minimum, another minimum in the early 16th century, and a period of anomalously high activity in the 12th and early 13th centuries. We can think of these as the grand minima and a grand maximum of the solar cycle, although we cannot judge from these data whether they are cyclic features.

The earlier minimum, which we might call the Spörer Minimum, persisted by our 10-parts-per-mil criterion from about 1460 through 1550. Its ^{14}C deviation is not quite as great as that during the Maunder Minimum, although that distinction is not a consistent feature of all representations of the ^{14}C history (58). We can presume that the Spörer Minimum was probably as pronounced as the Maunder Minimum and that during those years there were few sunspots indeed. It appears to have reached its greatest depth in the early 16th century when there were also very few aurorae reported.

We noted earlier the possibility of an intensification of solar activity in the 12th and 13th centuries, on the basis of naked-eye sunspot reports from the Orient. Evidence for the same maximum is found in the historical aurora record (Fig. 4): the number of aurorae in Fritz's catalog (37) is about constant for the 9th, 10th, and 11th centuries (23, 27, and 21 aurorae per century, respectively), rises abruptly for the 12th century (53 aurorae), and then falls for the next three centuries (16, 21, and 7 aurorae). The ^{14}C record (Fig. 5a) shows a similar anomaly in the same direction: a decrease in ^{14}C which could be attributed to a prolonged increase in solar activity.

We must take care in assigning any of the ^{14}C variations to a solar cause for there are other important mechanisms. The overwhelming long-term effects on ^{14}C production are ponderous changes in the strength of the earth's magnetic field (59, 60). Archeomagnetic studies have shown that in the past 10,000 years the earth's magnetic moment has varied in strength by more than a factor of 2, following an apparently sinusoidal envelope with a period of about 9000 years, on which shorter-term changes are impressed. The terrestrial moment reached maximum strength at about A.D. 100, at which time we would expect to find a minimum in ^{14}C production because of enhanced shielding of the earth against cosmic rays.

The good fit of the observed (smoothed) curve of geomagnetic change to the long-term record of fossil ^{14}C is shown in Fig. 5b, from a recent compilation (61), here replotted with increasing ^{14}C in the downward direction to display increasing solar activity and increasing geomagnetic strength as upward-going effects. Damon (57) has stressed that the long-term trends in the radiocarbon content of the atmosphere have been dominated in the past 8000 years by the geomagnetic effect, while the shorter-term fluctuations have probably been controlled by changes in solar activity. This point seems clear in Fig. 5b, where, near the modern end of the curve, the Maunder Minimum (M) and Spörer Minimum (S) stand out as obvious excursions from the long-term envelope of geomagnetic change. And at about 1200 we find a broad departure in the opposite direction, which might fit the 12th- and 13th-century maximum in sunspot and auroral reports. Whether the sun was indeed responsible is open to question, however, for Bucha (59) has pointed out that this ^{14}C decrease follows a similar short-term increase in the earth's magnetic moment (not shown in Fig 5b), which had its onset at about A.D. 900. Moreover, there is uncertainty in the fit of the smoothed archeomagnetic curve to the radiocarbon data, and a shift to the right or left will change the apparent contrast of these shorter-term excursions.

We should like to know how solar activity in a possible 12th-century Grand Maximum compares with the present epoch, but the present is an era of confusion in ^{14}C. The ^{14}C concentration has been falling steeply since the end of the 19th century, and the deviation ($\Delta ^{14}C$) is now about −25 parts per mil. Were this a solar effect, it would be evidence of anomalously high solar activity. In fact, the sharp drop is an effect of human activity—the result of fossil fuel combustion, which introduces CO_2 with different carbon isotopic abundance ratios—the so-called Suess Effect (47). If fossil fuel combustion is responsible for all of the modern ^{14}C trend, then during the 12th-century Grand Maximum (when industrial pollution was not significant), the natural ^{14}C deviation may have been much greater than at present and the sun may have been more active than we are accustomed to observing in the modern era. There were possibly more spots on more of the sun during the 12th-century Grand Maximum and, if the 11-year cycle operated then, there may have been higher maxima and higher minima than any we see in Fig. 1.

The shallow dip and rise in the 14th and early 15th centuries (Fig. 5a) suggest the presence of a subsidiary solar period of about 170 years, but these features seem for now too slight to warrant speculation; we may expect that additional ^{14}C data will clarify the case. The information available at present allows one to describe the history of the sun in the last millennium as follows: a possible Grand Maximum in the 12th century, a protracted fall to a century-long minimum around 1500, a short rise to "normal," and then the fall to the shorter, deeper Maunder Minimum, after which there has been a steady rise in the envelope of solar activity (25).

This last phase, which includes all detailed records of the sun and the sunspot cycle, does not appear in the ^{14}C history as very typical of the sun's behavior in the past, particularly if the *phase* of the long-term curve is important. During most of the last 1000 years the long-term envelope of solar activity was either higher than at present, or falling, or at grand minima like the Maunder Minimum. As with the present climate, what we think of as normal may be quite unusual. The possibility that solar behavior since 1715 was unlike that in the past has already been proposed to help explain the sudden auroral turn-on. Another piece of evidence comes from records of the sun's appearance at eclipse.

Absence of the Corona at Eclipse

Historical accounts of the solar corona at total eclipse offer another possible check on anomalies in past solar behavior. We know that the shape of the corona seen at eclipse varies with solar activity: when the sun has many spots, the corona is made up of numerous long tapered streamers which extend outward like the petals of a flower. As activity wanes, the corona dims and fewer and fewer streamers are seen. At a normal minimum in the solar cycle the corona seen by the naked eye is highly compressed and blank except for long symmetric extensions along its equator. We now believe that coronal streamers are rooted in concentrated magnetic fields on the surface of the sun, which, in turn, are associated with solar activity and sunspots. As sunspots fade, so do concentrated surface fields and associated coronal structures. Continuous, detailed, observations of the solar corona in x-ray wavelengths from Skylab have confirmed the association of coronal forms with loops and arches in the surface fields and have shown that in areas where there are no concentrated fields, loops, or arches there is no apparent corona (62).

Were there a total absence of solar activity, we would still expect to observe a dim, uniform glow around the moon at eclipse: the zodiacal light, or false corona, would remain, since it is simply sunlight scattered from dust and other matter in the space between the earth and the sun. At times of normal solar activity the corona seen at eclipse is a mixture of the true corona (or K corona) and the weaker glow of the zodiacal light (or F corona). The latter is a roughly symmetric glow around the sun which falls off in brightness from the limb and is distended in the plane of the planets where interplanetary dust is gravitationally concentrated. If the F corona were ever seen alone, we would expect it to appear as a dull, slightly reddish, eerie ring of light of uniform breadth and without discernible structure.

Fig. 6. Early 17th-century observation of a solar eclipse, by projection in a darkened room, as depicted in (80). Hevelius himself is depicted at the left, marking the obscuration of the sun by the lunar disk. [By permission of the Houghton Library, Harvard University]

In fact, firsthand descriptions of total solar eclipses during the Maunder Minimum seem entirely consistent with an absence of the modern structured corona, but proof seems blurred by the customs of observing eclipses in the past and by the fact that scientists seldom describe what is missing or what is not thought to be important. The solar origin of the corona was not established until the late 19th century; before that it seemed equally well explained as sunlight scattered in our own atmosphere, or on the moon. Solar eclipses were regularly and routinely observed throughout the 17th century but not to study the physical sun. They were occasions to test the then popular science of orbit calculation: careful measurement and timing of solar obscuration by the moon offered checks on lunar and terrestrial motions and opportunity to measure the relative sizes of solar and lunar disks. Such details are best obtained not at the eyepiece of a wide-field telescope in the open air but in a darkened room, by projection of the disks of the moon and sun upon a card, as we see in a contemporary drawing from Hevelius (Fig. 6). Under these restrictive conditions a corona, structured or not, could escape detection, particularly since it appeared so briefly and at just the time when undivided attention was demanded to observe the precise minutia of obscuration (63).

Nor was it so important to seek out geographic places on the central path of a total eclipse. The corona—K or F—is so faint that it cannot be seen except in exact totality. But if one's purpose were astronomical mensuration and timing, a partial or near-total eclipse was almost as good as a total eclipse and could be observed more accurately in the familiar conditions of permanent observatories. Since partial solar eclipses can be seen over large areas and thus occur frequently at any location, there was not the impetus of today to travel far and wide to set up camp for one-time tries in distant, hostile lands. Eclipse expeditions are a modern fad that did not take hold until about the 19th century (64).

These fundamental differences severely limit the number of cases we can test. There were 63 opportunities to see the sun eclipsed between 1645 and 1715 (65), but only eight of them passed through those parts of Europe where astronomers did their daily work (Fig. 7). Another case (1698) comes from the New World. Only a few of the European eclipses reached totality near any permanent observatory, and the three best observed occurred at the end of our period of interest—in 1706, 1708, and 1715, when spots had begun their return.

Nevertheless, from this list comes a handful of accounts which bear on the question and answer it consistently. They are descriptions of the corona from the eclipses of 1652, 1698, 1706, and 1708, the only contemporary firsthand descriptions of the sun eclipsed that I

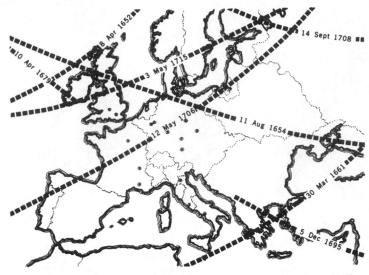

Fig. 7. Paths of totality for solar eclipses in Europe, from 1640 to 1715, from Oppolzer (65). Sites of observatories which reported eclipse observations in the period are shown as double circles.

can find (66). They were written, in general, by amateurs and nonconformists who watched the spectacle with eyes open to all of it. None describes the corona as showing structure. Not one mentions the streamers which at every eclipse in the present time are so easily seen with the naked eye to stretch as much as a degree or more above the solar limb. All describe the corona as very limited in extent: typically only 1 to 3 arc minutes above the solar limb. In each case the corona is described as dull or mournful, and often as reddish. No drawings were made. Every account is consistent with our surmise of what the zodiacal light would look like at eclipse, were the true corona really gone.

By 1715, the annual sunspot number had reached 26 and was climbing. At the eclipse of that year, at the end of the Maunder Minimum, the corona is fairly well described, and for the first time we have drawings of it. For the first time distinct coronal structures are described emanating from the sun. R. Cotes of Cambridge University described the corona (in a letter to Isaac Newton) as a white ring of light around the moon, its densest part extending about 5 arc minutes above the limb; he then added the following (67):

Besides this ring, there appeared also rays of a much fainter light in the form of a rectangular cross. . . . The longer and brighter branch of this cross lay very nearly along the ecliptic, the light of the shorter was so weak that I did not constantly see it.

We may presume that the light of the shorter branch was the polar plumes which we see today at times of sunspot minimum and that the longer, brighter branch was the familiar equatorial extensions seen at times of low sunspot activity. Thus by 1715 we find the corona described in modern terms and fitting a familiar form.

In her paper on the dearth of aurorae Clerke mentioned, without example, that it appeared to her probable that during the "prolonged sunspot minimum" the radiated structure of the solar corona was also "in abeyance" (17). Recently Parker has repeated Clerke's conjecture (68). The case for a disappearance of the structured corona during the Maunder Minimum might seem more solid were it not for the fact that the earliest description yet found for the rayed or structured corona at *any* eclipse is that of Cotes in 1715.

R. R. Newton has expressed the situation very explicitly, on the basis of his own researches for definite accounts of the corona as positive documentation of historical solar eclipses (23, p. 99):

The corona is mentioned in most modern discussions of total solar eclipses, and to most people it is probably the typical and spectacular sight associated with a total eclipse. In view of this, it is surprising to see how little the corona appears in ancient or medieval accounts. . . .

Newton continues (23, p. 601):

. . . there is no clear reference to the corona in any ancient or medieval record that I have found. The most likely reference is perhaps the remark by Plutarch . . . but the meaning of Plutarch's remark is far from certain.

I should add that here Newton is refer-

ring to *any* unambiguous description of the corona, K or F.

A misleading statement common in popular stories of eclipses is that the solar corona was seen in antiquity much as we would describe it today. Usually cited are two early accounts, one by Plutarch (about A.D. 46 to 120) and another by Philostratus (about A.D. 170 to 245). Both reports are ambiguous at best, and neither distinguishes between a structured or an unstructured appearance (69). The situation in all subsequent descriptions before the 18th century seems to be no different. At the eclipse of 9 April 1567 Clavius reported seeing "a narrow ring of light around the Moon" at maximum solar obscuration (although Kepler challenged this as possibly an annular eclipse). Jessenius at a total eclipse in 1598 reported "a bright light shining around the Moon." And Kepler himself reported that at the eclipse of 1604 (70): "The whole body of the Sun was effectually covered for a short time. The surface of the Moon appeared quite black; but around it there shone a brilliant light of a reddish hue, and uniform breadth, which occupied a considerable part of the heavens." None of these or any other descriptions that I can find fit a rayed or structured corona; in many are the words "of uniform breadth," and it seems to me most likely that we are reading descriptions of the zodiacal light, or of a K corona so weak that its radiance is overpowered by the glow of the F corona.

It could be that, until the scientific enlightenment of the 18th century, no one felt moved to describe the impressive structure of the solar corona at eclipse. Indeed, there are other examples from the history of eclipse observation where large and striking features were missed by good observers who were watching other things (71). Perhaps the rays of the corona at eclipse were thought to be so much like the common aureole around the sun that they were not deemed worthy of description. Other excuses could be offered. It will be hard for anyone who has seen the corona with the naked eye to accept these explanations and to believe that, of the thousands of observers at hundreds of total eclipses, not one would have commented on a thing so breathtaking and beautiful. It thus seems to me more probable that, through much of the long period of the Maunder Minimum and the Spörer Minimum, extending between perhaps 1400 and 1700, the sun was at such a minimum of activity that the K corona was severely thinned or absent altogether. The same may have been true for a much longer span before 1400 and for different reasons may apply as well to the Grand Maximum of the 12th and 13th centuries and possibly earlier. But here the records are so dim and scant that conclusions seem unwarranted. In any case the corona as we know it may well be a modern feature of the sun. It is an interesting question, and another important challenge for historians.

Summary and Conclusions

The prolonged absence of sunspots between about 1645 and 1715, which Spörer and Maunder described, is supported by direct accounts in the limited contemporary literature of the day and cited regularly in astronomy works of the ensuing century. We may conclude that the absence was not merely a limitation in observing capability because of the accomplishments in other areas of astronomy in the late 17th and early 18th centuries, and because drawings of the sun made at the time show almost all the sunspot detail that is known today. Major books by Scheiner and Hevelius, published just before the onset of the Maunder Minimum, describe wholly adequate methods for observing the sun and sunspots. We may assume that a fairly steady watch was kept, since the dearth of spots was recognized at the time and since the identification of a new sunspot was cause for the publication of a paper. We can discount the possibility of 70 years of overcast skies, since there is no evidence of such an anomaly in meteorological lore and since nighttime astronomy was vigorous and productive through the same period. Evidence which confirms the Maunder Minimum comes from records of naked-eye sunspot sightings, auroral records, the now-available history of atmospheric ^{14}C, and descriptions of the eclipsed sun at the time.

I can find no facts that contradict the Maunder claim, and much that supports it. In questions of history where only a dim and limited record remains and where we are blocked from making crucial observational tests, the search for possible contradiction seems to me a promising path to truth. I am led to conclude that the "prolonged sunspot minimum" was a real feature of the recent history of the sun and that it happened much as Maunder first described it.

Earlier in this article I reviewed the possible impact of a real Maunder Minimum on theories of the sun and the solar cycle. For some implications the distinction between no sunspots and a few (annual sunspot numbers of one to five) is crucial; it is important to know whether during the great depression of the Maun-

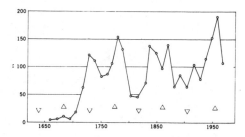

Fig. 8. Annual mean sunspot numbers at maxima in the 11-year cycle, from 1645 to the present, to demonstrate long-term trends in solar activity. Evident is the well-known 80-year cycle (extrema shown as triangles) imposed on a persistent rise since the Maunder Minimum. The 78- or 80-year cycle was first noted by Wolf (*81*) and later studied in detail by Gleissberg (*82*). The solar constant has also been slowly rising through the period during which it has been measured, since about 1908 (*25, 72*).

der Minimum the solar cycle continued to operate at an almost invisible level, with so few spots that they were lost in our fuzzy definition of "zero." Maunder held that there were enough instances of sunspot sightings through the period to make this case likely, and that the isolated times when a few spots appeared enabled one to identify the crests of a sunken spot curve "just as in a deeply inundated country, the loftiest objects will still raise their heads above the flood, and a spire here, a hill, a tower, a tree there, enable one to trace out the configuration of the submerged champaign" (*16*). This explanation seems to me unlikely, since the known, visible crests are not at regular spacings. We can hope that more thorough investigation of contemporary literature will enable us to make this important distinction which for now seems beyond the limit of resolution.

The years of the Maunder Minimum define a time in the ^{14}C record when the departure from normal isotopic abundance exceeded 10 parts per mil. If we take a ^{14}C deviation of this magnitude as a criterion of major change in solar behavior, we may deduce from ^{14}C history the existence of at least two other major changes in solar character in the last millennium: a period of prolonged solar quiet like the Maunder Minimum between about 1460 and 1550 (which I have called the Spörer Minimum) and a "prolonged sunspot maximum" between about 1100 and 1250. If the prolonged maximum of the 12th and 13th centuries and the prolonged minima of the 16th and 17th centuries are extrema of a cycle of solar change, the cycle has a full period of roughly 1000 years. If this change is periodic, we can speculate that the sun may now be progressing toward a grand maximum which might be reached in the 22nd or 23rd centuries. The overall envelope of solar activity has been steadily increasing since the end of the Maunder Minimum (Fig. 8), giving some credence to this view. Moreover, throughout the more limited span during which it has been measured, the solar constant ap-

pears to have shown a continuous rising trend which during the period from 1920 through 1952 was about 0.5 percent per century (*72*).

The coincidence of Maunder's "prolonged solar minimum" with the coldest excursion of the "Little Ice Age" has been noted by many who have looked at the possible relations between the sun and terrestrial climate (*73*). A lasting tree-ring anomaly which spans the same period has been cited as evidence of a concurrent drought in the American Southwest (*68, 74*). There is also a nearly 1 : 1 agreement in sense and time between major excursions in world temperature (as best they are known) and the earlier excursions of the envelope of solar behavior in the record of ^{14}C, particularly when a ^{14}C lag time is allowed for: the Spörer Minimum of the 16th century is coincident with the other severe temperature dip of the Little Ice Age, and the Grand Maximum coincides with the "medieval Climatic Optimum" of the 11th through 13th centuries (*75, 76*). These coincidences suggest a possible relationship between the overall envelope of the curve of solar activity and terrestrial climate in which the 11-year solar cycle may be effectively filtered out or simply unrelated to the problem. The mechanism of this solar effect on climate may be the simple one of ponderous long-term changes of small amount in the total radiative output of the sun, or solar constant. These long-term drifts in solar radiation may modulate the envelope of the solar cycle through the solar dynamo to produce the observed long-term trends in solar activity. The continuity, or phase, of the 11-year cycle would be independent of this slow, radiative change, but the amplitude could be controlled by it. According to this interpretation, the cyclic coming and going of sunspots would have little effect on the output of solar radiation, or presumably on weather, but the long-term envelope of sunspot activity carries the indelible signature of slow changes in solar radiation which surely affect our climate (*77*).

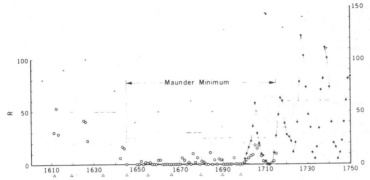

Fig. 9. Estimated annual mean sunspot numbers, from 1610 to 1750: open circles are data from Table 1; connected, closed circles are from Waldmeier (3); dashed lines (decade estimates) and crosses (peak estimates) are from Schove (8–11); triangles are Wolf's estimated dates of maxima for an assumed 11.1-year solar cycle (3, 6).

from the sun, with possible terrestrial effects. Our present understanding of the solar wind is that its flow is regulated by closed or open magnetic field configurations on the sun (78). We can only guess what effect a total absence of activity and of large-scale magnetic structures would have on the behavior of solar wind flow in the ecliptic plane. One possibility is that, were the sun without extensive coronal structure during the Maunder Minimum, the solar wind would have blown steadily and isotropically, and possibly at gale force, since high-speed streams of solar wind are associated with the absence of closed structures in the solar corona. During an intensive maximum, as is suggested for the 12th and 13th centuries, the solar wind was probably consistently weak, steady, and with few recurrent streams.

The existence of the Maunder Minimum and the possibility of earlier fluctuations in solar behavior of similar magnitude imply that the present cycle of solar activity may be unusual if not transitory. For long periods in the historic past the pattern of solar behavior may have been completely different from the solar cycle today. There is good evidence that within the last millennium the sun has been both considerably less active and probably more active than we have seen it in the last 250 years. These upheavals in solar behavior may have been accompanied by significant long-term changes in radiative output. And they were almost certainly accompanied by significant changes in the flow of atomic particles

The reality of the Maunder Minimum and its implications of basic solar change may be but one more defeat in our long and losing battle to keep the sun perfect, or, if not perfect, constant, and if inconstant, regular. Why we think the sun should be any of these when other stars are not is more a question for social than for physical science.

Table 1. Estimated annual mean sunspot numbers, R, from 1610 to 1715; X, sunspots noted but not counted; XX, unusual number of sunspots noted but not counted; (X), unusually small number of sunspots noted but not counted. Schove's values are for the maxima of each supposed cycle.

Year	R	Waldmeier (3)	Schove (9)	Year	R	Waldmeier (3)	Schove (9)	Year	R	Waldmeier (3)	Schove (9)
1610	X			1646				1681	2		
1611	30	Minimum		1647				1682	0		
1612	53			1648				1683	0		
1613	28			1649		Maximum	40	1684	11		
1614				1650	0			1685	0	Maximum	50
1615	X			1651	0			1686	4		
1616	X	Maximum	90	1652	3			1687	0		
1617	X			1653	0			1688	5		
1618	(X)			1654	2			1689	4		
1619		Minimum		1655	1	Minimum		1690	0	Minimum	
1620				1656	2			1691	0		
1621	X			1657	0			1692	0		
1622	X			1658	0			1693	0	Maximum	30
1623	X			1659	0			1694	0		
1624	X			1660	4	Maximum	50	1695	6		
1625	41			1661	4			1696	0		
1626	40	Maximum	100	1662	0			1697	0		
1627	22			1663	0			1698	0	Minimum	
1628				1664	0			1699	0		
1629	(X)			1665	0			1700	2		5
1630				1666	0	Minimum		1701	4		11
1631				1667	0			1702	6		16
1632	(X)			1668	0			1703	8		23
1633				1669	0						
1634	(X)	Minimum		1670	0			1704	9		36
1635	(X)			1671	6			1705	18		58	50
1636				1672	4			1706	15		29
1637				1673	0			1707	18		20
1638	X			1674	2			1708	8		10
1639	XX			1675	0	Maximum	60	1709	3		8
1640		Maximum	70	1676	10			1710	2		3
1641				1677	2			1711	0		0
1642	6			1678	6			1712	0		0
1643	16			1679	0			1713	2		2
1644	15			1680	4	Minimum		1714	3		11
1645	0	Minimum						1715	10		27

Appendix: Sunspot Numbers

I have used contemporary accounts of telescopic observation of the sun to reconstruct estimated annual mean sunspot numbers for the period from 1610 to 1715 (Table 1 and Fig. 9). Principal sources were Wolf's compilations (6) and (13–16, 19, 20, 28–32). The journal sources are, for the most part, the same as those that were used by LaLande, Spörer, and Maunder; thus, except for the direct numerical data from Wolf, Scheiner, and Hevelius, sunspot numbers given here are simply a literal quantification of Maunder's descriptive account. Full reliance has been placed on unchallenged statements in contemporary literature which specify periods in which no sunspots were seen, as, for example, between 1656 and 1660, 1661 and 1671, 1689 to 1695, 1695 to 1700, and 1710 to 1713.

Earlier I classified Wolf's historical sunspot data; by the same criteria the data in Table 1 should be given a reliability grade of "poor," since they come from largely discontinuous sets and since allowance for observer and site can only be guessed. The estimated annual sunspot numbers are uncertain to at least a factor of 2, and zero as an annual average means 0 to perhaps 5. The fact that the telescopes of Flamsteed and Cassini were in less than perfect observing sites could have caused these observers to miss a class of tiny, isolated spots which might be detected and counted by keen observers today. The more important point is that their sites and instruments were certainly adequate to detect any level of activity higher than that at the minima of the present solar cycle; they might have missed a few spots but they could not have missed a large number.

My sunspot numbers for the period 1700 to 1715 are somewhat lower than those given for the same period by Waldmeier (3), who took them from Wolf. Both values are shown in Table 1 and Fig. 9. The general agreement seems heartening, but the difference may be important since it is in the only span of overlap with other direct numerical compilations. It is also in the least reliable part of Wolf's data and the period of recovery from the Maunder Minimum, for which a more gradual rise seems reasonable. Auroral data and eclipse observations from the period of overlap seem to me to support the more suppressed sunspot curve (Fig. 9). I find it hard to justify Wolf's numbers for his first and possibly second cycles and suspect that his unusual-shaped maximum for 1705 was an artificiality of unrealistic correction factors. Wolf did not have confidence in most of

the data for 1700 to 1749 (6), and his numbers toward the beginning of that period may represent, more than anything else, a wishful extrapolation of normalcy. I also show in Fig. 9 and Table 1 Schove's estimates of decade-averaged and peak sunspot numbers from the STP (8–11), which we can also expect to be systematically high (79).

Numbers given for 1625 to 1627 and 1642 to 1644 (from Scheiner and Hevelius) are probably more reliable than any subsequent data in Table 1, since they are based on more nearly continuous daily drawings. Data for 1611 through 1613 come from the observations of Galileo. Waldmeier (3) and Schove (8–11) have apparently followed Wolf in assuming that these three islands of data before 1650 sample extrema of the sunspot cycle: Galileo and Scheiner at maxima, Hevelius at minimum. If these periods are all nearer maxima, as I suspect, they give some hint of the fall to the long minimum that followed. The nature of the fall suggests that the telescope was invented barely in time to "discover" sunspots before their numbers shrank to nearly zero. Had the invention of the telescope been delayed by as little as 35 years, the telescopic discovery and more thorough counting of sunspots could have been postponed a full century, burying forever the principal evidence for the Maunder Minimum.

References and Notes

1. The Wolf sunspot number (or sunspot relative number) is defined as $R = k(10g + f)$, where f is the total number of spots (irrespective of size), g is the number of spot groups, and k is a normalizing factor to bring the counts of different observers, telescopes, and sites into agreement.
2. R. B. Leighton, *Astrophys. J.* **156**, 1 (1969).
3. M. Waldmeier, *The Sunspot-Activity in the Years 1610–1960* (Schulthess, Zurich, 1961).
4. R. J. Bray and R. E. Loughhead, *Sunspots* (Wiley, New York, 1965), p. 1.
5. H. Schwabe, *Astron. Nachr.* **20** (No. 495) (1843). For an interesting discussion of Schwabe, his lonely work, and the prejudice against the idea of cyclic solar behavior before that time, see M. J. Johnson, *Mem. R. Astron. Soc.* **26**, 196 (1858).
6. R. Wolfe, *Sunspot Observations, 1610–1715*, facsimile of a typescript from Eidgen Sternwarte in Zurich (in G. E. Hale Collection, Hale Observatory Library, Pasadena, Calif.). The 11-page manuscript lists the days of each year on which spots were or were not seen, the numbers of spots (where known), and notes and references. Other more condensed accounts of the period by Wolf include: *Astron. Mitt. Zürich* **1**, viii (1856); *ibid.* **24**, 111 (1868).
7. In (3, p. 8), Waldmeier states that "Wolf intended to prove for a longer interval the sunspot-periodicity discovered shortly before by . . . Schwabe." In one of his papers [*Astron. Mitt. Zürich* **1**, viii (1856)] Wolf explained that, in periods where data were sparse, he assumed the continued operation of the 11.11-year cycle.
8. D. J. Schove, *Terr. Magn. Atmos. Electr.* **52**, 233 (1947); *J. Br. Astron. Assoc.* **71**, 320 (1961).
9. ———, *J. Geophys. Res.* **60**, 127 (1955).
10. ———, *Ann. N.Y. Acad. Sci.* **95**, 107 (1961).
11. ———, *J. Br. Astron. Assoc.* **72**, 30 (1962).
12. J. A. Eddy *et al.*, *Sol. Phys.*, in press.
13. F. W. G. Spörer, *Vierteljahrsschr. Astron. Ges. (Leipzig)* **22**, 323 (1887); *Bull. Astron.* **6**, 60 (1889).
14. E. W. Maunder, *Mon. Not. R. Astron. Soc.* **50**, 251 (1890).
15. ———, *Knowledge* **17**, 173 (1894).
16. ———, *J. Br. Astron. Assoc.* **32**, 140 (1922).
17. A. M. Clerke, *Knowledge* **17**, 206 (1894).
18. Late examples include: E. H. Burritt, *The Geography of the Heavens* (Huntington & Savage, New York, 1845), p. 180; R. A. Proctor, *The Sun* (Longmans, Green, London, 1871), p. 164.
19. W. Herschel, *Philos. Trans. R. Soc. London* **265** (1801). In this wide-ranging and oft-cited paper Herschel reveals his belief in the influence of solar fluctuations on weather, based on his own observation of a correlation between the price of wheat in London and the number of visible sunspots. In making his point, he uses the extreme periods of spot absence of the Maunder Minimum, during which time the price of wheat rose. Herschel attributes the connection to reduced rainfall when the sun was less spotted, and to the inexorable workings of the law of supply and demand. This paper reveals, among other things, that the quest for a solar-weather connection predated the discovery of the solar cycle. It was not Herschel's worst mistake: in the same paper he tells of his belief in a habitable and possibly inhabited sun.
20. J. LaLande, *Astronomie* (Desaint, Paris, 1792; and Johnson Reprint Corporation, New York, 1966), vol. 3, pp. 286–287. This encyclopedic work was probably the unacknowledged source of most of the 19th-century descriptions of past periods of prolonged sunspot absence. La-Lande's references included original journal reports and Jacques Cassini's *Eléments d'Astronomie* (Imprimerie Royale, Paris, 1740), pp. 81–82, 182. Jacques Cassini was the son of G. D. (Jean) Cassini, who discovered the sunspot of 1671 and the moons of Saturn.
21. G. E. Hale, "Photography of the solar prominences" [thesis, Massachusetts Institute of Technology (1890); reprinted in *The Legacy of George Ellery Hale*, H. Wright, J. Warnow, C. Weiner, Eds. (MIT Press, Cambridge, Mass., 1972), p. 117]; C. A. Young, *The Sun* (Appleton, New York, 1896), p. 193.
22. R. R. Newton, *Ancient Astronomical Observations and the Acceleration of the Earth and Moon* (Johns Hopkins Press, Baltimore, 1970), p. 39.
23. ———, *Medieval Chronicles and the Rotation of the Earth* (Johns Hopkins Press, Baltimore, 1972), pp. 99, 600–601.
24. C. M. Smythe and J. A. Eddy, *Bull. Am. Astron. Soc.*, in press.
25. J. A. Eddy, *Bull. Am. Astron. Soc.* **7**, 365 (1975); *ibid.*, p. 410.
26. Attributed to M. J. Rees, in *Project Cyclops*, J. Billingham, Ed. (NASA publication CR 114445, Stanford/NASA Ames Research Center, Moffett Field, Calif., 1973), p. 3.
27. H. C. King, *The History of the Telescope* (Sky Publishing, Cambridge, Mass., 1955), pp. 50–59.
28. A. H. Pingré (and M. G. Bigourdan), *Annales Célestes du Dix-Septième Siècle* (Gauthier-Villars, Paris, 1901).
29. This invaluable year-by-year diary (28) of astronomical advance in the 17th century was begun by Pingré in 1756 and completed by Bigourdan in 1901. It illuminates a most interesting century in astronomy and by length alone (639 pages) attests to the vigor of observational work at the time.
30. See, for example, G. D. Cassini, *Anc. Mem.* **10**, 727 (1688); J. Cassini, *Hist. Acad. R. Sci.* (*Amsterdam*) (1701), pp. 132, 356; *ibid.* (1702), pp. 185, 194; *ibid.* (1703), pp. 18, 141, 148, 151.
31. C. Scheiner, *Rosa Ursina sive Sol ex Admirando Facularum* (Apud Andream Phaeum Typographum Ducalem, 1630).
32. J. Hevelius, *Selenographia sive Lunae Descripto* (Gedani, Danzig, 1647).
33. A. Wilson, *Philos. Trans.* **64**, 6 (1774).
34. The *Rosa Ursina* (31), although large (25 by 36 by 8 cm) and beautifully set, has not enjoyed kind reviews; comments on the book range from "voluminous," "enormous," and "ovrage considérable renfermant plus de 2000 observations" to the less couched words of astronomer Jean Delambre: "There are few books so diffuse and so void of facts. It contains 784 pages; there is not matter in it for 50 pages" [*Histoire de l'Astronomie Moderne* (Imprimerie de Huzard-Courcier, Paris, 1821), vol. 1, p. 690; cited in (35)].
35. R. Grant, *History of Physical Astronomy* (H. and G. Bohn, London, 1852), p. 216.
36. The Maunder Minimum coincided with a prolonged period of distinct climatic anomaly—years of severe winters and abnormal cold, but there is no evidence of unbroken overcast. Astronomers are neither so mute nor so long-suffering that they would have kept quiet through year after year of continuous, frustrating cloud cover. The time was one of vigorous growth and

discovery in observational astronomy, as, for example, in the important revelations of Saturn already cited. Throughout the 70 years of the Maunder Minimum comets were regularly discovered and observed. We may conclude that during these years skies were at least tolerably clear, and certainly adequate to allow at least sporadic if not normal sampling of aurorae and sunspots, had they been there to see.

37. For this study I have used H. Fritz, *Verzeichniss Beobachter Polarlichter* (C. Gerold's Sohn, Vienna, 1873), which is still probably the most thorough published compilation of ancient aurorae. If criticized, it is more generally for sins of commission than omission; some of the ancient aurorae listed may not have been aurorae at all but meteors or comets [C. Stormer, *The Polar Aurora* (Oxford Univ. Press, New York, 1955), p. 14; (*38*, p. 20)].
38. S. Chapman, in *Aurora and Airglow*, B. M. McCormac, Ed. (Reinhold, New York, 1967).
39. E. H. Vestine, *Terr. Magn. Atmos. Electr.* **49**, 77 (1944).
40. E. Halley, *Philos. Trans. R. Soc. London* **29**, 406 (1716). Halley mentions that the aurora borealis had rarely been seen since the early 17th century.
41. Schove (*8–11*) has noted a tendency for auroral counts to alternate by century, with more in even centuries (such as the 16th and 18th) and fewer in odd, in which most of the Maunder Minimum took place.
42. Galileo and the other "discoverers" of sunspots were well aware of the existence of sunspots and naked-eye reports of them before they looked at the sun with telescopes [G. Abetti, in *IV Centenario della Nascita di Galileo Galilei* (Barbèra, Florence, 1966), p. 16].
43. For example, see S. Kanda, *Proc. Imp. Acad. (Tokyo)* **9**, 293 (1933). Kanda's compilation is more valuable in its own right than as a clue to past epochs of maxima, since large spots have been known to occur during years of minimum activity.
44. More recent studies of specific ancient oriental sunspot reports have been carried out by D. J. Schove and P. Y. Ho [*J. Br. Astron. Assoc.* **69**, 295 (1958); *J. Am. Orient. Soc.* **87**, 105 (1967)].
45. S. Nakayama [in *A History of Japanese Astronomy* (Harvard Univ. Press, Cambridge, Mass., 1969), pp. 12–23] has discussed the limitations of the "Institutional Framework of Astronomical Learning" in early Japan and the resultant repression of ideas and research. I have found no evidence that the Maunder Minimum was a unique period in this regard, however, and the almost precise coincidence with other evidences from Europe make the Far East sunspot gap seem real to me.
46. S. Matsushita, *J. Geophys. Res.* **61**, 297 (1956). I have taken from Matsushita's list only those auroral reports that he deemed "certain" or "very probable."
47. H. E. Suess, *J. Geophys. Res.* **70**, 5937 (1965).
48. P. E. Damon, A. Long, D. C. Grey, *ibid.* **71**, 1055 (1966).
49. I. U. Olson, Ed., *Radiocarbon Variations and Absolute Chronology* (Almqvist & Wiksell, Stockholm, 1970).
50. P. E. Damon (personal communication) has compiled radiocarbon data from five laboratories (University of Arizona; State University of Gronigen, Netherlands; University of California, San Diego; University of Pennsylvania; and Yale University).
51. M. Stuiver, *J. Geophys. Res.* **66**, 273 (1961); *Science* **149**, 533 (1965); J. R. Bray, *ibid.* **156**, 640 (1967); P. E. Damon, *Meteorol. Monogr.* **8**, 151 (1968); J. A. Simpson and J. R. Wang, *Astrophys. J.* **161**, 265 (1970).
52. H. E. Suess, *Meteorol. Monogr.* **8**, 146 (1968).
53. H. DeVries, *Proc. K. Ned. Akad. Wet. B* **61** (No. 2), 94 (1958).
54. J. C. Lerman, W. G. Nook, J. C. Vogel, in (*49*, p. 275). There are several available compilations of relative ^{14}C concentration; the most commonly cited is probably that of Suess (*47*) for Northern Hemisphere trees. P. E. Damon has kindly provided a compilation of ^{14}C data from five world radiocarbon laboratories (*50*), which has been very helpful in establishing real features. I have used the recent Groningen data cited here since they include a large sampling from trees of the Southern Hemisphere, where the larger ocean surface might be expected to bring about, in effect, faster tree response to real changes in atmospheric concentration. Fluctuations in ^{14}C atmospheric concentration are severely damped out in tree-ring concentrations because of the finite time of exchange between the atmosphere and the trees; the time constant is on the order of 10 to 50 years. The presence of absorbing oceans in the equilibrium process acts as an added sink, or leak, and, since the problem is analogous to that of determining changes in the rate of water flow into a bucket by noting its level, a leaky bucket makes a slightly more responsive system. In fact, there are only minor differences between the historical curve of Lerman *et al.* and that given by Suess and others; they show the same extrema at about the same times.
55. The use of ^{14}C data to deduce solar changes in the past and the possible relation of these changes to the history of the terrestrial climate have been the subject of numerous papers; for example, see (*51, 52, 56*); J. R. Bray, *Nature (London)* **220**, 672 (1968); P. E. Damon, A. Long, E. J. Wallick, *Earth Planet. Sci. Lett.* **20**, 300 (1973).
56. J. R. Bray, *Science* **171**, 1242 (1971).
57. P. E. Damon, in (*49*, p. 571).
58. The earlier compilations by Suess (*47*) and by Damon (*48, 50*) show that the deviation at 1500 is approximately equal to that of the Maunder Minimum period.
59. V. Bucha, *Nature (London)* **224**, 681 (1969); in (*49*, p. 501).
60. R. E. Lingenfelter and R. Ramaty, in (*49*), p. 513.
61. Y. C. Lin, C. Y. Fan, P. E. Damon, E. J. Wallick, *14th Int. Cosmic Ray Conf.* **3**, 995 (1975).
62. G. S. Vaiana, J. M. Davis, R. Giacconi, A. S. Krieger, J. K. Silk, A. F. Timothy, M. Zombeck, *Astrophys. J.* **185**, L47 (1973).
63. The 17th-century style of observing solar eclipses is well described throughout Pingré's compendium (*28*). A principal result from each eclipse was a table giving times of obscuration and the amount of the disk covered in "digits"—12 digits corresponding to the solar diameter and total obscuration.
64. A. J. Meadows, *Early Solar Physics* (Pergamon, London, 1970), p. 9.
65. T. R. von Oppolzer, *Canon of Eclipses* (reprinted by Dover Publications, New York, 1962).
66. V. Wing, *Astronomia Instaurata* (R. and W. (Leybourn, London, 1656), pp. 98–102; (*35*, pp. 364, 376–391; *28*, p. 570); J. Cassini, *Mem. Acad. Sci. (Amsterdam)* (1706), p. 322.
67. Cited in A. C. Ranyard [*Mem. R. Astron. Soc.* **41**, 503 (1879)]. Cotes might have given a more thorough account had he been free of a perennial eclipse nuisance, for, according to Halley, Cotes "had the misfortune to be oppressed with too much company" (*35*, p. 379). Halley's own description of the 1715 corona, from the same reference, follows: "a few seconds before the sun was all hid, there discovered itself round the moon a luminous ring . . . perhaps a tenth part of the moon's diameter in breadth. It was of pale whiteness . . . [and] concentric with the moon."
68. E. N. Parker, in *Solar Terrestrial Relations*, D. Venkatesan, Ed. (University of Calgary, Calgary, 1973), p. 6; *Sci. Am.* **233**, 42 (September 1975).
69. The Plutarch reference is to his account of the solar eclipse of 27 December A.D. 83; his description follows, as given by R. R. Newton (*22*, p. 114; *23*, pp. 99–100): ". . . [during a solar eclipse] a kind of light is visible about the rim which keeps the shadow from being profound and absolute." Newton feels that Plutarch's "kind of light" could be the rim of light visible during an annular eclipse or light from solar prominences, but that, if it is the corona, this is the earliest extant account. In any case it does not help us in answering whether the K corona was seen, since Plutarch's description could as well or better be the zodiacal light. The reference to Flavius Philostratus is from a passage in his fictional and controversial *Life of Apollonius of Tyana*, written about A.D. 210. Newton avoids it completely, but we should probably expose it to light: "About the time that [Apollonius] was busy in Greece a remarkable phenomenon was seen in the sky. A crown like a rainbow formed around the sun's disk and partly obscured its light. It was plain to see that the phenomenon portended revolution and the Governor of Greece [the tyrant Domitian] summoned Apollonius . . . to expound it. 'I hear, Apollonius, that you have Science in the supernatural' " [translation of J. S. Phillimore (Clarendon, Oxford, 1912) of book VIII, chap. 23]. In Philostratus's story the "crown" (Στεφανος) portends the name of Stephanus who later murdered Domitian. The use of the word is thus couched in symbolism and gives no evidence that Philostratus had ever seen either a total solar eclipse or the structured corona.
70. R. Grant (*35*, p. 377–378) gives the Clavius, Jessenius, and Kepler accounts.
71. J. A. Eddy, *Astron. Astrophys.* **34**, 235 (1974).
72. E. Öpik, *Irish Astron. J.* **8**, 153 (1968); see also (*25*). A change in solar luminosity of 0.5 percent per century corresponds to 0.005 stellar magnitude per century and is thus outside the limits of practical detection in other G stars.
73. For example, see G. Manley, *Ann. N.Y. Acad. Sci.* **95**, 162 (1961); Suess (*52*); Bray (*56*); *Adv. Ecol. Res.* **7**, 177 (1971); S. H. Schneider and C. Maas, *Science* **190**, 741 (1975).
74. A. E. Douglass, *Climatic Cycles and Tree Growth* (Publication 289, Carnegie Institution of Washington, Washington, D. C.), vol. 1, p. 102 (1919); vol. 2, pp. 125–126 (1928). Douglass found that from 1660 to 1720 the curve of Southwest tree growth "flattens out in a striking manner," and, before knowing of Maunder's work, he described the end of the 17th century as a time of unusually retarded growth in Arizona pines and California sequoias.
75. A good review of past climate history is given in (*76*), from which the climate incidents cited here were derived. The Little Ice Age lasted roughly from 1430 to 1850; it was marked by two severe extremes of cold, roughly 1450 to 1500 and 1600 to 1700, if we take H. H. Lamb's index of Paris-London Winter Severity as a global indicator.
76. W. L. Gates and Y. Mintz, Eds., *Understanding Climate Change* (National Academy of Sciences, Washington, D.C., 1975), appendix A.
77. If changes in the solar constant are reflected in the envelope of solar activity, and if the rate of change has held to the 0.5 percent per century rate cited earlier (*72*), then we can estimate that during the Maunder Minimum the solar flux was about 1.4 percent lower than at present—a number not inconsistent with temperature estimates during that coldest period of the Little Ice Age (*76*).
78. A. Hundhausen, *Coronal Expansion and the Solar Wind* (Springer-Verlag, Berlin, 1972); A. S. Krieger, A. F. Timothy, E. C. Roelof, *Sol. Phys.* **29**, 505 (1973).
79. The solar emphasis of the Spectrum of Time Project (STP) was first directed at fixing the epochs of presumed 11-year maxima of the past solar cycle (*10*). Amplitudes of past cycles (10-year averages) and of past maxima of the cycle were estimated on the basis of the best information available: auroral counts and other unspecified data, with an arbitrary correction for what fraction of aurorae was recorded in a given century (*11*). Moreover, in the STP there was a built-in constraint to generate nine solar cycles in each 100 years, regardless of whether there was evidence for them or not (*9*). These and other assumptions tend to dilute possible drastic changes in the past (like the Maunder Minimum) and to nullify possible long-term drifts in the amplitude of solar activity. The Maunder Minimum shows up as a significant drop in the number of sunspots in the STP, but with $R_{max} = 30$ at its weakest "maximum," in 1693, which falls in the middle of a 5-year period for which direct accounts from the contemporary literature report that no spots were seen. It is unfair to press the comparison since the STP covers a much longer span than the Maunder Minimum and, more to the point, it should be noted that the STP shows Maunder's "prolonged sunspot minimum" in figure 1 and table 2 of (*9*).
80. J. Hevelius, *Machina Coelestis* (Simon Reiniger, Danzig, 1673).
81. R. Wolf, *Astron. Mitt. Zürich* **14** (1862).
82. W. Gleissberg, *Publ. Istanbul Univ. Obs. No. 27* (1944).
83. I am indebted to the libraries of Harvard College, the U.S. Naval Observatory, and the Hale Observatories for the privilege of access. I thank O. Gingerich, H. Zirin, T. Bell, J. Ashbrook, D. MacNamara, G. Newkirk, M. Stix, M. Altschuler, L. E. Schmitt, and P. E. Damon for help and suggestions. I am most indebted to E. N. Parker for calling my attention to Maunder's papers, and for personal encouragement in all the work reported here. This research was funded entirely by NASA contract NAS5-3950. The National Center for Atmospheric Research is sponsored by the National Science Foundation.

23

JUNK IN SPACE: WHAT GOES UP

Nine Saturdays from today, on June 30th, the largest man-made object in space, America's 75-tonne Skylab, will tumble ignominiously back to earth. Or so Britain's Royal Aircraft Establishment (RAE) at Farnborough now predicts. Give or take roughly a week either way.

America's National Aeronautics and Space Administration (Nasa) thinks it could come down even earlier, by June 19th. But the RAE was predicting 1979 as the time of Skylab's decay five years ago, when Nasa was confident it would last out till 1983.

Skylab's demise will be the second dramatic reminder in 18 months of the apparent dangers posed by all the hardware man has put into space since the first Sputnik in 1957. Unlike the Russian spy satellite that came down over Canada in January, 1978, Skylab carries no nuclear material. However, it is expected to spew 400-500 bits and pieces, weighing anything from under one to several hundred pounds, over its long, dying trail.

An embarrassed Nasa is saying:

● The public should not be alarmed. With any luck, Skylab's debris will do no harm—coming down in the 70% of the earth's surface covered by water or in sparsely populated areas.

● On the other hand, Nasa cannot say for certain exctly where the debris will fall. Nor will it be able to do so even during the last days and hours of Skylab's decay. It can only cross its fingers—and pay up if, unluckily, there is damage or injury.

[*Editor's Note:* Material has been omitted at this point. Skylab fell in a deserted spot in Australia on 11 July 1979.]

The trick is to know the air density at the satellite's closest approach to earth, the point of maximum drag. Scientists can say how density varies with height on average. The snag is that there are enormous variations around the average. And these are devilishly difficult to predict.

The first pitfall is caused by variations in the sun's 11-year sunspot cycle: this was the source of Nasa's boob. These variations can be huge. At a height of 300 miles, for example, air density at the maximum of the cycle can be 10 times greater than at the minimum. That means a satellite at such a height could have a life of five years if launched two years before a solar minimum, but one of only six months if launched at the maximum. Moreover, no two sunspot cycles are exactly the same.

Part V

LONGER CYCLES

Editor's Comments
on Papers 24 Through 28

24   **HALE**
     The Law of Sun-Spot Polarity

25   **GLEISSBERG**
     The Eighty-Year Solar Cycle in Auroral Frequency Numbers

26   **KIRAL**
     Excerpts from *Autocorrelation and Solar Cycles*

27   **COLE**
     Excerpts from *Periodicities in Solar Activity*

28   **DAMON**
     Excerpt from *Solar Induced Variations of Energetic Particles at One AU*

The 22-Year Cycle

The *sunspot cycle,* as we now know, is magnetically not an 11-year period; it is a 22-year period. The first hint that there were systematic differences between odd and even 11-year cycles came from Wolf (1883), who was followed by Turner (1913). Meanwhile, Hale (1868–1938), born in Chicago and an engineering graduate of Massachusetts Institute of Technology, investigated the magnetism of the sun and discovered (Hale, 1908) the strong magnetic field centered on a sunspot. The *leader spot* of a sunspot pair had a characteristic polarity and its partner had an opposite polarity. After the sunspot minimum of 1912, in addition to Carrington's *zone leap,* Hale observed a *sign leap* (1913) as the polarity patterns reversed. Near the next minimum, in 1922, a further sign leap led Hale to announce his law (Paper 24). Americans at this time entered a field that previously had been a special province of the Swiss, the Germans, and the British. The structure of the 22-year cycle has, however, been studied since then especially in the USSR (see Rubashev, 1964, p. 19; Vitinsky, 1965, pp. 12, 24).

The odd cycles since 1850 have been stronger than the even cycles and Gnevyshev and Ol' (1948) claimed that, unlike even cycles, their strength could be predicted from the preceding cycle. They used the following:

$$R_{Odd} = 0.94 \, R_{Even} + 32.4$$

Cycles 19 and 21 roughly confirmed this formula and Vitinsky (1965, p. 25) claimed that the rule was likewise confirmed in sixty-five percent of the pre-Zurich pairs of Paper 20. Nevertheless, the *Hale minima* as defined in the U.S. (see Fig. 16 in our Introduction), follow the even cycles, for example, 20 to 21 in 1976. Before 1850 the 22-year cycle does not show up in spectral analyses. The 22-year cycle is then nearly a heliomagnetic cycle, the 80-year cycle affects sunspot phase, and the 200-year cycle affects both phase and magnitude.

The 80-Year Cycle

The term *sunspot cycle* normally denotes the familiar cycle of 11 years but in 1862 Wolf (1862 and cf. Paper 24) was already aware that there were medium and long secular cycles, and in 1872 he tentatively suggested (cf. Stewart, 1864; Wolfer, 1891, p. 553) 55 and 160 years respectively. Meanwhile Fritz (1881; see Siscoe, 1980, for earlier ideas) suspected that there were cycles of 56 years in the aurora borealis, somewhat similar to the 50-year period considered by Pilgram in 1788 (cf. our Appendix D). Zurich data since 1700 show a vague 80-year cycle with two medium peaks at 55 to 59 and especially 88 to 94 (Paper 27 and cf. Fig. 2 in Wittmann, 1978). An independent set of sunspots from medieval China showed a main cycle of 62 years with lesser peaks at 81 and 91 years (Table 5 in Yunnan Observatory, 1977). A large set of mainly auroral dates from A.D. 299 to 1604 analyzed by Bain (unpublished) recorded a significant period of 92 years, with some cyclicity at periods of 53, 62, and 72 years (and also at 116 and 133 years).

The spectral analysis of sunspot magnitudes (Paper 20) given in Fig. 9 of our Introduction shows periodicities at 42 and 58 years with diffuse power in the 80 to 100 range, whereas a Maximum Entropy Spectral Analysis (MESA) chart of auroral numbers (Paper 19) also analyzed by Bain (personal communication) shows similar peaks at 51, 58, and 65 years and a broad peak in the 80 to 130-year range. Auroral maxima and minima (Paper 39) suggest that a 95-year wave train persisted from A.D. 950 to 1980 but that it was not present in the first millennium; likewise they suggest that the 60-year or 59-year wave train commenced only about A.D. 1200. This double structure explains why the 80-year cycle does not appear in spectral analyses (Fig. 9 in our Introduction) and why it does appear when Gleissberg uses secular smoothing (cf. Gleissberg, 1965, part of which appears as Paper 25; cf. Gleissberg, 1952, 1966, 1972, 1973).

In *sunspot phase* (or length) the 80-year cycle is clearer—it has been assumed to be a multiple of the 11.1 cycle—but to call it a 78-year cycle (as done in Papers 20 and 38) is unjustified and reminiscent of Biblical numerology. In Papers 26 (cf. Paper 21) and 27 it appears—again with a double structure—with cycles at both 78½ and 94½ years. Personal bias in paper 20 must explain some of the power at 7 cycles in Papers 26 and 27, and Appendix A (which has no built-in 80-year cycle) should be used in the future.

Many solar parameters are supposed to reflect the 80-year cycle. Thus the geomagnetic aa-index (see Part II; Dickc, 1978, 1979*a,* 1979*b;* Feynman and Crooker, 1978) has shown a progressive increase from 1900 to 1960. Some of the changes noted may relate to the longer 200-year cycle. In Radiocarbon ΔQ (see Appendix C) Stuiver and Quay find only a cycle of 135 years. The inconsistency of all the so-called cycles in this chapter is partly explained by their reinterpretation as temporary wave-trains, and apparent durations are noted in Appendix D.

The 200-Year Cycle

In a survey (Schove, 1947, 1948) of the early attempts of Fritz (1881) and Kanda (1933) to determine the dates of sunspot and auroral maxima I was struck by the much greater success rate in even centuries. The sunspot-auroral minima of odd centuries since A.D. 200–all odd cycles except the ninth–were equally clear, and in Paper 20 a solar cycle of 200 to 205 years was taken for granted. This cycle, like the ordinary 11-year cycle, seems to have a *double maximum* (see p. 143 in Paper 20) a weaker *mid-century maximum* thus came between 1950 and 1970 as had been anticipated in 1955. The structure was investigated by Cole (Fig. 3*a* in Paper 27), who believed that the short solar cycles of 10.45 (which characterized even active centuries such as our own), came into phase again with the average cycle length at the spectral peak of about 200 years. In Paper 26 (Fig. 2) the power spectrum shows the same peak in cycle *lengths* between 16 and 20 cycles (177–222 years). In cycle *strengths* the spectral density has a peak (Fig. 9 in our Introduction at 201 years; some power also at 143 to 149 years is confirmed by the radiocarbon cycles (Suess, 1980*a,* 1980*b*) of 200 and 150 years.

Some of the apparent cycle strengths at 200 years derived from my own tabulations (Papers 19 and 20) could have originated in my personal bias, but a 202-year cycle (200 years in Houtermans et al.,

1967; 182 years in Paper 28; and 180 years in de Jong et al., 1979, and Fig. 2 in Schove, 1980) has also been found in radiocarbon dates (see Appendix D) through most of the last 8000 years. However, the 202-year cycle does not show regularity in Chinese sunspots, and the Yunnan group (1977, Fig. 3) suggested double peaks, the stronger at 240 to 270 and the weaker at 165 to 210. From 4000 B.C. to 1 B.C. a radiocarbon cycle of 286 years is found by Damon (Paper 28); this roughly fits Aaby's cycle of 270 years (cf. p. 255 in Paper 39; Svenonius and Olausson, 1979). Cole (Paper 27) notes a possible 280-year cycle in the sunspot-cycle phase since A.D. 300, but throughout the eight millennia of the Bristlecone period Suess (1980a, 1980b) finds only the 203-year cycle significant. In the period since A.D. 700 Stuiver (Pepin et al., eds., 1980), using decadal records of radiocarbon, reports cycles of 46, 133, and 202 years. We conclude that the *long cycle* of about 200 years (170 to 270 years) is more stable and less complex than the *medium cycle*. The multiple nature of these cycles probably reflects an origin in currents beneath the surface of the sun.

Radiocarbon Irregularities

The first radiocarbon dates for the end of the last glacial stage came out at 10,000 years ago or 8000 years B.C. exactly as the Swedish varve specialists had expected. However, when it became clear that radiocarbon dates differed irregularly from tree-ring dates, the conventions b.p., b.c., and a.d. were suggested (Schove, 1967) to distinguish the ^{14}C dates, and we now know that 10,000 b.p., that is 8050 b.c., may be nearer 9000 B.C. (Paper 39) than 8000 B.C. The apparent (a.d.) dates since about 1950 were affected by the atomic bomb, and those since about 1850 by the *Suess* effect, caused by the burning of fossil-fuels such as coal that had long since lost all their active radiocarbon. Before 1850, however, there were other short-term effects noted by De Vries (1958): these were thought at first to be due to climatic oscillations but are now known (see Appendix D; Stuiver and Suess, 1966; Suess, 1980b; Stuiver, 1980) to be due to solar oscillations. Back to about 1690 in the Maunder Minimum this was demonstrated by Stuiver (1961, 1965, 1978), and Houtermans and coworkers (1967; see Fig. 10 in our Introduction) showed that coherence existed with the sunspot curve of Paper 20 back to at least A.D. 900; they found also that the 200-year cycle was present in both data sets. Gray and Damon (1970) confirmed that a curve based on lagged sunspot numbers after Lingenfelter's model (see Lingenfelter and Ramarty, 1970) fitted radiocarbon changes well since 1200 (see Fig. 3 in Paper 29).

The latest analysis by Neftel (Fig. 2 in Suess, 1979; cf. Neftel et al., 1981) of the 8000-year radiocarbon record indicated a mean value of 204 years. Cyclicity was also apparent (Suess, personal communication) at about 89, 104, 143, 154, 305, 499, about 955, and perhaps about 2289 years. Like the multiples of 155 (cf. p. 140 in Paper 20 for phase irregularities and cf. Fig. 3 in Anderson, 1954) and of 450 to 500 (claimed as a sunspot periodicity in Fig. 2 in Yunnan Observatory, 1977) all these periods can be checked for significance when sufficient annual and decadal radiocarbon evidence becomes available; meanwhile, Appendix D includes those cycles significant in the past millennium.

Radiocarbon minima and maxima revealed by 30-year running means (see Stuiver, 1982; cf. Fig. 3 in Paper 28) often follow corresponding auroral *maxima* and *minima* (as given in Papers 25 and 39) and are dated approximately as follows. C14 Minima: *A.D. 210, 310, 385, 525, 625, 760, 855, 965, 1145, 1380, 1475,* and *1515* (both relative), *1616, 1787,* and *1877.* C14 Maxima: *A.D. 75, 265, 355* (relative), *445, 610* (relative), *685* and *725, 795, 905, 1055, 1185* and *1235* (both relative), *1335, 1460, 1535, 1705* and *1815.* The typical delay is 15 or 20 years (due to atmospheric and biogenic mixing). Using this relationship in reverse, Eddy (1978) has calculated from the ^{14}C irregularities the pattern of solar activity through the centuries back to 5320 B.C. His curve is reproduced as Fig. 1 in Paper 36 and an update table of the pattern is included in our Conclusion.

REFERENCES

Anderson, C. N., 1954, Notes on the Sunspot Cycle, *Jour. Geophys. Research* **59:**455–461. (Fig. 3 is still valid.)

de Jong, A. F. M., W. G. Mook, and B. Becker, 1979, Confirmation of the Suess Wiggles: 3200–3700 B.C., *Nature* **280:**48.

De Vries, H., 1958, Variation in Concentration of Radiocarbon with Time and Location on Earth, *Koningl. Neerland. Akad. Wetenschap. Proc.*, ser. B, **61**(2):94–102.

Dicke, R. H., 1978, Is There a Chronometer Hidden Deep in the Sun? *Nature* **276:**676–680.

Dicke, R. H., 1979a, The Clock Inside the Sun, *New Scientist* **83:**12–13.

Dicke, R. H., 1979b, Solar Luminosity and the Sunspot Cycle, *Nature* **280:**24–27.

Eddy, J. A., 1978, Climate and the Changing Sun, in *Encyclopedia Britannica Yearbook of Science and the Future,* Encyclopedia Britannica Inc., Chicago, pp. 263, 282.

Feynman, J., and N. U. Crooker, 1978, The Solar Wind at the Turn of the Century, *Nature* **275:**626–627. (*Cf. Jour. Geophys. Research* **85:**2991–2997, 1980.)

Fritz, H., 1881, *Das Polarlicht,* Brockhaus, Leipzig, Germany, 348p.

Gleissberg, W., 1952, *Die Haufigkeit der Sonnenflecken,* Akademie-Verlag, Berlin, 91p.

Gleissberg, W., 1965, The Eighty-Year Solar Cycle in Auroral Frequency Numbers, *British Astron. Assoc. Jour.* **75:**227–231.

Gleissberg, W., 1966, Ascent and Descent in the Eighty-Year Cycles of Solar Activity, *British Astron. Assoc. Jour.* **76:**265–268.

Gleissberg, W., 1972, The Eighty-Year Solar Cycle and Its Use for Solar-Activity Forecasting, *Jour. Interdisciplinary Cycle Research* **3:**391–394.

Gleissberg, W., 1973, The Eleven-Year and Eighty-Year Cycles in the Frequency of Sunspots Easily Visible to the Naked Eye, *Jour. Interdisciplinary Cycle Research* **4:**313–318. (See Fig. 1 in our Introduction.)

Gnevyshev, M. N., and A. I. Ol', 1948, On the Twenty-two Year Cycle of Solar Activity, *Astron. Zhur.* **25:**18–20. (In Russian.)

Gray, D. C., and P. E. Damon, 1970, Sunspots and Radiocarbon Dating in the Middle Ages, in *Scientific Methods in Medieval Archaeology,* R. Berger, ed., University of California Press, Berkeley, pp. 167–182.

Hale, G. E., 1908, On the Possible Existence of a Magnetic Field in Sunspots, *Astrophys. Jour.* **28:**315–343.

Hale, G. E., 1913, Preliminary Results of an Attempt to Detect the General Magnetic Field of the Sun, *Astrophys. Jour.* **38:**27–98.

Houtermans, J., H. E. Suess, and W. Munk, 1967, Effect of Industrial Fuel Combustion on the Carbon-14 Level of Atmospheric CO_2, in *Radioactive Dating and Methods of Low-Level Counting,* International Atomic Energy Agency, ST1-PUB-152, Wien, Austria, pp. 57–68.

Kanda, S., 1933, Ancient Records of Sunspots and Aurorae in the Far East and the Variation of the Period of Solar Activity, *Imperial Academy Japan Proc.* **9:**293–296. (Sine curve solution.)

Lingenfelter, R. E., and R. Ramaty, 1970, Astrophysical and Geophysical Variations in C14 Production, in *Nobel Symposium 12: Radiocarbon Variations and Absolute Chronology,* J. U. Olsson, ed., Wiley, Chichester, England, pp. 513–535. (See also other papers in this volume.)

Neftel, A., H. Oeschger, and H. E. Suess, 1981, Secular Non-random Variations of Cosmogenic Carbon-14 in the Terrestrial Atmosphere, *Earth and Planetary Sci. Letters* **56:**127–147.

Pepin, E. O., J. A. Eddy, and A. B. Merill, eds., 1980, *The Ancient Sun: Fossil Record in the Earth, Moon, and Meteorites,* Pergamon, Oxford, England, 581p.

Pilgram, A., 1788, *Wetterkunde,* vols. 1 and 2, Wien, Austria.

Rubashev, B. M., 1964, *Problemy solnechnoy aktivnosti,* Nauka, Moscow and Leningrad, 362p. (*Problems of Solar Activity,* translated in NASA TTF 244.)

Schove, D. J., 1947, The Sunspot Cycle Before 1750, *Terrestrial Magnetism* **52:**233–238.

Schove, D. J., 1948, Sunspot Epochs, A.D. 188–1750, *Popular Astronomy* **56:**247–252.

Schove, D. J., 1967, World Climate, *Roy. Meteorol. Soc. Proc. Internat. Symp. World Climate.*

Schove, D. J., 1980, The 200-, 22- and 11-year Cycles and Long Series of Climatic Data, Mainly Since A.D. 200, in *Sun and Climate,* Centre National d'Etudes Spatiales, pp. 87–100.

Siscoe, G. L., 1980, Evidence in the Auroral Record for Secular Solar Variability, *Rev. Geophysics and Space Physics* **18:**647–658.

Stewart, B., 1864, On the Large Sun-spot Period of About 56 Years, *Royal Astron. Soc. Monthly Notices* **24:**197.

Stuiver, M., 1961, Variation in Radiocarbon Concentration and Sunspot Activity, *Jour. Geophys. Research* **66:**273–278.

Stuiver, M., 1965, Carbon 14 Content of Eighteenth and Nineteenth Century Wood: Variations Correlated with Sunspot Activity, *Science* **149:**533–535.

Stuiver, M., 1978, Radiocarbon Timescale Tested Against Magnetic and Other Dating Methods, *Nature* **273:**271.

Stuiver, M., 1980, Solar Variability and Climatic Change During the Current Millenium, *Nature* **285:**868–871.

Stuiver, M., 1982, A High-Precision Calibration and the A.D. Radiocarbon Time Scale, *Radiocarbon* **24:**1–26.

Stuiver, M., and H. E. Suess, 1966, On the Relationship Between Radiocarbon Dates and True Sample Ages, *Radiocarbon* **8:**534–540.

Suess, H. E., 1979, The Radiocarbon Record in Tree-rings of the Last 8000 Years, in *Proceedings of the 10th International Radiocarbon Conference,* Bern and Heidelberg, Germany.

Suess, H. E., 1980*a,* The Radiocarbon Record in Tree-rings of the Last 8000 Years, Proceedings of the 10th International Radiocarbon Conference, Bern and Heidelberg, *Radiocarbon* **22:**200–209.

Suess, H. E., 1980*b,* Solar Activity. Cosmic Ray Produced Carbon-14 and the Terrestrial Climate, in *Sun and Climate,* Centre National d'Etudes Spatiales, pp. 307–310.

Svenonius, B., and E. Olausson, 1979, Cycles in Solar Activity, Especially of Long-Period and Certain Terrestrial Relationships, *Palaeogeography, Palaeoclimatology, Palaeoecology* **26:**89–97.

Turner, H., 1913, Sunspots and Meteor Swarms, *Royal Astron. Soc. Monthly Notices* **7:**82–109.

Vitinsky, Y. I., 1965, *Solar Activity Forecasting,* Israel Program for Scientific Translations, Jerusalem, 129p. (*Prognozy solnechnoi aktivnosti,* Akad. Nauk Izd. SSSR, Leningrad, 151p.) (Standard work.)

Wittmann, A., 1978, The Sunspot Cycle Before the Maunder Minimum, *Astronomy and Astrophysics* **66:**93–97.

Wolf, R., 1862, Mitteilungen über die Sonnenflecken, No. 15, *Astronomische Mitt.,* pp. 151, 190. (Long cycles.)

Wolf, R., 1872, Mitteilungen über die Sonnenflecken, *Astronomische Mitt.*

Wolf, R., 1883, Mitteilungen über die Sonnenflecken, Nos. 54–56., *Astronomische Mitt.* (Hints of 22-year cycle.)

Wolfer, A., 1891, *Les Travaux de M. Wolf,* Royal Astronomical Society Pamphlet.

Yunnan Observatory, 1977, A Recompilation of our Country's Records of Sunspots Through the Ages and an Inquiry into Possible Periodicities in Their Activity, *Chinese Astronomy* **1:**347–359. (First published in Chinese in *Acta Astron. Sinica* **17:**217–227, 1976.)

24

Reprinted from *Natl. Acad. Sci. (USA) Proc.* **10**:53–55 (1924)

THE LAW OF SUN-SPOT POLARITY

By George E. Hale

Mount Wilson Observatory, Carnegie Institution of Washington

Communicated, December 6, 1923

In a paper on "The Direction of Rotation of Sun-spot Vortices," published in the *Proceedings of the National Academy of Sciences* in June, 1915, I described the general reversal of the magnetic polarity of sun-spots observed at the minimum of solar activity in 1912. After the usual interval of about eleven years another minimum has occurred, and the incoming spots of the new cycle following it show another reversal of polarity. It therefore becomes possible to formulate a polarity law based on observations by Hale, Ellerman, Nicholson, Joy, Pettit and others of the Zeeman effect in 2136 sun-spot groups during the period 1908–1923.*

About 60 per cent of all sun-spots are bipolar, consisting of two spots or groups of spots of opposite magnetic polarity. In classifying the observations we give chief weight to these, and treat single spots followed by flocculi as the preceding members of incomplete bipolar groups, and single spots preceded by flocculi as the following members of such groups. Before the minimum of 1912 the polarity of preceding spots in the northern hemisphere was S (south seeking pole) or negative, and that of following spots N or positive. The spots of the southern hemisphere gave opposite polarities, i.e., N for the preceding and S for the following members.

As is well known, the last spots of a cycle occur in low latitudes, ranging from 0° to about 18°. The first spots of a new cycle appear in high lati-

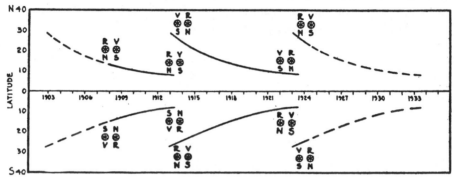

FIGURE 1

tudes, sometimes exceeding 40°. As the cycle progresses, the average latitude of all spots decreases steadily, as shown by the curves in Fig. 1.

To our surprise, the high latitude spots of the new cycle, which began in 1912 when spots were very few, were opposite in polarity to the low latitude spots of the previous cycle. As this cycle advanced and the spots became more and more numerous, the new polarities were found to characterize all spots observed, with only 4 per cent of exceptions. The average latitude of the spots gradually decreased, in the customary way, and the recent spots marking the end of the cycle have been near the equator, though several of them observed in 1923 reached latitudes as high as 15°.

The first spot of the next (third) cycle appeared on June 24, 1922, at latitude 31° N. After a long interval several other spots of this cycle have been observed, including a number of bipolar groups. They again show a reversal of polarity, back to the conditions existing in the first cycle before the minimum of 1913. The results are shown graphically in Figs. 1 and 2.

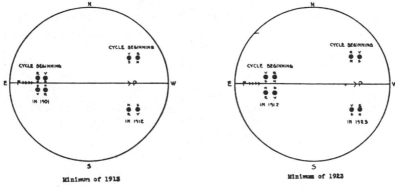

FIGURE 2

It thus appears that near the time of minimum solar activity four spot zones, characterized by distinct magnetic polarities, may co-exist on the sun. This condition lasts only two or three years, after which the last low-latitude spots of the old cycle disappear.

We cannot yet determine with certainty the sign of the dominant electric charge in the spot vortex. If it is always the same, the vortices of the preceding and following spots of bipolar groups must whirl in opposite directions. Moreover, the reversed polarities of corresponding spots (preceding or following) in the same hemisphere of successive cycles must also mean opposite directions of whirl. A series of observations of the Evershed effect at low levels in bipolar spots is needed to settle this question and thus to determine the sign of the dominant charge.

The sun-spot period, if defined in the usual way as representing the variation in the number or total area of all spots on the sun, is about 11.1 years. But if we regard the period as the interval between successive appearances of spots of the same magnetic polarity, the present results indicate that it is twice as long.

The details of this investigation will be published in the *Astrophysical Journal*.

* For the methods employed, see "The Magnetic Polarity of Sun-spots," Contributions from the Mount Wilson Observatory No. 165, *Astrophys. J., Chicago*, **49**, 1919 (153–178).

25

Reprinted from pages 227–228 and 230–231 of British Astron. Assoc. Jour.
75:227–231 (1965)

THE EIGHTY-YEAR SOLAR CYCLE IN AURORAL FREQUENCY NUMBERS

W. GLEISSBERG
(Communicated by DR D. J. SCHOVE)

Auroral frequency numbers were published recently by Schove (1) for each decade since A.D.290, and also for numerous decades between 500 B.C. and A.D. 200. In his discussion of these numbers, which may be denoted by N_A, Schove stated that the 80-year solar cycle "did not seem to be clearly

[*Editor's Note:* Table 1, which appears on page 229 in the original, has been omitted. The unsmoothed decadal values are reprinted in Paper 9 (see also our Appendix C, column *g*) and Gleissberg's smoothed values are charted in Figure 8 in our Introduction.

In a follow-up paper entitled "Ascent and Descent in the Eighty-year Cycle of Solar Activity" (1966, *British Astron. Assoc. Jour.* **76**:265–268), Gleissberg cited Rubashev's *Problemy solnechnoy aktivnosti* (1964, Nauka, Moscow and Leningrad, p. 362; translated in *NASA TTF 244*) and concluded that the following rules apply to cycles of about 80 years: the longer the period of ascent, the higher the ascent; the longer the period of descent, the deeper the descent; and the mean lengths of the periods of ascent and descent are 38 years and 41 years respectively.]

marked" in the auroral frequency pattern in medieval times. This result seems to be surprising, for, as I have shown (2,3), the 80-year cycle is well marked in the series of the intervals D_M between the epochs of consecutive auroral maxima during 16 centuries, epochs which had been published by Schove.[4] Moreover, Link[5] found the 80-year cycle also in auroral intensity since A.D. 400. Thus, it appears worthwhile to investigate whether the 80-year cycle could be traced out also in Schove's auroral frequency numbers N_A since A.D. 290.[1]

Secular smoothing has proved to be a suitable method for exhibiting the 80-year cycle.[6] As I explained earlier in this JOURNAL[3] secular smoothing of a series of given quantities consists of taking moving averages of every five consecutive quantities weighted so that half weight is given to the first and the last ones. There is also another way of obtaining secularly smoothed values: one can take moving averages of every four consecutive quantities and then take moving averages of every two consecutive averages; these latter averages are the secularly smoothed values of the given series. It is easy to see that both methods of secular smoothing lead to the same result.

If secular smoothing is applied to the auroral frequency numbers N_A, as given by Schove[1] from A.D. 290 onwards, the data in Table 1 are obtained. The arrangement of Table 1 is the same as that of Schove's table in his paper.[1] As each secularly smoothed value $\overline{N_A}$ corresponds to the central of the five quantities N_A used in forming the averages, the series of the secularly smoothed values $\overline{N_A}$ in Table 1 starts only from A.D. 310 and ends in 1930, whereas the uninterrupted series of auroral frequency numbers N_A in Schove's table runs from A.D. 290 until A.D. 1950.

All the maxima of $\overline{N_A}$ which appear in Table 1 are marked with an asterisk. To avoid any arbitrariness, each value $\overline{N_A}$ being larger than both of its neighbours was considered as a maximum. In two cases, viz. in 760–780 and 1440–1450, the maximum extends over three or two equal values, respectively. There are 19 maxima in Table 1, the first one occurring in 370 and the last one in 1840. This means that, in the maxima of $\overline{N_A}$ are interpreted as peaks of a cyclic variation, 18 cycles correspond to 1470 years; thus, one cycle, on the average, equals about 82 years. There can be little doubt that this variation is identical with the 80-year solar cycle.

The presumable identity of the variation of the values $\overline{N_A}$ with the 80-year cycle, is supported by a comparison with other data concerning the location of the peaks of the 80-year cycle. As mentioned above, I have traced out this cycle in the series of the intervals D_M between the years of auroral maxima determined by Schove.[4] From the last few centuries, when reliable sunspot data are available, it is known that, in the 80-year cycle, minima of the intervals D_M nearly coincide with maxima of solar activity. Thus, if the variations of $\overline{N_A}$ in Table 1 really are produced by the 80-year cycle one should expect that the maxima of $\overline{N_A}$ would occur, at least approximately, in the same decades as the minima of $\overline{D_M}$, the secularly smoothed values of D_M. Column

(a) of Table 2 contains the decades into which, according to Table 1, the maxima of $\overline{N}_A$ have fallen. The decades where minima of $\overline{D}_M$ occurred can be taken from my previous papers; see Table 1 in (2) and Fig. 2 in (3). These decades are given in column (b) of Table 2. The corresponding data as deduced by Link[5] from his own aurora catalogue for the interval 400–1600 are added in column (c).

Taking into consideration the uncertainty and incompleteness of medieval aurora records, we can say that the data of column (a) of Table 2 are in rather good agreement with the data in columns (b) and (c). It appears, however, that one maximum, at about A.D. 910–930, is missing in column (a); on the other hand, there seem to be two gaps, at about A.D. 500–520 and 1300–1310, in column (c).

TABLE 2

DECADES OF MAXIMA OF SECULARLY SMOOTHED AURORAL FREQUENCY:

(a) as taken from Table 1,
(b) as deduced from Schove's list of auroral maximum years (2, 4),
(c) as given by Link[5] for A.D. 400–1600.

(a) Auroral Activity	(b) Cycle length	(c)
370	350	—
440	440	440
500	520	—
570	580	590
660	670	670
770	740	740
840	840	850
—	910	930
980	990	1 000
1 110	1 110	1 130
1 190	1 180	1 190
1 260	1 230	1 260
1 310	1 300	—
1 370	1 380	1 370
1 450	1 460	1 450
1 560	1 530	1 570
1 610	1 600	—
1 730	1 700	—
1 770	1 760	—
1 840	1 840	—

204

The above discussion leads to the conclusion that the auroral frequency numbers established by Schove[1] imply the 80-year solar cycle. For, it can be produced from these numbers by the method of secular smoothing.

Acknowledgement

It is a pleasure to me to thank Dr D. J. Schove for revising this paper with respect to the language and for communicating it to the Association.

References

1 Schove, D. J., *J. British Astron. Assoc.*, **72**, 30, 1962.
2 Gleissberg. W., *Publ. Istanbul Univ. Obs.*, No. 57, 1955.
3 Gleissberg. W., *J. British Astron. Assoc.*, **68**, 148, 1958.
4 Schove, D. J., *J. Geophys. Research*, **60**, 127, 1955.
5 Link, F., *Bull. Astron. Inst. Czechoslovakia*, **14**, 226, 1963.
6 Gleissberg, W., *Terr. Magn. and Atmosph. Electr.*, **49**, 243, 1944.

26

AUTOCORRELATION AND SOLAR CYCLES

A. Kiral

These excerpts were translated expressly for this Benchmark volume by D. J. Schove, St. David's College, England, from Istanbul Univ. Obs. Publ. No. 70, pp. 12-21, 1961. Copyright ©1979 by D. J. Schove. In the original, the following summary appeared in English. Table 4, Figures 1 and 2, and the Bibliography have been reprinted from the original.

Summary: Using the data of "The Sunspot Cycle, 649 B.C. to A.D. 2000" which was published by D. J. Schove in the *Journal of Geophysical Research,* 60, 127–146, (1955), I was able to confirm the existence of a long solar cycle of 80 years (7 cycles of eleven years) with the aid of the method of autocorrelation and the power spectrum which W. Gleissberg had already established using his method of secular smoothing.

This study has also proved the existence of the periods of 2, 9, 18 cycles of eleven years of the solar activity and gave the possibility to reject all other cycles.

It must be noted that in the application of this method the numbers of cycles of eleven years were taken as independent variable instead of time.

Solar activity, and solar frequency in particular, has for a long time been known to be regulated by two fundamental cycles: the 11-year cycle discovered by Schwabe, and the 80-year cycle, postulated as early as 1862 by Wolf (1), which has been the subject of several investigations by Gleissberg (2) and others. Research workers have associated other cycles with the two well-known cycles mentioned here, the reality of which is to a greater or lesser extent doubtful. The period of 22 years relates to the reversal of magnetic polarity rather than the frequency of sunspots. However, specific correlations with various solar phenomena have been the subject of several papers (3).

By harmonic analysis and successive approximation Kimura (4) managed to find, among others, a period of 82.2 years, little different from the 81-year period of Wolf (5). Among the 29 periods found by Kimura [Table I] and those found by Wolf [Table II], the 80-year period occupies an important position next to the 11-year cycle because of its high amplitude . . . [The outmoded analyses of Turner (6) and Michelson (7) are then mentioned.]

[*Editor's Note:* Table I, which includes Turner's periods, has been omitted at this point.]

In view of the impossibility of representing the series of observations as a Fourier series, Gleissberg has applied to the lists of Schove (8) an

Table II.   Periods and Corresponding Amplitudes of Wolf *(Astronomische Mitt.,* 1862)

Period (in yrs.)	Amplitude
11	21.2
81	17.7
10	17.1
8	

arithmetical method—sufficiently orthodox for statistical calculations—defined and christened *secular smoothing* (9, 10). [See Paper 25.] Thanks to this method he proved the existence of the long period of 80 years, and he was also able to determine its value as 78.8 ± 3.3 years (i.e., 7.1 eleven-year cycles). . . .

[*Editor's Note:* Material including Table III has been omitted at this point.]

We will try for our part to apply to the Schove series the method of autocorrelation that proves so fruitful in periodicity analysis. We limit ourselves to the period of observation between A.D. 290 and 1947, the variables being the same as in Table IV—the numbers *(k)* of the cycles and the intervals *(uᵢ)* between successive maxima.

The method of autocorrelation we use here has been described by Kendall (11) . . . who shows that the correlogram has indeed an advantage over the periodogram, the latter yielding generally unreliable results for the values of the periods . . . [e.g., where cycles are damped, nonpersistent, or subject to phase change. Statistical methods at that time in use (12, 13, 14) are then discussed further.]

A glance at the correlogram (Fig. 1) enables us to infer from the maxima at $K = 6, 15, 22, 26, 35, 41,$ and 48 cycles, a mean period of seven 11-year cycles. In addition to this period secondary maxima for $K = 3, 12, 19,$ and 30 are characteristic. The [corresponding] latter period is of the order of nine 11-year cycles. The reality of these cycles is confirmed by the *power spectrum* (Fig. 2). . . .

The power spectrum clearly reveals the secondary periods of two, nine, and eighteen 11-year cycles [i.e., 22, 100, and 200 years] with the fundamental period of seven 11-year cycles [i.e., 80 years]. Their respective amplitudes reveal their relative importance.

These cycles, in addition to the 11-year cycle, may be supposed to affect our weather (see Figs. 1 and 2). Indeed, in research on climatic periodicities we do meet cycles of 24 years (15), 89 years (16), and Méméry's 100-year period (17). Moreover, the periodicity of aurorae of 200–205 years is established. These (meteorological) cycles correspond respectively to our (solar) cycles.

The effective existence of periods of two, seven, nine, and eighteen 11-year cycles (i.e., approximately 22, 78, 100, and 200 years) of solar activity can thus be confirmed and we believe that any other period between these values is imaginary.

Tableau IV

No. du cycle	Époque d. Max.	u'_i	u_i
−182	290		
131	302	12	+1
−130	311	9	−2
129	321	10	−1
128	330	9	−2
127	342	12	+1
126	354	12	+1
125	362	8	−3
124	372	10	−1
123	387	15	+4
122	396	9	−2
121	410	14	+3
−120	421	11	0
119	430	9	−2
118	441	11	0
117	452	11	0
116	465	13	+2
115	479	14	+3
114	490	11	0
113	501	11	0
112	511	10	−1
111	522	11	0
−110	531	9	−2
109	542	11	0
108	557	15	+4
107	567	10	−1
106	578	11	0
105	585	7	−4
104	597	12	+1
103	607	10	−1
102	618	11	0
101	628	10	−1
−100	642	14	+3
99	654	12	+1
98	665	11	0
97	677	12	+1
96	689	12	+1
95	699	10	−1
94	714	15	+4
93	724	10	−1
92	735	11	0
91	745	10	−1
−90	754	9	−2
89	765	11	0
88	776	11	0
87	787	11	0
86	798	11	0
85	809	11	0
84	821	12	+1
83	829	8	−3
−82	840	11	0
81	850	10	−1
−80	862	12	+1
79	872	10	−1
78	887	15	+4
77	893	11	0
76	907	8	−3
75	917	11	0
74	926	9	−2
73	938	12	+1
72	950	12	+1
71	963	13	+2
−70	974	11	0
69	986	12	+1
68	994	8	−3
67	1003	9	−2
66	1016	13	+2
65	1027	11	0
64	1038	11	0
63	1052	14	+3
62	1067	15	+4
61	1078	11	0
−60	1088	10	−1
59	1098	10	−1
58	1110	12	+1
57	1118	8	−3
56	1129	11	0
55	1138	9	−2
54	1151	13	+2
53	1160	9	−2
52	1173	13	+2
51	1185	12	+1
−50	1193	8	−3
49	1202	9	−2
48	1219	17	+6
47	1228	9	−2
46	1239	11	0
45	1249	10	−1
44	1259	10	−1
43	1276	17	+6
42	1288	12	+1
41	1296	8	−3
−40	1308	12	+1
39	1316	8	−3
38	1324	8	−3
37	1337	13	+2
36	1353	16	+5
35	1362	9	−2
34	1372	10	−1
33	1382	10	−1
−32	1391	9	−2
31	1402	11	0
−30	1413	11	0
29	1429	16	+5
28	1439	10	−1
27	1449	10	−1
26	1461	12	+1
25	1472	11	0
24	1480	8	−3
23	1492	12	+1
22	1505	13	+2
21	1519	14	+3
−20	1528	9	−2
19	1539	11	0
18	1548	9	−2
17	1558	10	−3
16	1572	14	+3
15	1581	9	−2
14	1591	10	−1
13	1604,5	13	+2
12	1615,5	11	0
11	1626,0	10,5	−0,5
−10	1639,5	13,5	+2.5
9	1649,0	9,5	−1,5
8	1660,0	11,0	0
7	1675,0	15,0	+4,0
6	1685,0	10,0	−1,0
5	1693,0	8,0	−3,0
4	1705,5	12,5	+1,5
3	1718,2	12,7	+1,7
2	1727,5	9,8	−1,7
−1	1738,7	11,2	+0,2
0	1750,3	11,6	+0,6
+1	1761,5	11,2	+0,2
2	1769,7	8,2	−2,8
3	1778,4	8,7	−2,8
4	1788,1	9,7	−1,8
5	1805,2	17,1	+6,1
6	1816,4	11,2	+0,2
7	1829,9	13,5	+2,5
8	1837,2	7,3	−3,7
9	1848,1	10,9	−0,1
+10	1860,1	12,0	+1,0
11	1870,6	10,5	−0,5
12	1883,9	13,3	+2,3
13	1894,1	10,2	−0,8
14	1907,0	12,9	+1,9
15	1917,6	10,6	−0,4
16	1928,4	10,8	−0,2
17	1937,4	9,0	−2,0
18	1947,5	10,1	−0,9

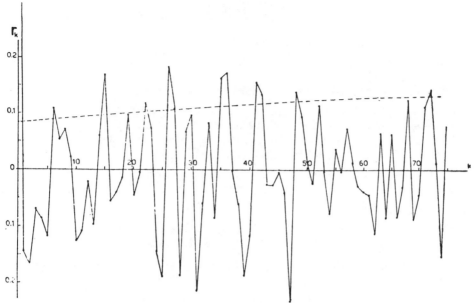

Fig. 1. Le corrélogramme

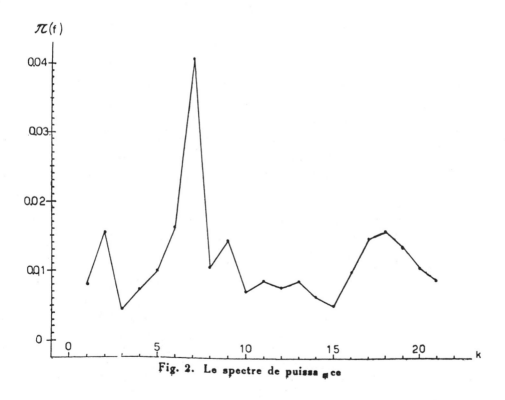

Fig. 2. Le spectre de puissance

Bibliographie

(1) R. Wolf, Astr. Mitt. Zürich, No. 14, 1862.

(2) Voir la littérature de W. Gleissberg, Publ. Istanbul University Observatory, No. 26, 1944 et No. 57, 1955.
W. Gleissberg, Die Häufigkeit der Sonnenflecken, Berlin, 1952.

(3) L. Taffara, Mem. Soc. Astr. Ital., nouveau sér., 4, 395, 1929;
B. Meyermann, AN 269, 114, 1939; 279, 45, 1950;
M. Cimino et G. Armellini. Proc. Nat. Acad., Rome, 1, 624, 1946;
W. Grotrian, Zs. f. angew. Phys., 2, 376, 1950.

(4) H. Kimura, On the harmonic analysis of sun-spot relative numbers, (MN 73, 543, 1918).

(5) R. Wolf, AN 2463.

(6) H. H. Turner, On the harmonic analysis of Wolf's sun-spot numbers, with special reference to Mr. Kimura paper, (MN 73, 549, 1918).

(7) A. A. Michelson, Determination of periodicities by the harmonic analyzer with an application to the sun-spot cycle, (ApJ 38, 264, 1913).

(8) D. J. Schove, The sunspot cycle, 649 B. C. to A. D. 2000, (Journal of Geophysical Research, 60, 127, 1955).

(9) W. Gleissberg, A table of secular variations of the solar cycle, (Terrestrial Magnetism and Atmospheric Electricity, 49, 243, 1944).
W. Gleissberg, Die Häufigkeit der Sonnenflecken, Berlin, 1952.

(10) W. Gleissberg, Vorläufige Bestimmung der mittleren Länge des 80 jährigen Sonnenfleckenzyklus, (Naturwissenschaften, 42, 410, 1955).

(11) M. G. Kendall, The Advanced Theory of Statistics, II (3. éd.), 402, 1951.

(12) J. Ashbrook, R. L. Duncombe et A. J. J. van Woerkom, A statistical analysis of the light curve of the variable star μ Cephei, (AJ 59, 12, 1954).

(13) M. S. Bartlett, J. R. Statist. Soc., Suppl. 8, 27-41, 1946.

(14) W. Feller, General Theory of Probability. Princeton, 1951.

(15) F. B. Groissmayr, Eine 24 jährige Witterungsperiode, (Annalen der Hydrographie und maritimen Meteorologie 65, 118, 370, 1937; 68, 200, 1940; 69, 145, 1941; 70, 80, 1942).

(16) C. Easton, Afwijkingen en periodiciteit der wintertemperaturen in West-Europa sedert het jaar 760, (Versl. Kon. Acad. v. Wet. Amsterdam,25, 1119-1134, 1917); C. Easton, Periodicity of winter temperatures in western Europe since A. D. 760, (Proc. Roy. Ac. Sc. Amsterdam, 20, 1092-1107, 1917); C. Easton, Les hivers dans l'Europe occidentale, Leyden, 1928; W. Köppen, Das Gesetz in der Wiederkehr strenger Winter in Westeuropa, (Meteorologische Zeitschrift, 47, 205, 1930)

(17) F. Kratoschwill. Über kalte und strenge Winter in Mitteleuropa, (Meteorologische Zeitschrift, 57, 420, 1940).
H. Ménéry, Les bases de l'influence des phénomènes solaires en météorologie, 1936.
O. v. Myrbach, Der kalte Winter 1939/40 im hundert jährigen Wetterrhytmus und seine Beziehung zum Sonnenfleckenverlauf, (Metedrologische Zeitschrift, 57, 442, 1940).

27

Reprinted from pages 103, 104, 105–106, 106–107, 108, and 110 of *Solar Physics*
30:103–110 (1973)

PERIODICITIES IN SOLAR ACTIVITY*

T. W. COLE

Division of Radiophysics, CSIRO, Sydney, Australia

Abstract. The techniques of power spectral analysis are used to determine significant periodicities in the annual mean relative sunspot numbers. The main conclusion is that a period of 10.45 yr is very basic and can be associated with an excitation of new solar cycles. When combined with a period of 11.8 yr, associated here with the free-running length of a solar cycle, the mean cycle length of 11.06 yr and a phase variation of 190 yr are explained. Similarly the amplitude variations with periods 88 and 59 yr (previously described as the 80-yr cycle) are due to an amplitude modulation of the solar cycle by a period of 11.9 ± 0.3 yr. The results dispute several associations of planetary position and solar activity.

* Radiophysics Publication RPP 1647, January, 1973.

[*Editor's Note:* Introductory material and an explanation of the Blackman-Tukey method (see Paper 28) have been omitted.]

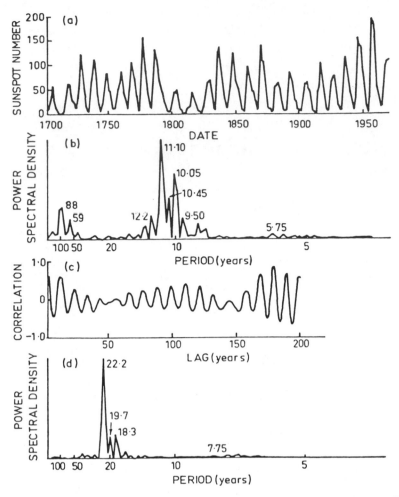

Fig. 1.   (a) The mean annual Wolf relative sunspot numbers plotted from 1700 until 1969. (b) The power spectral density of the annual relative sunspot numbers from 1700 until 1969 plotted against the periodicity in years. (c) The autocorrelation of the series in (a) plotted against lag in years up to a maximum lag of 200 yr. (d) The power spectrum of the 22-yr solar magnetic cycle obtained by inverting the sign of alternate peaks in (a).

By considering the polarity of the solar magnetic field, the sunspot cycle can also be represented as a 22-yr cycle in which every second peak in Figure 1(a) is reversed in sign (Bracewell, 1953). The power spectrum of this 22-yr cycle is shown in Figure 1(d), and in this case the strongest and significant periodicities are the group near the 22-yr period and another near the 7.75-yr period.

3. The Phase of the Sunspot Cycle

The phase of the solar cycle can be studied by comparing the dates of the solar cycle maxima or minima with those expected for a constant period of 11.06 yr, the mean spacing of the cycles. The residual phase so obtained is plotted in Figure 2(a) for solar cycle maxima between 300 AD and 1968 (Waldmeier, 1961; *Astronomische Mitteilungen*; Schove, 1955).

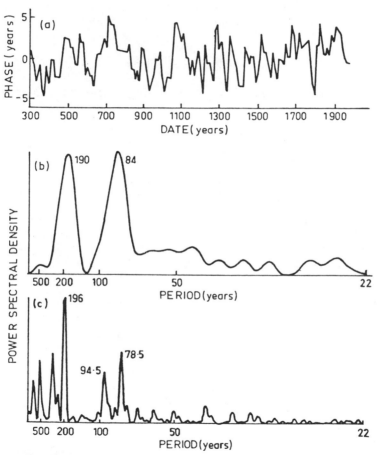

Fig. 2. (a) The relative phase of the dates of solar cycle maxima relative to a constant 11.06-yr period, plotted from 300 AD until 1968 using the data of Schove (1955) for maxima before the year 1600. (b) The power spectrum of the section of the curve in (a) between 1626 and 1968. (c) The power spectrum of the residual phase for the whole 152 cycles of Figure 2(a).

213

The power spectrum of this residual phase over the period between 1626 and 1968 (where the data are more accurate) is shown in Figure 2(b); it reveals two significant peaks at 190- and 84-yr periods. Both peaks are broadened, indicating that they consist of components which cannot be resolved by the 340-yr length of data. Some extra information can be obtained from the less accurate data of Schove (1955), and Figure 2(c) is a plot of the power spectrum of the residual phase for the complete 1668-yr interval of Figure 2(a). It indicates three main periodicities in the phase of periods 196, 94.5 and 78.5 yr. The phase also contains periodicities of 280, 560 and 1050 yr but it is difficult to place accuracy limits on these longer-term variations derived from the data of Schove (1955).

The periodicities can now be interpreted, but first they will be used to predict the dates of the next few solar cycle maxima and minima.

[*Editor's Note:* Material has been omitted at this point. The next minimum is thus given by Cole as 1988 with a maximum of 70 in 1993.]

5. The 190-yr Periodicity

The strongest feature in the phase of the 11-yr cycle over the last 270 yr is a 190-yr periodicity. The component of period 94.5 yr is most likely a second harmonic of the 190-yr component. The form of the 190-yr variation is therefore not sinusoidal. Further understanding of this periodicity can be gained through the spectra of Figure 1.

Given a 190-yr phase modulation, the power spectrum in Figure 1(b) should show this as a series of peaks about the basic frequency of the cycle with a separation corresponding to a 190-yr period. Indeed, the peaks about the 10.45-yr peak are separated, on average, by this amount. Further, modulation theory (see, for example, Everitt and Anner (1956)) indicates that the ratio of the various peaks in Figure 1(a) is due to a phase modulation of $\pm 90°$ and the non-symmetry about the centre of the group of peaks at the 10.45-yr period indicates a non-sinusoidal form to the phase modulation.

The autocorrelation function and the spectrum of Figure 1(d) add to the argument in support of the real existence of a 190-yr phase modulation of the solar cycle.

Figure 3(a) illustrates the mean behaviour of the 190-yr phase variation. It is obtained by superimposing 190-yr segments of the residual phase of Figure 2(a). Also drawn on the diagram are solid lines which are those which would be obtained from a periodicity of 10.45 yr in the data. A discussion of the 10.45-yr period is left to a later section, but it can be seen that the phase is strongly modulated and that for almost half of the time the sunspot cycle is, on average, closer to 10.45 yr than the mean, 11.06-yr period.

The form of the 78.5-yr phase modulation is apparently sinusoidal, as is shown in the superimposed residual phase plotted in Figure 3(b).

[*Editor's Note:* Material has been omitted at this point.]

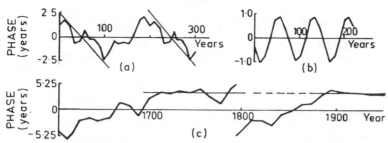

Fig. 3. (a) The mean shape of the 190-yr variation in phase derived by superimposing 190-yr segments of Figure 2(a) for the interval 300 AD until 1968 AD. The sloping lines represent the phase of a 10.45-yr periodicity. Note that the section between 190 and 300 yr is merely a repeat of the section from 0 to 110 yr. (b) The mean 78.5-yr variation in phase derived by superimposing 78.5-yr segments of Figure 2(a). (c) The relative phase between dates of solar cycle minima and a 10.45-yr period for the interval 1626 to 1968. The section from 1790 onwards has been displaced by one cycle. The two sections shown by a solid, thin line are those where the period is close to 10.45 yr.

[*Editor's Note:* Material has been omitted at this point. Cole concludes that the mean cycle length of 11.06 years and the 190-year phase modulation are a result of two periods of 11.8 and 10.45 years representing respectively a characteristic free-running time for the solar cycle, and a well-defined periodicity that excites the new solar cycle.]

References

Astronomische Mitteilungen der Eidgenössischen Sternwarte, Zürich, No. 273, 279, 283, 284.
Bigg, E. K.: 1967, *Astron. J.* **73**, 463.
Blackman, R. B. and Tukey, J. W.: 1959, *The Measurement of Power Spectra*, Dover, N.Y.
Bracewell, R. N.: 1953, *Nature* **171**, 649.
Everitt, W. L. and Anner, G. E.: 1956, *Communication Engineering*, McGraw-Hill, N.Y.
Gleissberg, W.: 1952, *Die Häufigkeit der Sonnenflecken*, Akademie-Verlag, Berlin.
Jose, P. D.: 1965, *Astron. J.* **70**, 193.
Schove, D. J.: 1955, *J. Geophys. Res.* **60**, 127.
Waldmeier, M.: 1957, *Z. Astrophys.* **43**, 149.
Waldmeier, M.: 1961, *The Sunspot-Activity in the Years 1610–1960*, Schulthess & Co. AG, Zürich,
Waldmeier, M.: 1966, *Astron. Mitteilungen der Eidgenössischen Sternwarte*, Zürich, No. 274.
Wood, K. D.: 1972, *Nature* **240**, 91.
Zhukov, L. V. and Muzalevskii, Yu. S.: 1969, *Astron. Zh.* **46**, 600. (English translation in *Soviet Astron AJ* **13**, 473.)

28

Reprinted from pages 434–441 and 446–448 of *The Solar Output and Its Variation,*
O. R. White, ed., Colorado Associated University Press, Boulder, 1977, 526p.

SOLAR INDUCED VARIATIONS
OF ENERGETIC PARTICLES AT ONE AU

Paul E. Damon
Laboratory of Isotope Geochemistry,
Department of Geosciences,
University of Arizona
Tucson, Arizona

[*Editor's Note:* In the original, material on solar-induced particles precedes this excerpt.]

III. ATMOSPHERIC ^{14}C AND SOLAR ACTIVITY

The most fundamental assumption of ^{14}C dating is that the ratio of ^{14}C to ^{12}C for atmospheric CO_2 remains constant with time. Willard Libby, in his classic book *Radiocarbon Dating*, thoroughly analyzed this assumption and concluded that it had not changed significantly in historic times (Libby, 1955). This conclusion was based on the measurement of biogenic organic materials of historically known age ranging from A.D. 1072 to 2950 B.C., i.e., encompassing ~5,000 y of historic time. It soon became apparent that combustion of fossil fuels (Suess, 1955) and atomic bomb tests (Rafter and Ferguson, 1957) were changing the ^{14}C content of the atmosphere, but, except for these artificial perturbations, constancy was assumed prior to the twentieth-century technological explosion. However, shortly after the demonstration of artificial perturbations, de Vries (1958) was able to show that there had been measurable natural variations of the atmospheric ^{14}C concentration during the Little Ice Age (sixteenth through nineteenth centuries). This was accomplished by determining the ^{14}C concentration and stable carbon isotopic composition of tree rings and charred wheat of known age. These natural fluctuations in the atmospheric ^{14}C concentration are known as the de Vries effect.

Since the initial demonstration of the de Vries effect, a number of laboratories have systematically determined the ^{14}C concentration of tree rings in order to evaluate the fluctuation of the concentration of ^{14}C in the atmosphere. This work has been greatly aided by the establishment of a 7,484-y chronology for bristlecone pine (Ferguson, 1970). To determine the fluctuation of atmospheric radiocarbon, measurements are made of the ^{14}C concentration and stable isotopic concentration of dendrochronologically dated tree rings. The results are expressed as Δ values which can be thought of as the per mil difference between the ratio of ^{14}C to ^{12}C in the atmosphere at any time, t, relative to the atmospheric ratio of ^{14}C to ^{12}C for the year 1890, which is chosen as a standard prior to technological perturbation. I have discussed the basic assumptions and equations, methods of analyses, and evaluation of errors elsewhere (Damon, 1970).

Figure 1 shows the results from five laboratories of Δ measurements for dated tree rings supplied primarily by the University of Arizona Laboratory for Tree-Ring Research (Damon, Long, and Grey, 1970; Damon, Long and Wallick, 1972; Lerman, Mook, and Vogel, 1970; Ralph and Michael, 1970; Stuiver, 1969; Suess, 1965, 1967; and data of Suess in Houtermans, 1971). All samples that included

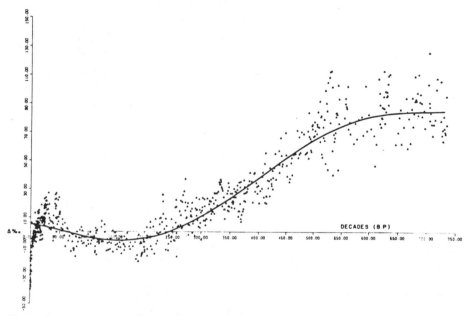

Fig. 1   Per mil variation (Δ) in the ^{14}*C activity of the atmosphere relative to 1890 A.D. as the reference time vs. dendrochronologic age in decades B.P. (before 1950 A.D.). The Δ values are determined by measuring the* ^{14}*C activity and stable carbon isotope composition of dendrochronologically dated tree rings. The trend curve is a third-order orthogonal polynomial. The relatively large change represented by the trend curve results from a quasi-sinusoidal change in the intensity of the earth's dipole magnetic field. Carbon 14 production is also modulated by solar activity as can be seen for the Little Ice Age (100-500 years B.P.) during which the Maunder and Spörer minima occurred. Twentieth-century Δ values have been affected by combustion of fossil fuels.*

217

more than 25 annual rings were rejected. Most of the remaining samples contained ten rings or less. The overall trend of the data shown by the solid line in Figure 1 is thought to be caused by a quasi-sinusoidal variation of the earth's magnetic field intensity with a period around 8,000 to 10,000 y. The reader may refer to the proceedings of the Twelfth Nobel Symposium for a thorough discussion, by various workers, of the geomagnetic field effect (Olsson, 1970). The Suess effect due to combustion of fossil fuels is evident after 1890, and the fluctuations observed by de Vries during the sixteenth to nineteenth centuries are also quite evident.

It can be seen that the de Vries effect consists of fluctuations around the trend curve that continue with greater or lesser intensity during the entire 7,484-y span of time. By using an electrical analog model proposed by de Vries (1959) and also by simply summing each solar cycle independently and comparing the resulting histogram, inverted, with the variation curve, Stuiver (1961, 1965) was the first to show a convincing relationship between the de Vries effect fluctuations and solar activity. The nature of the modulating mechanism of the galactic cosmic ray flux has been discussed in § I. Following Stuiver, Grey and others (Grey, Damon, and Long, 1966; Grey, 1969; Grey and Damon, 1970) used a computer model to demonstrate the relationship between Δ and solar activity. The model assumed that the ^{14}C content of the atmosphere during the jth year, n_{aj}, would be equal to the amount produced during that year, Q_j, plus the residuum from the previous year, $n_{a(j-1)}$, where Q_j was given by Lingenfelter's equation (Lingenfelter, 1963) relating Q_j to the sunspot number, R. The residuum was assumed to be lost from the atmospheric reservoir at a rate τ, the mean life of a disturbance in the ^{14}C content of the atmospheric reservoir. The equation, which was evaluated, year by year, by an iterative integration using an IBM 7092 computer, was as follows:

$$n_{aj} = n_{a(j-1)} e^{-1/\tau} + Q_j \quad . \tag{5}$$

Figure 2 shows a comparison between the theoretical curve derived from the above model and a smoothed curve through the measured Δ data where the sunspot number is derived from measurements by the Zurich Observatory (Waldmeier, 1961). The less accurate data of Schove (1955), derived from historical accounts of aurorae, which vary with solar activity, were used to test the agreement between measured and calculated values for the thirteenth through the nineteenth centuries (Figure 3). It seemed to us that the agreement could hardly be a matter of chance. The model predicts that the atmosphere will act as a low-pass filter attenuating 11-y solar-cycle-induced changes in Δ by two orders of magnitude (see Figure 2 for the theoretical magnitude of the 11-y solar cycle fluctuations). Damon, Long, and Wallick (1973) confirmed this great attenuation (see Figure 4). Further confirmation has been provided by Stuiver (1974, 1975).

Despite the good agreement between theory and measurement provided by the simple model discussed above, Ekdahl and Keeling (1973) point out that more

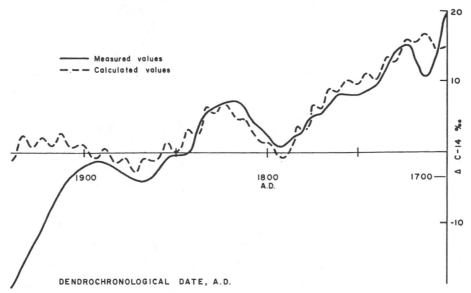

Fig. 2   *A comparison of per mil* ^{14}C *variations* (Δ) *as determined by measurements of tree rings and as calculated from a simple solar modulation model based on Lingenfelter's (1963) relationship between sunspot number and* ^{14}C *production (after Grey and Damon, 1970). The theoretical curve shows the predicted 11-year* ^{14}C *cycle (3°/oo peak to trough) which is near the limit of detection for precise gas counting measurements (see Fig. 4). It also shows the decline in the* ^{14}C *content of the atmosphere after the Maunder solar activity minimum and a distinct peak between 1800 and 1850 A.D. The measured data follow the theoretical curve closely until 1890 A.D. after which there is a divergence resulting from the combustion of fossil fuels.*

complicated ^{14}C reservoir models, e.g., their five- and six-reservoir models, predict a significantly greater attenuation of changes in ^{14}C production with a period greater than 20 y, although the simple model does predict about the same attenuation for the 11-y solar cycle ^{14}C production variations. The disparity in predicted attenuations could be a fault of the reservoir models or an underestimation by Lingenfelter of the effect of solar cycle activity on ^{14}C production which was compensated in the iterative integration by choosing a value of τ that produced excellent agreement. Values of τ between 2 and 100 y were programmed. The model was not very sensitive for values greater than 40 years, but a τ of 100 years was used in the calculations shown in Figures 2 and 3. Two- and three-box reservoir models give similar atmospheric residence times, i.e., circa 50 y (Damon, 1970; Houtermans, Suess, and Oeschger, 1973).

Using the data set shown in Figure 1, the average value of Δ was computed (Table 4) for the Maunder and Spörer minima (Eddy, 1976) and preceding, intermediate, and succeeding intervals of time. It can be seen that values of Δ increase by 10°/oo during both the Spörer and Maunder minima. There is a 180-y interval

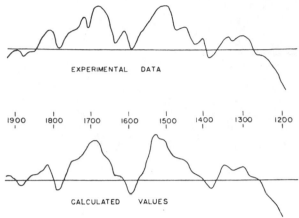

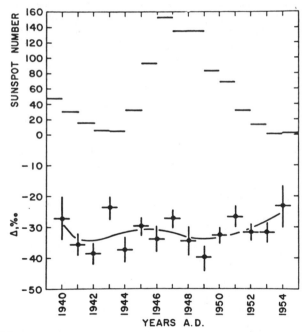

Fig. 3 *Comparison of experimental and calculated values of Δ vs. time (years A.D.) from the simple solar modulation model based on Lingenfelter's (1963) relationship between sunspot number and ^{14}C production for the Little Ice Age using sunspot data from the Zurich Observatory (Waldmeier, 1961) and Schove's (1955) sunspot estimates based on historical accounts of aurorae (after Grey and Damon, 1970). The Δ values for the Spörer (1450-1550 A.D.) minimum and Maunder (1645-1715 A.D.) minimum are about 15 °/oo (see Table 4).*

Fig. 4 *Per mil ^{14}C measurements (Δ values) for annual rings from the Radio Ridge tree compared with sunspot numbers (Damon, Long, and Wallick, 1973). Note the small variation in ^{14}C (Δ) indicated by the orthogonal polynomial trend curve.*

TABLE 4. Average relative atmospheric ^{14}C concentrations (Δ) during the Maunder and Spörer minima.

Time (A.D.)	Epoch	Δ	$\bar{\sigma}$	σ**
1716-1791	Following 75 years	6.1	±1.3	(±7.7)
1645-1715	Maunder minimum	16.8	±1.3	(±5.6)
1551-1644	Intermediate Period	6.3	±1.7	(±6.6)
1450-1550	Spörer minimum*	14.3	±1.6	(±8.4)
1374-1449	Preceding 75 years	2.9	±1.9	(±4.2)

*The Spörer minimum, according to Eddy (1976), extends from A.D. 1460-1550. Since its extent is not precisely known, I have taken a full century.

**The standard deviation (σ) claimed by the various authors who have provided the measurements is ± 5°/oo. It is interesting that somewhat more than half of the variance is due to measurement errors. The standard deviation of the mean ($\bar{\sigma}$) leaves no doubt as to the validity of the high Δ values during the Maunder and Spörer minima.

between the midpoints of the two minima. Damon, Long, and Grey (1966) showed that the best sinusoidal fit to the Δ data for the Little Ice Age had a period of 200 y and an amplitude of 10°/oo. Table 5 shows an analysis of the Δ data using the Blackman-Tukey (1958) Fourier analysis. For the time from 0 to 2,000 y, a 182-y periodicity is observed with a signal-to-noise ratio of 3.0. According to Schove (1955), average sunspot numbers vary from about 20 for the Maunder minimum to about 90 for the intermediate period. For a 180- to 200-y periodicity, the six-box model of Ekdahl and Keeling (1973) predicts an attenuation of 18. Using the most recent estimates by Lingenfelter and Ramaty (1970) of the dependence of ^{14}C production on sunspot number would have yielded a variation of about 100°/oo in ^{14}C production for the fluctuation from R = 20 to R = 90. An attenuation of 18 yields a Δ amplitude of 5.5°/oo within a factor of two of the observed value. A variation of R from 0 to 150 is required for perfect agreement. According

TABLE 5. Significant ^{14}C periodicities (Blackman and Tukey, 1958) for 2,000-y intervals.

Interval (years, B.P.)	Period	Signal to noise ratio
0-2,000	400	2.5
	182˙	3.0
	69	3.6
	56	1.8
2,000-4,000	500	3.0
	286	6.1
4,000-6,000	1000	1.2
	286	1.7
	100	1.9

to Eddy's (1976) analysis, there were virtually no sunspots during the Maunder Minimum. Is it possible that the sunspot number for the intermediate period was greater than previous estimates would suggest? In any case, agreement within a factor of two is adequate, although analysis of the eighteenth- and nineteenth-century sunspot and Δ data suggest to me that the more complicated reservoir models are predicting too great an attenuation for frequencies between 100 and 200 y.

·Eddy (1976) points out that the 11-y cycle and longer cycles may not be continuous and, in fact, solar activity may be more erratic than many workers would like to believe. The Fourier power-spectrum analysis of Table 5 confirms his suggestion. The 180-y periodicity is quite evident from 0 to 2,000 years B.P. (B.P. = before present), but does not appear for the interval from 2,000 to 6,000 B.P. Irregular fluctuations do persist, as can be seen from Figures 5, 6, and 7. Next, we shall take up the question of whether or not these irregular fluctuations

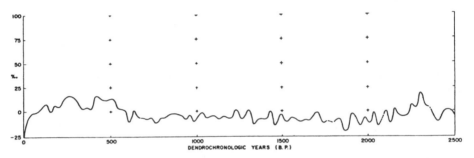

Fig. 5 Smooth curve drawn through the 25-y-interval averages of ^{14}C data (Δ) vs. dendro-chronologic age (0-2,500 B.P.). Note that the solar activity modulation of ^{14}C production is relatively subdued during the earlier part of the twentieth century when the geomagnetic field intensity is most intense (dashed lines indicate extrapolation for intervals without data).

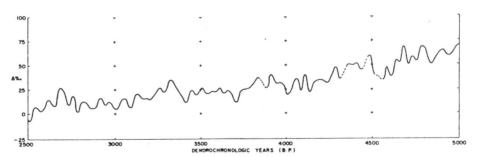

Fig. 6 Smooth curve drawn through the 25-y-interval averages of ^{14}C data (Δ) vs. dendro-chronologic age (2,500-5,000 B.P.). Note that the solar activity modulation of ^{14}C is more intense as the Δ values rise near the geomagnetic field intensity minimum. (Dashed lines indicate extrapolation for intervals without data.)

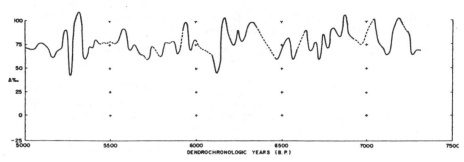

Fig. 7 Smooth curve drawn through the 25-y-interval averages of ^{14}C data (Δ) vs. dendro-chronologic age (5,000-7,500 B.P.). Note that the solar activity of ^{14}C is markedly more intense during this period of low geomagnetic field intensity. (Dashed lines indicate extrapolation for intervals without data.)

vary greatly in amplitude. In order to accomplish this, we will use the scatter of measurements around the trend curve in Figure 1 as a measure of the intensity of these fluctuations.

[*Editor's Note:* In the original, the physics is further discussed and acknowledgments are given. Only those references cited in this excerpt have been included.]

REFERENCES

Blackman, R.B., and Tukey, J.W., 1958, *The Measurement of Power Spectra*, Dover Press, New York.

Damon, P.E., 1970, *ibid.*, p. 571.

_____ , Long, A., and Grey, D.C., 1966, *J. Geophys. Res. 71*, 1055.

_____ , Long, A., and Grey, D.C., 1970, *Radiocarbon Variations and Absolute Chronology* (Proceedings 12th Nobel Symposium), ed. I.U. Olsson, John Wiley & Sons, New York, 615.

_____ , Long, A., and Wallick, E.I., 1972, *Proceedings Eighth International Conference on Radiocarbon Dating*, Wellington, New Zealand, 44.

_____ , Long A., and Wallick, E.I., 1973, *Earth Planet. Sci. Lett. 20*, 300.

_____ ; Ferguson, C.W.; Long, A.; and Wallick, E.I., 1974, *Am. Antiq. 39*, 350.

de Vries, Hl., 1958, *Proc. K. Ned. Akad. Wet.*, B61.

_____ , 1959, *Researches in Geochemistry*, ed. P.H. Abelson,, John Wiley & Sons, New York, 169.

Eddy, J.A., 1976, *Science 112*, 1189.

Ekdahl, C.A., and Keeling, C.D., 1973, *Carbon and the Biosphere* (Proceedings 24th Brookhaven Symposium in Biology), ed. G.M. Woodwell and E.V. Pecan, Technical Information Services, U.S. Atomic Energy Commission, CONF-720510, 51.

Ferguson, C.W., 1970, *Radiocarbon Variations and Absolute Chronology* (Proceedings 12th Nobel Symposium), ed. I.U. Olsson, John Wiley & Sons, New York, 237.

Grey, D.C., 1969, *J. Geophys. Res. 74*, 6333.

_____ and Damon, P.E., 1970, *Scientific Methods in Medieval Archaeology*, ed. R. Berger, University of California Press, Berkeley, 167.

———— , Damon, P.E., and Long, A., 1966, paper presented at Sixth Western National Meeting, American Geophysical Union, Los Angeles, 7-9 Sept.

Houtermans, J.C., 1971, Ph.D. thesis, University of Bern, Switzerland (see appendix for data of H.E. Suess).

———— , Suess, H.E., and Oeschger, H., 1973, *J. Geophys. Res. 78*, 1897.

Lerman, J.C., Mook, W.G., and Vogel, J.C., 1970, *Radiocarbon Variations and Absolute Chronology* (Proceedings 12th Nobel Symposium), ed. I.U. Olsson, John Wiley & Sons, New York, 275.

Libby, W.F., 1955, *Radiocarbon Dating*, 2d ed., University of Chicago Press, Chicago.

Lingenfelter, R.E., 1963, *Rev. Geophys. 1*, 35.

———— and Ramaty, R., 1970, *Radiocarbon Variations and Absolute Chronology* (Proceedings 12th Nobel Symposium), ed. I.U. Olsson, John Wiley & Sons, New York, 513.

Olsson, I.U. (ed.), 1970 *Radiocarbon Variations and Absolute Chronology* (Proceedings 12th Nobel Symposium), John Wiley & Sons, New York.

Rafter, T.A., and Fergusson, G.J., 1957, *Science 126*, 557.

Ralph, E.K., and Michael, H.N., 1970, *Radiocarbon Variations and Absolute Chronology* (Proceedings 12th Nobel Symposium), ed. I.U. Olsson, John Wiley & Sons, New York, 619.

Schove, D.J., 1955, *J. Geophys. Res. 60*, 127.

Stuiver, M., 1961, *ibid. 66*, 273.

———— , 1965, *Science 149*, 533.

———— , 1969, *Radiocarbon 11*, 545.

———— , 1974, paper presented at Symposium on Quaternary Dynamics, Geological Society of America National meeting, Miami Beach, Fla., 18-20 Nov.

———— , 1975, paper presented at First Miami Conference on Isotope Climatology and Paleoclimatology, Key Biscayne, Fla., 16-22 Nov.

Suess, H.E., 1955, *Science 122*, 415.

———— , 1965, *J. Geophys. Res. 70*, 5937.

———— , 1967, *Radioactive Dating and Methods of Low-Level Counting*, International Atmospheric Energy Agency, Vienna, 143.

Waldmeier, M., 1955, *Ergebnisse und Probleme der Sonnenforschung*, 2d ed., Geest und Portig, Leipsig, East Germany.

———— , 1961, *The Sunspot Activity in the Years 1610-1960*, Schulthess, Zurich, Switzerland.

Part VI

SUNSPOTS IN HISTORY AND THEIR EFFECT ON CLIMATE

Editor's Comments
on Papers 29 Through 36

29   HERSCHEL
Excerpts from *Observations tending to investigate the Nature of the Sun, in order to find the Causes or Symptoms of its variable Emmission of Light and Heat; with Remarks on the Use that may possibly be drawn from Solar Observations*

30   JEVONS
The Periodicity of Commercial Crises, and its Physical Explanation

31   SCHOVE
Excerpts from *Solar Cycles and the Spectrum of Time Since 200 B.C.*

32   SCHOVE
Tree Rings and Climatic Chronology

33   SCHOVE
The Biennial Oscillation, Tree Rings and Sunspots

34   SCHOVE
Biennial Oscillations and Solar Cycles, AD 1490–1970

35   SCHOVE
Excerpts from *African Droughts and the Spectrum of Time*

36   SISCOE
Excerpt from *Solar-Terrestrial Influences on Weather and Climate*

Trade Cycles and Revolutions

Sunspots are often blamed by the press for bad weather. Sunspots and weather were associated with each other in Italy in 1651 by Riccioli (1651 and 1653), who recalled the *warm* and *dry* September 1632 when there were *no* spots and the *cold* June 1642 when there were *many*. In England, on the other hand, Herschel (Paper 29) noted that the cold, wet weather of 1795 coincided with an absence of spots and concluded that the sun occasionally had an *indisposition,* the failure to produce spots being the obvious symptom. Using LaLande's 1771 historical references (which have been reprinted, see LaLande, 1966) to what we now term sunspot minima of about 1680, 1687, and 1711, Herschel found that the price of wheat was higher at those periods. These views have indeed been extended today by Eddy (Paper 28), who connects the cold of the seventeenth century as a whole with the Maunder sunspot minimum. King (in pp. 109–125 of Shapley et al., 1975) has noticed that the world wheat production reached peaks in 1958 and 1968, approximately at sunspot maxima, although a simplistic view of the solar-weather relations is no longer acceptable. The concept of the pressure parameter (see Fig. 4.6 in Paper 35 and Fig. 4 in Paper 38) nevertheless leads us to expect that, on the average, in many *temperate* regions the weather should be drier at sunspot maxima as Herschel had supposed, and that in many *low latitude* regions such as India the correlation should be the opposite.

The *trade cycle* was discovered (Clarke, 1838) just before Schwabe (see Paper 7) discovered the *sunspot cycle*. Both cycles were originally thought to be about ten years, a period slightly lengthened later: the trade cycle was placed at ten or eleven (Clarke, 1847) *before* Wolf corrected Schwabe's sunspot period from ten to eleven years. Trade crises were studied back to about 1700 by Jevons (Paper 30, 1878, cf. 1909, p. 195) who slightly revised the list of Clarke (1847)—1793, 1804, 1815, 1826, 1837, and 1847—for Jevons (Paper 30) added 1857 and 1866 but cast doubt on the crisis Hyde had dated 1805. Jevons connected the trade cycle with the sunspot cycle and successfully predicted the crash of 1878. Charts of subsequent business cycles (see Williams in Fairbridge, 1961, Fig. 2, p. 84) have indicated later crises in 1885, 1893 to 1900, 1909, 1921, 1931 to 1934, 1949, and 1958. Jevons found that eighteenth-century trade in India and North America reflected these crises and suspected tropical droughts (known to have occurred at sunspot minima) as the cause (see Paper 35 and Wood, 1977). The connection between the two cycles is usually dismissed as accidental (Sparkes, 1974; Rostow, 1978; and Kindleberger, 1978) although the dates cited show some clustering soon after sunspot minima; the importance of weather naturally increases as we go backward through the centuries.

In the Middle Ages, Jevons argued (1878) that English prices depended on local weather and he found that in the period from 1259 to 1400 prices were consistent with a cycle of 11.1 years, with maxima in years of form $1262 + (11.1)\,n,$ and minima in years of form $1268.2 +$

(11.1) *n,* where *n* is an integer. Jevons had no chi-squared statistical tests and he did not know the dates of medieval sunspot maxima so that he did not have much confidence in that result. In our "remainder" terminology (Papers 11 and 20) this corresponds to 7 and 2 respectively, that is, $X - 1$ and $N + 1$, as we do now know that sunspot maxima and minima at that time had remainders averaging 8 and 1. This suggests high prices at sunspot *maxima,* but further study is needed. In English wheat prices since 1500, Granger and Hughes (1971) could find no evidence of any 11.1-year cycle but a study of European famines (Schove and Dobson, in progress; cf. Paper 34 and Schove, 1974a) shows a significant distribution within the 11-year cycle for years $X - 5$ to $X + 5$: 45, 43, 72, 79, 73, 87, 79, 77, 77, 67, 57. In northeastern Europe the famine peak is later (X to $X + 4$) than in the Mediterranean ($X - 3$ to $X + 2$). In the Tropics famines are more frequent at sunspot minima (N), again as Herschel and Jevons would have expected; King (1975, p. 19) finds that twenty-nine of the sixty-eight major famines since A.D. 1700 came in the years $N - 1$ and $N,$ and these were usually low-latitude famines.

Revolutions since 1500 in Sorokin's list (see Paper 33), 172 total, were put into 11 classes depending on the year in relation to sunspot maximum, from $X - 5$ to $X + 5$. Expressed as percentages the results were: 7, 6, 9, 11, 11, 10, 11, 9, 11, 8, 8. The revolution years 1789, 1830, 1848, 1870, 1906, and 1917 were all, like 1979 (Iran), near sunspot maxima and this association is especially pronounced in weak solar cycles as might be expected from a similar pattern in weak cycles for famines. As soon as a large sample of revolts is available it should be possible to separate different varieties (e.g., peasant and urban) and to make a geographical analysis for each century (see Parker and Smith, 1978, Fig. 1).

Global Pressure Patterns

A crop of papers appears annually on sunspot-weather correlations (see Bandeen and Maran, 1975; Gnevyshev and Ol', 1977) but most well-known claims have been shown to be unsound statistically (Pittock, 1978). Recent papers (e.g., McCormac and Seliga, 1979; Herman and Goldberg, 1978; Markson, 1978) are noted in working documents by Shapley and colleagues (1975, 1977, 1979). Five-year sunspot data are given by Miles and Gildersleeves (1978).

Global pressure maps have been connected with sunspots by many different authors (e.g., Hanslik, 1930, 1931; Willett, 1961, 1969; Sleptsov-Shevelich, 1972; Rubashev, 1978). A common feature of the maps is a resemblance in low latitudes to what I term (Fig. 4 in paper 35) the pressure parameter pattern. Even for individual years such as 1893, 1906, and 1917, this is evident in the anomaly maps of Clayton (1939, and cf. 1923, 1934). Global maps for these years, completed by Schove (unpublished), confirm this. Patterns for the period within two years of sunspot maximum again show similar features.

The months of January and July have been investigated by Parker (Figs. 11 and 12 in our Introduction; cf. Lamb, 1972, Appendix IV, pp. 496–498). His map for January shows significantly low pressure over Australia, Mauritius, and northeastern Brazil and high pressure over New Zealand and Easter Island. However, a new and unexpected feature is the low pressure over Iceland at spot maximum; this negative anomaly at 65° N migrates from 90° W at year $X - 2$ to 30° E (Parker, 1976, Figs. 1a and 1b) one year before the minimum $(N - 1)$. The July maps (Fig. 12 in our Introduction) show unexpected features at 65° S, which may be similarly interpreted.

There is, however, a distinction in high and temperate latitudes between strong and weak cycles shown by Willett (Figs. 13 and 14 in our Introduction), who continued Hanslik's useful studies (1930, 1931) and contrasted the odd (usually strong) and even cycles. The contrast leads to a 22-year weather cycle especially near the zero isopleth in Fig. 4 of Paper 36 and in corresponding seasonal maps. Willett's (1961, 1969) maps exemplify the parameter and area rules of Paper 11. The 22-year cycle in English temperature since 1659 was effectively demonstrated by Bain (see Fig. 15 in our Introduction) who has since proved that in the other months, especially in Spring (Bain, personal communication, using Maximum Entropy Spectral Analysis), there is a similar cycle of about 22 years.

Subcycles and Pressure Parameter

In my principal papers (1949, 1950, 1954, 1962, and 1966) and theses (1953, 1958) on European climatic history I ignored sunspots. When suggesting (1951) that hail was more frequent in the period 1640 to 1670 I did not think of the Maunder Minimum as a possible cause of a cold upper troposphere. In my subsequent papers on solar-terrestrial connections I nevertheless found that:

The magnitude and even the *sign* of the sunspot effect often depends on the precise year and season in relation to the year (X) of sunspot maximum.

Meteorological series since 1749 are thus normally too short to specify the solar effects (but see Schove, 1972) and long series of proxy data (e.g., tree-rings) are necessary.

229

Strong and weak sunspot cycles often cause opposite changes in the
pressure parameter, the geography of this eigenvector explaining
many successes and failures of sunspot forecasting and also the
success of the 22-year cycle near the zero isopleth of the map in
Paper 35 (Fig. 4.6).

Only in the equatorial Indian Ocean do all amplitude classes have
similar effects on barometric pressure.

Strong solar cycles (often 10 years) are associated with 2-year weather
cycles and weak solar cycles (often 12 years) are associated with
3-year weather cycles, but lag effects recur up to 40 years afterward.

The effects of the 80-year cycle are not yet clear but decades of low
and rapidly falling sunspots (1590s, 1690s, 1790s, 1870s, 1920s, and
perhaps 1990s) are often cyclonic with cool summers in northwestern
Europe despite the usual effect of the 11-year cycle.

Papers 31 and 32 come from the 1961 New York symposium
edited by Professor Fairbridge (1961). In Table 1 of Paper 31 the
relative anomalies should be defined as *floating decadal anomalies
relative to encircling 30-year means;* the best glacial evidence now
comes from the Lower Grindelwald glaciers (see Groveman, 1979, for
global changes) of the Alps, where the main glacial maxima are dated
1593 to 1640, 1777 to 1778, and 1820 to 1860 with short advances in
about 1670 and 1720 and with warmer interglacials in about 1686 and
1730 (Zumbühl in Messerli et al., 1978). In the western half of the USA
Harris (Schove and Fairbridge, eds., 1983), combining series studied
separately by LaMarche and Stockton (1974), finds summer temperature minima in 30-year moving means centered at about *698,* about
913, about *1349,* about *1471, 1616* to *1655, 1762,* and *1846* with
shorter cold spells at *1668* to *1682* and *1703* to *1711,* and with intervening 30-year maxima centered at about *977, 1166, 1728,* and *1794.*
The chart (Fig. 19 in our Introduction) supplied by King shows that
using periods of 10 to 70 years there is some positive correlation
between sunspots and temperature.

Dendroclimatology, considered in Paper 32, has made great
strides since 1961 by Fritts (see Fritts, 1976), LaMarche (see LaMarche
and Stockton, 1977), and others at Tucson, Arizona. Although the
tropical results in Paper 32 seem valid, other climatic connections
with the 11-year cycle have not been generally accepted. The USA
climatic pattern (Paper 34) nevertheless shows a 22-year cycle (see
Fig. 16 in our Introduction) not appreciated when I wrote the article; in
the western plains dry periods at about 1780, about 1800, 1840s, 1860,
1880, 1900, 1930s, 1950s, and 1970s are confirmed by current research.

Thirty-year moving barometric means were studied further in
1962 (Figs. 17 and 18 in our Introduction and cf. Paper 32). In the
period 1876 to 1905 (designated as c. *1890*) sunspots were at a

minimum and in the period 1906 to 1935 (designated as c. *1920*) they were relatively strong. The trends in pressure and rainfall suggest possible solar effects; if genuinely solar, these trends should have been resumed in the sunspot-rich period 1951 to 1980, although on the other hand, the lull in sunspots in the 1960s and the tropical droughts of 1963 to 1972 confirm the predictions made (Schove, 1961, p. 255). Tree-ring studies over long periods are now needed.

Indian Ocean pressure data, reversed to yield Southern Oscillation (S.O.) and Pressure Parameter (PP) indices (Wright, 1975), have been compared with sunspot data (Schove, 1980; Paper 35, Fig. 4.6). The S.O. anomaly by half-years (in hundredths of a mb) from the summer of $X - 5$ to the winter of $X + 5\frac{1}{2}$ (where X is the year of sunspot maxima) is thus: 0, +13, +5, +12, −16, −20, −11, +3, +9, +17, +7 (sunspot maximum summer), +5, +7, +3, +1, +17, +27, +5, −13, −5, +3, −2 (following winters for the same 11 years, i.e., $X - 4\frac{1}{2}$ to $X + 5\frac{1}{2}$). In the *equatorial* Indian Ocean the pattern is similar for *different* amplitude classes, and the solar cycle divides into three pressure cycles of 3 to 4 years in that region. In Ethiopia, Wood (1977) found that droughts since A.D. 1066 have usually occurred near sunspot minima (year $N - 2$ to $N + 2$).

In Fairbridge, ed. (1961, p. 185), Brier explains that the harmonic dial method is of special value since it can be used to illustrate differences between odd and even, or weak and strong solar cycles. Papers 33 and 34, inspired by the work of Berson and Kulkarni (1968), link the quasibiennial oscillation to sunspots (cf. Lamb, 1972, p. 210). Computer studies (made by the Meteorological Office, Bracknell) reveal what I interpret as a north-south wobble of the Upper High biennially at sunspot maximum (see Fig. 4 in Paper 38).

Ol's charts (1969) can be used to show that in the Alps, for example, pressure is higher in the odd years such as $X + 1$ than in the even years such as $X + 2$. Pressure rises north of the Indian Ocean in Siberia and in southeastern Asia but falls in the region of the Davis Strait and north Pacific troughs. The southward "nods" of the Upper High are illustrated unintentionally even in Pittock's (1978, Fig. 8a) analysis of the seasonal swing of the latitude of the surface high-pressure belt in eastern Australia (with maxima $X, X + 2, X + 4, X + 6$). Many other curious relationships, often with long lags, can be found—thus, for example, the triennial cycles in the USA low-level trees seem to lose phase by one year in twenty-six. After a strong solar cycle such as the present, droughts in the western USA seem more likely in years $X + 1$, $X + 4$ and $X + 8$ or 9 (i.e., 1988); similarly we can predict that, statistically, counting X as 1979, two-year pulses are slightly more likely about 1983 to 1984 in Ethiopian rainfall, about 1988 and 2000 in USA winter-sensitive trees of the western USA, and about 1990 to about 1996 in upper timberline summer-sensitive trees

231

(see also Paper 39). The effects in English summer rainfall and temperatures are shown by an oddness index (Schove, 1972, pp. 352–353; 1974*b*, p. 171).

The explanation of the 2.2-year cycle is still obscure; its mean value is about 2.17 and thus not exactly one-fifth of the *average* sunspot cycle. Nevertheless, the stronger cycles are shorter than the average of 11.1 years and they may be more effective.

Harmonics of the solar cycle may exist. We should thus expect 5-year, 3.3-year and 2.5-year cycles to be associated with runs of 10-year solar cycles. Tabony's (1979, Fig. 6) chart of the varying amplitude of the 5-year cycle in Wales and England's rainfall since 1727 shows peaks near the 1760s, 1820s, 1870, and 1925 to 1965, and it so happens that these were generally periods when the sunspot cycle was 10 years and not 11 years. However, a global study is needed.

African fluctuations are considered in Paper 35 and the excerpts in Paper 38 (p. 458) include the map of the first eigenvector of pressure changes by 5-year periods—the pressure parameter map. A corresponding temperature chart (using information from a manuscript left by the late Dr. G. S. Callendar to Schove) shows that the pressure fall in the Indian Ocean coincides with a cooling throughout the Tropics except for a warm crescent extending northwest from New Zealand where the temperature changes are controlled by the 5-year pressure parameter changes. Temperature is lowered also in the Davis Strait, the west coast of North America, and in temperate South America (see p. 608 in Paper 32); it is warmer in northern and eastern Europe and especially in Newfoundland.

A pressure map for the summer season (unpublished) shows a rise in the western Mediterranean and in southeastern Asia, but a fall in northern Europe especially in northern Norway; an *upper ridge* normally lies over Scandinavia in the late summer season (July to September) and this is eroded like the Upper High.

Nile floods over the period A.D. 641 to 1522 showed only weak cycles about 11 years and 22 years (22.12 as estimated by Brooks, 1928); the latter cycle is perhaps due to the amplitude effect shown in Fig. 4.8. of Paper 35. The effects of the 22-year cycle have been considered by Ol' (1969; see also Rubashev, 1978), but the pressure pattern change when all seasons are combined is very weak (see Miles, 1974).

The 22-Year Cycle

For meteorological comparisons it is convenient to plot alternate sunspot cycles with a negative sign (Anderson, 1939; Bracewell, 1953; Dewey, 1966; Stetson, 1947, p. 155).

Important papers on the terrestrial effects of the Hale cycle appeared in 1979 (see Svenonius and Olausson, 1979). At present the papers have three separate interpretations, which are described here— *solar, magnetic,* and *planetary.* The introductory papers of Dicke (1978, 1979*a*, 1979*b*) explains a 22-year rhythm in heavy *hydrogen* in the ice layers as the climatic response to a regular solar rhythm reaching us from a layer deeper than that of the irregular sunspot cycles. The high *remainders* or lateness of weak cycles (see Paper 20, p. 139) in the arrival of sunspots on the sun's surface from such a deep layer may (I suggest) reflect a weak solar vertical wind. Once the precise length of this cycle is known the numbering of the cycles in Papers 20 and 26 can be checked, which imply back to A.D. 300 a double cycle of 22.17 years.

A 22-year cycle in North American droughts near the Hale minima is clearly demonstrated by Mitchell and his colleagues (1979) and this can be traced back to A.D. 1600 as Siscoe (Paper 36) explains. Moreover, as its amplitude varies with the amplitude of the solar cycle (see Fig. 16 in our Introduction), a direct magnetic effect has been suspected. In his study of famines in Ethiopia, Wood (1977) does not mention the 22-year cycle but his destructive droughts are listed as 1543, 1561, 1890, 1913, 1957, and about 1974, curiously similar to the Hale minima in the American sense.

In the Greenland ice-cores, summer temperature is reflected in the *melt features,* which at the *Dye* 3 site show (Hibler and Langway, 1977) some response to an 11-year cycle. Annual temperatures, reflected in the Oxygen-18 δ-index, show very clear evidence of a cycle of 20 years. In the period A.D. 1244 to 1971 Hibler and Johnsen (1979) confirmed this cycle for the mean of three sites. They also found a cycle of 7½ years (22½÷3) and they wondered whether the 20-year cycle (Mock and Hibler, 1976) could be the effect of Jupiter-Saturn conjunctions on the center of gravity of the sun in the solar system. Such planetary conjunctions have led to a solution of Maya chronology (Schove, 1977), but they have not been used before for meteorological chronology and further investigations are in progress (see Schove and Fairbridge, 1983).

The number of 22-year cycles with instrumental meteorological data is too few either for satisfactory barometric comparisons (see Ol', 1969; Miles, 1974; Hancock and Yarger, 1979; Wagner, 1971) or for a calculation of its precise length.

Longer Cycles

The meteorological effects (Paper 38) of *Medium* (80-year) and *Long* (200-year) cycles are still uncertain and difficult to distinguish.

Chinese evidence is inconclusive (see Paper 33). The Nile Flood data and various series of tree-rings suggest (see Schove 1980, pp. 87–88) that there is a barometric superoscillation between the Indian and Pacific Oceans at 30-year dates centered as follows: *818, 865, 925, 970, 1045, 1092, 1220, 1350, 1380, 1488, 1560, 1615, 1635, 1720, 1790, 1839, 1920,* and *1955* for the Indian Ocean low and about *773, 841, 942, 1022, 1065, 1210, 1235, 1273, 1308, 1400, 1460, 1650 to 1665, 1750,* and *1890* for the Indian Ocean high. Some dates in the second set fit periods of weak sunspot activity but the solar connection is again unreliable. Tentative charts of relationships between sunspots and weather since 3500 B.C. given in *Sun and Climate* (Schove 1980, pp. 95–100), should not be taken too seriously.

On longer time scales likewise (see Schove, 1980, especially on Fig. 1 on p. 99) we can recognize Indian and Pacific rainfall oscillations such as on short-time scales (Takahashi and Yoshino, 1980) can be due to solar influence. The Indian Ocean was thus wet about 9600 to 9000 b.p. and dry about 8300 to 8000 b.p. On the true B.C. time scale it was wet about the sixty-first, fifty-first, forty-fourth/forty-first, and twenty-seventh centuries B.C., but dry about 7000 B.C. and 2300 B.C., and fairly dry in the fourteenth and fifth centuries B.C. When these phases are more precisely dated it will be possible to compare them with the solar excursions tabulated in the Conclusion.

Longer cycles (from Pisias et al., 1973) in the range of 400 to 3000 years are discussed in Paper 39 but they are generally uncertain. Others are discussed in Schove and Fairbridge (1983 and cf. Anderson, 1939). Cycles over 7000 years are real (see Appendix D) and determine the main climatic changes of the Quaternary (Imbrie and Imbrie, 1979), but these arise from the dynamics and geometry of the earth-sun relationships and there is no need to involve solar cycles.

Other Phenomena

Comets and meteors are supposed to vary in brightness with the solar cycle but quantified comet data (Schove, 1975) do not yield a significant correlation. Earthquakes and the sunspot cycle were incorrectly associated with each other by Davison (1938), although new studies might now be made; certainly his list of earthquakes in England since A.D. 975 (Davison, 1924) does show some agreement with the maxima dates in Paper 20.

Epidemics often occur in cycles and, in particular, plague tends to follow a cycle of eleven years. Such plague cycles recurred in the West in the period A.D. 540 to 750 and A.D. 1347 to about 1840 (Biraben, 1975; and cf. Schove, 1974a). However, the phase relation between sunspots and the plague is not constant. Polio pandemics in the twentieth century have occurred near auroral maxima, but the history of polio is too short for a significant test and the same applies to other epidemics often discussed (e.g., Gnevyshev and Ol', 1977).

Locusts respond to the pressure parameter and thus there appears to be a real relationship with sunspots judging from a list of medieval locust outbreaks in China and modern outbreaks in Africa.

I have always assumed that any sunspot correlations found with famines, revolutions, epidemics, or other biological cycles would be associated only indirectly with the sun, the weather being the intermediate factor. The hypothesis of direct effects of radiation has, however, been suggested by Tchijevsky (1971) and some other writers, and further investigations are desirable.

REFERENCES

Anderson, C. N., 1939, A Representation of the Sunspot Cycle, *Terrestrial Magnetism and Atmospheric Electricity* **44:**175-179. (The 22-year cycle.)

Bandeen, W. R., and S. P. Maran, eds., 1975, Possible Relationships Between Solar Activity and Meteorological Phenomena, *NASA Spec. Pub. 366, Goddard Space Flight Center, Greenbelt, Maryland,* 443p.

Berson, F. A., and R. N. Kulkarni, 1968, Sunspot Cycles and the Quasibiennial Stratospheric Oscillation, *Nature* **217:**1133-1134.

Biraben, J-N., 1975, *Les hommes et la peste en France et dans les pays européens et mediterranéens,* La peste dans l'histoire, vol. 1, Mouton, Paris, 455p.

Bracewell, R. N., 1953, The Sunspot Number Series, *Nature* **171:**649-650. (The 22-year cycle.)

Brooks, C. E. P., 1928, Periodicities in the Nile Floods, *Royal Meteorol. Soc. Mem.* **2:**9-26. (The leading authority on solar-weather relationships, 1920-1950.)

Clarke, H., 1838, On the Political Economy and Capital of Joint Stock Banks, *Railway Mag.* **27:**288-293.

Clarke, H., 1847, *Railway Mag.*

Clayton, H. H., 1923, *World Weather,* Macmillan, New York, 263p.

Clayton, H. H., 1934, World Weather and Solar Activity, *Smithsonian Misc. Colln.* **89:**1-51.

Clayton, H. H., 1939, The Sunspot Record, *Smithsonian Misc. Colln.* **98:**1-19.

Davison, C., 1924, *History of British Earthquakes,* Cambridge University Press, Cambridge, England, 416p.

Davison, C., 1938, *Studies in the Periodicity of Earthquakes,* Murby, London, 108p.

Dewey, E. R., 1966, The 22-year Cycle in Sunspot Numbers with Alternate Cycles Reversed, *Cycles* **17:**221-225.

Dicke, R. H., 1978, Is There a Chronometer Hidden Deep in the Sun? *Nature* **276:**676-680.

Dicke, R. H., 1979a, The Clock Inside the Sun, *New Scientist* **83:**12-13.

Dicke, R. H., 1979*b,* Solar Luminosity and the Sunspot Cycle, *Nature* **280:**24–27.

Fairbridge, R. W., ed., 1961, *Solar Variations, Climatic Change and Related Geophysical Problems,* New York Academy of Science Annals, vol. 95(1), New York Academy of Science, 740p.

Fritts, H. C., 1976, *Tree-rings and Climate,* Academic, London, 562p.

Gnevyshev, M. N., and A. I. Ol', eds., 1977, *Effects of Solar Activity on the Earth's Atmosphere and Biosphere,* Israel Program for Scientific Translations, Jerusalem, 290p.

Granger, C. W. J., and A. D. Hughes, 1971, A New Look at Some Old Data. The Beveridge Wheat Price Series, *Royal Statistical Soc. Jour.,* ser. A, **134:**413–428.

Groveman, B. S., 1979, *Reconstruction of Northern Hemisphere Temperature 1579–1880,* M. S. thesis, University of Maryland, College Park, 78p.

Hancock, D. J., and D. N. Yarger, 1979, Cross-spectral Analysis of Sunspots and Monthly Mean Temperature and Precipitation for the Contiguous United States, *Jour. Atmos. Sci.* **36:**746–753.

Hanslik, S., 1930, The Effect of the Sunspot Periods on Air Pressure, *Gerlands Beitr. Geophysik* **28:**114–125. (In German.)

Hanslik, S., 1931, The Effect of the Sunspot Periods on Air Pressure, *Gerlands Beitr. Geophysik* **29:**138–155. (In German.)

Herman, J. R., and R. A. Goldberg, 1978, *Sun, Weather and Climate,* NASA SP-426, Washington, 372p. (Popular report.)

Hibler, W. D., and S. Johnsen, 1979, The 20-year Cycle in Greenland Ice-core Records, *Nature* **280:**481–483.

Hibler, W. D., and C. C. Langway, Jr., 1977, *Polar Oceans,* M. J. Dunbar, ed., Arctic Institute of North America, Calgary, Canada, pp. 589–602.

Imbrie, J., and K. P. Imbrie, 1979, *Ice Ages. Solving the Mystery,* Macmillan, London, 224p. (Long astronomical cycles.)

Jevons, W. S., 1878, Commercial Crises and Sun-spots, *Nature* **19:**33–37.

Jevons, W. S., 1909, *Investigations in Currency and Finance,* Macmillan, London, 347p.

Kindleberger, C. P., 1978, *Manias, Panics and Crashes,* Macmillan, London, 271p.

King, J. W., 1975, Sun-Weather Relationships, *Aeronatique et l'Astronautique* **13:**10–19.

LaLande, J., 1966, *L'Astronomie,* vol. 3, Johnson Reprint Corporation, New York, pp. 286–287.

LaMarche, V. C., Jr., and C. W. Stockton, 1977, Chronologies from Temperature-Sensitive Bristlecone Pines at Upper Treeline in the Western United States, *Tree-Ring Bull.* **34:**21–45.

Lamb, H. H., 1972, *Climate: Present, Past and Future,* vol. 1, Methuen, London, 613p.

McCormac, B. M., and T. A. Seliga, eds., 1979, *Solar-Terrestrial Influences on Weather and Climate,* Reidel, Dordrecht, Holland, 340p.

Markson, R., 1978, Solar Modulation of Atmospheric Electrification and Possible Implications for the Sun-Weather Relationship, *Nature* **273:**103–109.

Messerli, B., P. Messerli, C. Pfister, and H. T. Zumbühl, 1978, Fluctuations of Climate and Glaciers in the Bernese Oberland, Switzerland, and Their Geological Signficance, 1600 to 1975, *Arctic and Alpine Research* **10:**247–260.

Miles, M. K., 1974, The Variation of Annual Mean Surface Pressure Over the

Northern Hemisphere During the Double Sunspot Cycle, *Meteorol. Mag.* **103:**93–99. (Pressure in even maxima is higher in zone 20°–35°.)

Miles, M. K., and P. B. Gildersleeves, 1978, A Statistical Study of the Likely Influence of Some Causative Factors on Temperature Changes Since 1665, *Meteorol. Mag.* **107:**193–204. (Sunspot means by 5-year periods.)

Mitchell, J. M., C. W. Stockton, and D. M. Meko, 1979, Evidence of a 22-year Rhythm of Drought in the Western United States Related to the Half Solar Cycle Since the 17th Century, in *Solar Terrestrial Influences on Weather and Climate,* B. M. McCormac and T. A. Seliga, eds., Reidel, Dordrecht, Holland, pp. 125–143.

Mock, S. J., and D. J. Hibler III, 1976, The Twenty-year Oscillation in Eastern North America Temperature Records, *Nature* **261:**484–486.

Ol', A. I., 1969, Manifestation of the 22-year Solar Activity Cycle in the Earth's Climate, *Arkt. Antarkt. Nauč̌ Issled Inst. Leningrad* **289:**116–131.

Parker, B. N., 1976, Global Pressure Variation and the 11-year Solar Cycle, *Meteorol. Mag.* **105:**33–44.

Parker, G., and L. M. Smith, 1978, *The General Crisis of the Seventeenth Century,* Routledge & Kegan Paul, London, 283p.

Pisias, N. G., J. P. Dauphin, and C. Sancetta, 1973, Spectral Analysis of Late Pleistocene-Holocene Sediments, *Quaternary Research* **3:**3–9.

Pittock, A. B., 1978, A Critical Look at Long-term Sun-Weather Relationships, *Reviews Geophysics and Space Physics* **16:**400–420.

Riccioli, G. P., 1651 and 1653, *Almagestum novum astronomiam veterem novamque complectens, observationibus aliorium et propis, novisque theorematibus problematibus ac tabulis promotam,* vols. 1 and 2, Bologna, Italy.

Rostow, W. W., 1978, *The World Economy: History and Prospect,* Macmillan, London, 833p.

Rubashev, B. M., 1978, *Effects of Long Period Solar Activity Fluctuation on Temperature and Pressure of the Terrestrial Atmosphere,* NASA TM75345, Washington, D.C., 25p.

Schove, D. J., 1949, European Raininess and European Temperatures, A.D. 1500–1950, *Royal Meteorol. Soc. Quart. Jour.* **75:**175–179, 181.

Schove, D. J., 1950, The Climatic Fluctuations Since A.D. 1850 in Europe and the Atlantic, *Royal Meteorol. Soc. Quart. Jour.* **76:**147–165.

Schove, D. J., 1951, Hail in History, A.D. 1630–1680, *Weather* **6:**17–21.

Schove, D. J., 1953, *Climatic Fluctuations in Europe in the Late Historical Period (especially A.D. 800–1700),* M.Sc. thesis in geography, University of London, 397p.

Schove, D. J., 1954, Summer Temperatures and Tree-rings in North Scandinavia, A.D. 1461–1950, *Geog. Annaler* **36:**40–80.

Schove, D. J., 1958, *The Preliminary Reduction of Wind and Pressure Observations in N. W. Europe, 1648–1955,* Ph.D. thesis, University of London, 305p.

Schove, D. J., 1961, Major Pressure Oscillation 1875/1960, *Geofisica Pura e Aplplicata* **49:**255–263.

Schove, D. J., 1962, The Reduction of Annual Winds in N. W. Europe, A.D. 1635–1960, *Geog. Annaler* **44:**303–327.

Schove, D. J., 1966, Fire and Drought, A.D. 1600–1700, *Weather* **21:**311–314.

Schove, D. J., 1972, The Biennial and Triennial Cycles, Varve Cycles, Solar Cycles and Terrestrial Oscillations, *Jour. Interdisciplinary Cycle Research* **3:**349–354, 361–363, 409–411.

Schove, D. J., 1974a, Chronology and Historical Geography of Famine,

Plague and Other Pandemics, *Internat. Congress History and Medicine* (*23rd*) *Proc.*, pp. 1265–1272.

Schove, D. J., 1974*b*, Dendrochronological Dating of Oak from Old Windsor, Berkshire, A.D. 650–906, *Medieval Archaeology* **18:**165–172. (Dates should be 231 years earlier.)

Schove, D. J., 1975, Comet Chronology in Numbers A.D. 200–1802, *British Astron. Assoc. Jour.* **85:**401–407.

Schove, D. J., 1977, Maya Dates A.D. 352–1296, *Nature* **268:**670. (*Cf.* earlier erroneous result in *Nature* **261:**471–473.)

Schove, D. J., 1980, The 200-, 22-, and 11-year Cycles and Long Series of Climatic Data, Mainly Since A.D. 200, in *Sun and Climate,* Centre National d'Etudes Spatiales, Toulouse, pp. 87–100.

Schove, D. J., and M. Dobson, in progress, The Famine Project of the Spectrum of Time, A.D. 1–1980, in *Famine in History Colloquium.*

Schove, D. J., and R. W. Fairbridge, eds., 1983, *Ice-cores, Varves and Tree-rings,* Balkema, London.

Shapley, A. H., and H. W. Kroehl, 1977, *Solar-Terrestrial Physics and Meteorology: Working Document II,* Special Committee for Solar-Terrestrial Physics, Washington, D.C. 146p.

Shapley, A. H., H. W. Kroehl, and J. H. Allen, eds., 1975, *Solar-Terrestrial Physics and Meteorology: Working Document I,* Special Committee for Solar-Terrestrial Physics, Washington, D.C., 142p.

Shapley, A. H., C. D. Ellyett, and H. W. Kroehl, 1979, *Solar-Terrestrial Physics and Meteorology: Working Document III,* Special Committee for Solar Terrestrial Physics, Washington, D.C., 132p.

Sleptsov-Shevlevich, B. A., 1972, Manifestation of Solar Activity in Long-period Pressure Field Variations in the Northern Hemisphere of the Earth, *Geomagnetism and Aeronomy* **12:**285–287.

Sparkes, J. R., 1974, Sunspots and the Business Cycle, *Nature,* **252:**520.

Stetson, H. T., 1947, *Sunspots in Action,* Ronald Press, New York, 252p. (Popular book.)

Svenonius, B., and E. Olausson, 1979, Ring Widths of Trees, Solar Activity, and Weather Conditions in Sweden in the Period 1756–1975, *Geologiska Föreningens i Stockholm Forhandlinger* **100:**95–100.

Tabony, R. C., 1979, A Spectral Filter Analysis of Long-Period Rainfall Records in England and Wales, *Meteorol. Mag.* **108:**97–118.

Takahashi, K., and M. M. Yoshino, eds., 1980, *Climate Change and Food Production,* University of Tokyo Press, Tokyo.

Tchijevsky, A. L., 1971, Physical Factors of the Historical Process, *Cycles* **22:**11–27. (Other works available in the British Museum, London.)

Wagner, A. J., 1971, Long-period Variations in Seasonal Sea-level Pressure Over the Northern Hemisphere, *Monthly Weather Rev.* **99:**49–66.

Willett, H. C., 1961, The Pattern of Solar Climatic Relationships. The Double Solar Cycle, *N.Y. Acad. Sci. Annals* **95:**89–106.

Willett, H. C., 1969, *Geog. Annaler* **31:**301–307.

Wood, C. A., 1977, Preliminary Chronology of Ethiopian Droughts, in *Drought in Africa 2,* D. Dalby, R. J. Harrison-Church, and F. Bezzaz, eds., International African Institute, London, pp. 68–73.

Wright, P. B., 1975, *An Index of the Southern Oscillation,* Climatic Research Unit, University of East Anglia, Norwich, England, 22p.

29

Reprinted from pages 306, 310, and 315 of *Royal Soc. London Philos. Trans.*
91:265–318 (1801)

XIII. Observations tending to investigate the Nature of the Sun, in order to find the Causes or Symptoms of its variable Emission of Light and Heat; with Remarks on the Use that may possibly be drawn from Solar Observations. By William Herschel, *L.L.D. F.R.S.*

[*Editor's Note:* Introductory material has been omitted.]

SIGNS OF SCARCITY OF LUMINOUS MATTER IN THE SUN.

Visible Deficiency of empyreal Clouds.

July 5, 1795. 1ʰ 6′. The appearance of the sun is very different from what I have ever seen it before. There is not a single opening in the whole disk; there are no ridges or nodules; there are no corrugations.

A perfect Calm in the upper Regions of solar Clouds.

Dec 9, 1798. 12ʰ 33′. The sun has no openings of any kind; nor can I perceive any places that look disturbed, like those where openings have lately been.

[*Editor's Note:* The excerpt above illustrates the prolonged sunspot minimum now dated 1798.3. The next excerpt describes some openings or spots of the new cycle in 1801. The Wolf formula $10g + f$ would lead to a number $10(2) + (10 + 2 + 4) = 36$.]

March 2, 1801. There are six different sets of openings in the sun. One of them consists of ten; another of two; the rest are single.

[*Editor's Note:* Material has been omitted at this point.]

From these two last sets of observations, one of which establishes the scarcity of the luminous clouds, while the other shews their great abundance, I think we may reasonably conclude, that there must be a manifest difference in the emission of light and heat from the sun. It appears to me, if I may be permitted the metaphor, that our sun has for some time past been labouring under an indisposition, from which it is now in a fair way of recovering.

[*Editor's Note:* Herschel considers prices in earlier periods in which sunspots were either frequent or infrequent. The next excerpt indicates his approach.]

The fourth period on record, is from the year 1695 to 1700; in which time no spot could be found in the sun. This makes a period of 5 years; for, in 1700 the spots were seen again. The average price of wheat, in these years, was £3. 3s. 3⅕d. the quarter. The 5 preceding years, from 1690 to 1694, give £2. 9s. 4⅘d. and the 5 following years, from 1700 to 1704, give £1. 17s. 11⅕d. These differences are both very considerable; the last is not less than 5 to 3.

[*Editor's Note:* Material has been omitted at this point.]

30

Reprinted from *British Assoc. Adv. Sci. Rep.*, pp. 666–667, 1878

THE PERIODICITY OF COMMERCIAL CRISES, AND ITS PHYSICAL EXPLANATION

W. S. Jevons

In this paper the author took up again the inquiry into the relation between the solar spot period and commercial phenomena, which he had previously treated, but, as he now believes, unsuccessfully, at the Bristol Meeting of the British Association (see Report for 1875, Transactions of Sections, p. 217). Observing that commercial collapses have occurred in 1866, 1857, 1847, 1837 (in United States), 1825, and 1815, the author adopts the opinion of Dr. Hyde Clarke and others, that such phenomena tend to recur in a period of about ten years.

He then points out that the series, although apparently broken (a crisis occurring in 1809–10, rather than in 1805, when there was no great increase of bankruptcy), may probably be traced back through the 18th century. There were great crises in the years 1793 and 1763, and very distinct ones also in 1772–3 and 1783. The principal part of the paper was occupied with an attempt to show that the great crisis of 1721, which followed the South Sea Bubble, falls into the series, as it was preceded by "stock jobbing" manias about the years 1701 & 1711, and was followed by one in 1732–33. There remains, in order to complete the whole series, two periods, 1742 and 1752, when there were certainly no great manias, or crises. But the author adduces evidence of a certain weight to show that in those years there were remarkable rises in the price of wool, and about 1752–53, there was a marked rise in the price of tin, which forms one of the best indications of commercial activity and good credit. He believes, then, that trade reached maxima of activity in or about the years 1701, 1711, 1721, 1732, 1742, 1753, 1763, 1772, 1783, 1793 (1805 ?), 1815, 1825, 1837, 1847, 1857, 1866. The intervals vary from 9 to 12 years; the average interval is 10·3 years. But, as the earlier dates, 1701 and 1711, are not well established, and the panic of 1866 was probably premature, as shown by comparison with the three preceding panics, the author prefers to compare the undoubted collapse of 1721 with that of 1857, giving a mean interval of 10·46 years. If we take the year 1763, which was also a year of well marked crisis, and compare it with 1857, we obtain 10·444. The mean length of the commercial period is thus almost identical with Mr. J. A. Broun's estimate of the sunspot period, namely, 10·45 years, which again is almost exactly the same as Lamont's earlier estimate.

The author infers that the phenomena are probably connected; but, as he and other inquirers have failed in discovering a like cycle in the price of corn, he believes that the connection must be sought rather in the varying produce of the crops in the tropical or other countries, where the meteorological influence of the solar period has already been detected. In corroboration of this opinion he points out that the exports of goods from England to India, when graphically represented, display a certain appearance of decennial variation, especially in the earlier part of the 18th century.

In conclusion he alludes to the contemporary state of trade, quoting a statement of the numbers of bankruptcies in England and the United States, to show that the latest crisis must be assigned either to 1877, or, it may be, to 1878. In either case the theory of decennial variation will be verified as closely as can be expected.

[*Editor's Note:* The crash came as predicted in 1877 to 1878.]

31

Reprinted from pages 107, 109, 110, 115–118, 119–120, 122–123 of *New York Acad. Sci. Ann.* **95**:107–123 (1961)

SOLAR CYCLES AND THE SPECTRUM OF TIME SINCE 200 B.C.

Derek J. Schove

St. David's College, Beckenham, Kent, England

Introduction

The so-called Spectrum of Time project arose from appeals at international conferences of various kinds (meteorological, geographical, archaeological, and historical) for information about datable climatic and meteorological events. The items received were cross-dated and diagnosed, and those relating to sunspots and the auroras were singled out for a separate study of the sunspot cycles in the historical period. The method was described in various journals (Schove, 1947, 1951*e*, 1950*d*, and 1951*c*), and the detailed results have been tabulated in the *Journal of Geophysical Research* (Schove, 1955*b*). The meteorological information was likewise separated in order to survey the climatic fluctuations of the past 2000 years. Although it was essential to keep these two investigations apart—in order to avoid even unconscious bias—some of the main results will be summarized here. In my two papers included in this monograph, the detailed evidence that has been or is being described elsewhere will be omitted, and the emphasis will be on oscillations and on decadal data of potential significance in the American continent.

[*Editor's Note:* Material dealing with climatic fluctuations (e.g., the south-steering rule) that are not sunspot related has been omitted.]

Climatic Fluctuations in Europe since 1100 A.D.

The Spectrum of Time information has already proved useful for the study of historical meteorology both in Europe and in Asia. In Europe it is now possible to estimate changes of pressure and pressure gradient through the centuries (Schove, 1961, 1953*e* and *f*). From 1800 to 1950 the barometric evidence is sufficient (Schove, 1958, 1961) to determine the pressure (top curve of FIGURE 2), at a fixed central position ("The Wash") in eastern England, and to determine also the changes in the strength (central curve) and direction (lower curve) of the prevailing wind.

The pressure chart for the modern period shows maxima about 1830, 1860, 1900, and 1945, as if the barometer were in step with the mathematical series 3, 6, 10, and 15. These pressure peaks were studied geographically (FIGURE 3), and were found to have moved from north to south according to the principle of south steering (Schove, 1950*a* and *b*). Temperature patterns by 30-year periods moved southward in the same way (FIGURE 4), the cold phase (T–min.) preceding the pressure peak (P–max.) by about a decade (see FIGURE 5).

Terms such as P–min. and T–min. will be used again in the course of this paper; the following typical definitions are therefore listed here for convenience: S–max., 30-year period of maximum sunspot/auroral activity (dated by the 15th year); t–max. (cf. p–max., r–max.), 10-year period of maximum temperature (dated by the 5th year); and f(R)–max. (cf. f(T)–max.), 30-year period of maximum tree growth, where rainfall is the presumed major factor.

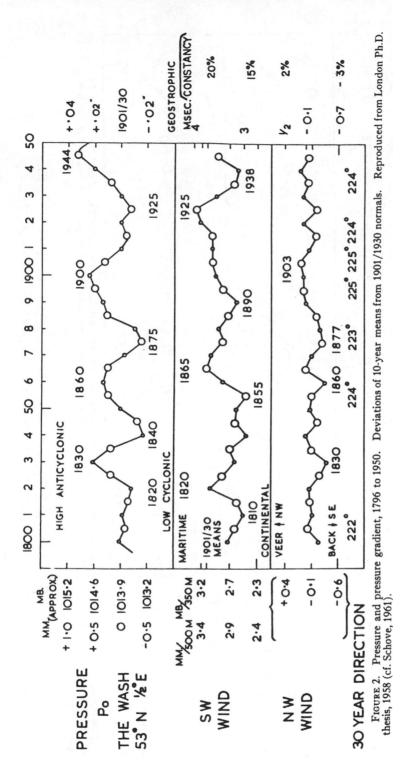

FIGURE 2. Pressure and pressure gradient, 1796 to 1950. Deviations of 10-year means from 1901/1930 normals. Reproduced from London Ph.D. thesis, 1958 (cf. Schove, 1961).

[*Editor's Note:* Figures 1, 3, 4, 5, and 8, and Table 2 are not included in these excerpts. In England principal decadal r-min are dated c. 1804, c. 1858–1859, c. 1891 and 1897, and c. 1944. Principal r-max are dated c. 1825–1827, c. 1843–1847, c. 1876–1878, and c. 1926–1927. There is a relative r-min at c. 1833 corresponding to the p-max of c. 1830.]

243

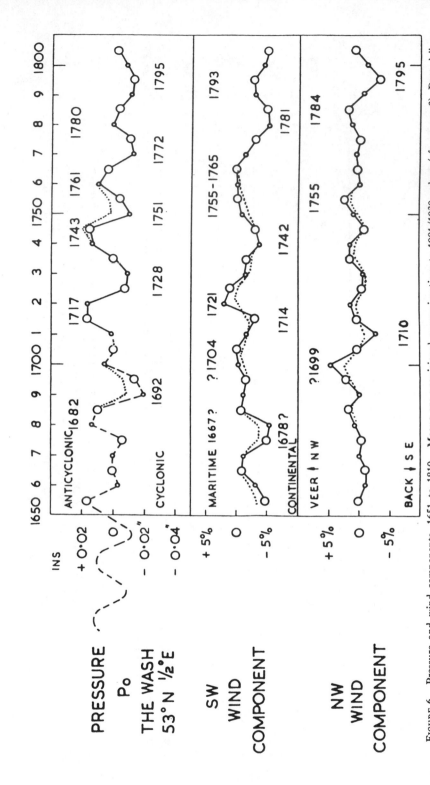

FIGURE 6. Pressure and wind components, 1651 to 1810. Means are provisional approximations to 1901/1930 values (cf. FIGURE 2). Dotted lines illustrate plausible alternatives based on the evidence of the weather patterns. Reproduced from London, Ph.D. thesis, 1958 (cf. Schove, 1961).

[*Editor's Note:* In England the principal r-min are dated c. 1741.5, c. 1783, and c. 1804. The principal r-max are dated c. 1771–1772 and c. 1793.]

TABLE 1

Decades*	Auroral numbers†	American tree rings	Winter temperature, Europe (F)	Pressure, Europe (C)
1100s	50	Wet, Snake; dry, Colorado	(+5)	−1
1111/20	70	Wet both areas	(−2)	+1
1120	80	Still wet, Colorado	(−3)	−1
1130	80		+3	+3
1140	40	Long dry, Colorado, 1131 to 1161	−7	−3
1150s	50	Long dry, Southern California 1140 to 1190	+1	−1
1160	45	Dry Snake area 1157 to 1180	+1	0
1170	65		+2	+2
1180	60	Wet phase begins generally	+10	+2
1190	50	Wet	+6	−3
1200s	70	Wet	−2	+4
1210	50	Wet phase ending	−7	−1
1220	60	Dry, Colorado; Wet, Snake and Missouri	(+1)	−4
1230	30		(−1)	−3
1240	50	Dry phase spreads south	+6	+5
1250s	50	Droughts reach Rio Grande (Smiley, this monograph)	(+1)	−3
1260	45		+6	+5
1270	45		+4	−4
1280	40	Drought ends, Missouri	+9	+2
1290	30	Last decade of drought elsewhere	+3	+4
1300s	60	Long wet phase begins, Colorado	−1	0
1310	40		−3	−5
1320	45	Wet spreads to Snake, Mo., Southern California	+1	+1
1330	45	Wet everywhere	+1	+1
1340	30	Drier Colorado	+7	−3

Decades*	Auroral numbers†	N.W.	S.W.	N.E.	S.E.	Mex.	Winter temperature, Europe (F)	Pressure, Europe (C)
1350s	50	Drier Snake also.					0	+2
1360	60	dr‡	d	·	·		+5	−2
1370	80	·	r	·			+4	+1
1380	65	·	r		r		+2	+1
1390	40	·	·		r		−1	+1
1400s	40	d	d	d	r		+1	−1
1410	25	r	·	r	d		+5	+2
1420	25	·	·	·	·		+5	+1
1430	35	d	d	r	·		−6	−3
1440	40	r	dd	·	r		+1	+2

TABLE 1—*Continued*

Decades*	Auroral numbers†	American tree rings					Tree rings.						
		N. W.	S. W.	N. E.	S. E.	Mex.	(A)	(B)	(C)	(D)	(E)	(F)	(G)
1450s	40	d	d	·	d					−4		−5	−1
1460	35	·	d	r	d					+1		−2	−1
1470	25	·	r	dr	d		−5			−2		+9	+3
1480	30	r	rr	dr	dd		+6			+4		+2	−2
1490	30	·	d	dd	d		−1			−5		−1	−3
1500s	35	·	·	dr	dr		−1	(−8)	−2	−4		+1	+1
1510	40	d	dr	rr	r		+2	−4	0	+2		+7	+2
1520	60	d	d	r	r		−3	0	+2	−1		−1	−5
1530	65	dr	·	dr	r		+4	+8	+5	+3	(+1)	+6	+6
1540	60	d	dr	dr	·		−5	−2	−12	−2	−1	+4	−3
1550	65	dr	rrd	r	r		−1	+4	+10	+5	0	−2	+2
1560	70	·	r	r	·		+3	−4	+6	+1	−2	+1	−1
1570	75	r	dd	d	d		−1	−2	−6	−3	0	−4	−1
1580	60	dr	dd	dd	d		0	−1	−8	−6	+3	−3	+3
1590	45	r	dr	r	d		−4	0	−3	−4	+2	−4	−1
1600s	50	r	rr	d	d		+7	−1	+10	+2	−2	+2	(0)
1610	50	rd	r	r	drr		−7	+6	+9	+5	0	+2	(+3)
1620	60	dd	dd	d	d		−1	−5	−9	−3	+2	−5	(−4)
1630	50	d	d	d	drr		+3	+5	−8	+2	−2	0	(+4)
1640	25	·	r	rd	r		−10	−1	+13	+3	0	+1	(−3)
1650	30	d	d	d	r		+9	−2	−4	+1	−5	0	(+3)
1660	25	r	d	d	rd		+1	−3	−8	−4	−3	−5	(−1)
1670	30	r	d	·	d		−7	+3	+9	+2	−3	−4	(−2)
1680	25	d	r	·	d		0	0	−3	−2	+3	−1	(+4)
1690	20	·	r	·	·		+4	+3	+3	+1	+4	−8	(−4)
1700s	25	r	·	·	·	d	+2	−4	−3	−3	+4	(+5)	(0)
1710	35	·	·	rd	·	dd	−9	−3	+1	0	+4	−2	+5
1720	60	·	·	·	rd	r	+4	+7	+4	+2	−4	+2	−4
1730	60	·	d	r	dd	r	+3	−2	−12	−6	−1	+4	−1
1740	40	·	rr	·	dr	·	−7	−3	+8	+4	+1	−4	+4
1750	36	rrd	d	d	·	dr	+6	+7	−5	0	−3	−2	−3
1760	57	·	·	r	r	·	−3	−7	+9	+1	0	−4	+2
1770	70	d	rdd	·	d	dd	+6	+3	−6	−3	+5	−2	−2
1780	72	dd	dr	rr	d	·	−9	−5	−7	0	+3	−7	+2
1790	28	dd	rd	d	r	r	+7	+3	+6	+3	0	−4	−2
1800s	26	·	rd	r	rdd		−1	+3	−2	−2	0	−4	+2
1810	22	r	d	·	·	rr	+7	−3	+3	0	−4	−4	−2
1820	33	·	ddr	dr	r	dd	+11	+3	−10	0	+4	+6	+2
1830	67	r	r	r	rr	r	−4	−6	+18	+7	−3	−2	+2
1840	57	dd	d	rrd	·	r	+1	+1	−16	−1	−2	+2	−5
1850	46	·	·	d	d	·	+1	+3	+5	−1	−1	−4	+3
1860	53	r	d	dr	dr	·	−6	−1	+7	−1	+3	+4	+3
1870	40	dr	·	dr	dd	·	+4	−1	−9	−2	0	0	−5
1880	35	rrd	dr	rd	d	r	+1	+4	+5	−1	0	0	+2
1890	45	ddr	dd	dr	dd	d	+3	−5	−7	−2	+1	+2	+2
1900s	37	rr	rr	dr	·	·	−7	+7	+4	+1	−3+	+4	+1
1910	41	rd	rr	rdd	rr	r	+5	−8	+5	+7	−2+	+9	−1
1920	42	dd	r	·	r	·	(−)	+3	−2	−3	−2+	+6	−2
1930	54	·	rd	d	·	rdd	(+)	( )	(−3)	(0)	( )	+9	−1
1940	74	( )	d	d	( )	( )	( )	( )	( )	( )	( )	(+)	(+)

* Decades are of the form 1101–10.

† Auroral numbers to 1700 are intended to correspond to sunspot numbers. Thereafter, sunspot numbers are as in Schove (1955), with approximately 10 per cent increase to 1740.

(Footnotes concluded on p. 118.)

TABLE 1—*Concluded*

‡ Droughts and moisture as reflected in the semiarid zone, from unpublished maps largely based on data in Schulman (1956). The north-south division is the Rockies. The NW is north of 42° N, but the NE is north of 39° N. The letters dr mean dry then wet. Mex. signifies west central Mexico.

§ Columns (*A*) through (*C*) are keyed as follows:

(*A*) North Scandinavian trees (adapted from Schove, 1954*a*). Relative anomalies. Selected to indicate summer temperature.

(*B*) Alaskan trees (adapted from Schove, 1953*e*). Relative anomalies. Selected to indicate summer temperature.

(*C*) Western United States trees (adapted from Schove, 1953*e*, columns 3 and 4). Relative anomalies.

(*D*) Southern United States trees (Schulman, 1956, Tables 78, 70, and 50, and standardized curves, pp. 86, 87).

(*E*) Java minus "Pacific." Differences between tree growth in Java and Argentine-plus-west central Mexico. Unreliable at present as a measure of pressure parameter. Values from 1900 affected also by disease in Argentine trees.

(*F*) Winter temperature in northern Europe (45° to 60° N). Based mainly on Schove, 1953*e*, but with revisions. Estimated from ice data and documentary evidence. Winter in this column is one half a year, for example, November to April, and latitude is 45°/55° N to 1700, 53° N after 1700. From 1700, Manley's (1957) temperature series (November to April) (C × 10) and length in quarter days of Baltic ice-free season at Leningrad, Union of Soviet Socialist Republics (Sokolov, 1955), have been given equal weight.

(*G*) Pressure in northwest Europe (for example, 50° N 5 E) estimated from documentary data to 1650. Relative anomalies. Adapted from Schove, 1953*e*, with revisions. Pressure to 1600: 45°–55° N, based on documentary evidence of dryness and similar factors. After 1600: "The Wash," 53° N ½° E (Schove, 1953*e*, with revisions, and 1958).

[*Editor's Note:* Material relating to pressure changes, geological latitudes, and climate has been omitted at this point.

A certain Roger about A.D. 1100 followed "three angels clothed in white garments and stoles, each bearing in his hand a cross over which there were the same number of burning tables . . ." (p. 81). He travelled from Windsor to Markyate, eight miles northwest of St. Albans, England. This direction is 20°E of true north. More certain auroral displays suggest that the magnetic meridian in northwestern Europe was indeed about 15°E. The story is told in *The Life of Christina of Markyate,* C. H. Talbot, ed. and trans., Clarendon Press, Oxford, England, 1959, 156p.]

Sunspots and Climatic Fluctuations

Sunspots and climatic fluctuations may often run in parallel but, as will be explained (Schove, elsewhere in this monograph), the 11-year sunspot cycle is

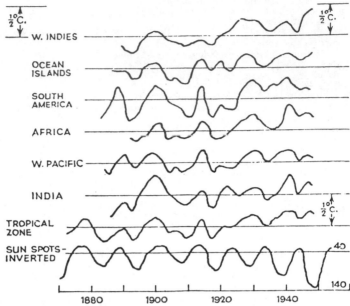

FIGURE 7. Temperature fluctuations in tropical regions, 5-year annual mean departures from mean 1901 to 1930. Reproduced by permission of the *Quarterly Journal of the Royal Meteorological Society* (Callendar, 1961).

often only an indirect cause. An instance of this is the well-known inverse relationship between sunspots and tropical temperatures. A world survey of temperature fluctuations has just been published (Callendar, 1961). The results are consistent with those found by Mitchell (elsewhere in these pages), but the two surveys illustrate different aspects. FIGURE 7 reproduces part of one of Callendar's diagrams, according to which the usual sunspot relationship was evidently reversed in the period 1921 to 1950. A pressure parameter is determined (Schove, elsewhere in this monograph) that may reflect more accurately than sunspot numbers the effective solar radiation.

The significance of the variation of decadal means (or, for example, 11-year overlapping means) of auroral numbers (TABLE 1) from cycle to cycle may be mentioned here: the cold waves in the tropics and in the United States often correspond to active solar cycles, although the present snowy winter of 1961 could hardly have been predicted from the great sunspot activity of the period 1955 to 1960.

References

CALLENDAR, G. S. 1961. Temperature fluctuations and trends over the earth. Quart. J. Roy. Meteorol. Soc. **87**: 1–12.

MANLEY, G. 1957. Climatic fluctuations and fuel requirements. Scott. Geogr. Mag. **73**: 19–28.

SCHOVE, D. J. 1947. The sunspot cycle before 1750. Terrestrial magnetism. **52**: 233–238. Williams & Wilkins. Baltimore, Md.

SCHOVE, D. J. 1948a. Sunspot epochs, A.D. 188–1610. Popular Astronomy. **56**: 247–262.

SCHOVE, D. J. 1949a. European raininess and European temperatures, A.D. 1500–1950. Quart. J. Roy. Meteorol. Soc. **75**: 175–179; 181.

SCHOVE, D. J. 1950a. The climatic fluctuation since A.D. 1850 in Europe and the Atlantic. Quart. J. Roy. Meteorol. Soc. **76**(328): 147–165. *Cf.* Meteorol. Abst. & Bibl., **1950**: 362 and Geogr. Rev., 1951. **41**: 656–660.

SCHOVE, D. J. 1950d. Visions in north-west Europe and dated auroral displays (A.D. 400–600). J. Brit. Arch. Assoc. 3rd ser. **13**: 34–49.

SCHOVE, D. J. 1951c. The earliest dated sunspot. J. Brit. Astron. Assoc. **61**: 22–24.

SCHOVE, D. J. 1951e. Sunspots, aurorae and blood rain. Isis. **42**: 33–138.

SCHOVE, D. J. 1953e. M.Sc. Thesis. London, England.

SCHOVE, D. J. 1953f. The preliminary reduction of early barometric and wind data. Paper 10, in U.G.G.I. Publication AIM No. 9/c. Proces-verbaux des seances de l'Association de Meteorologie, Memoirs and Discussions. : 187–193.

SCHOVE, D. J. 1954a. Summer temperatures and tree-rings in north-Scandinavia, A.D. 1461–1950. Geogr. Ann. Stockholm. **36**(H 1–2): 40–80.

SCHOVE, D. J. 1955b. The sunspot cycle, 649 BC–AD 2000. J. Brit. Astron. Assoc. **60** (2): 139.

SCHOVE, D. J. 1958. Ph.D. Thesis. London, England.

SCHOVE, D. J. 1961. Wind and pressure in N.W. Europe since 1630. Geogr. Ann. Stockholm, Sweden.

SCHOVE, D. J. & P. Y. HO. 1958. Chinese aurorae, AD 1048–AD 1070. J. Brit. Astron. Assoc. **69**: 295–304.

SCHULMAN, E. 1956. Dendroclimatic Changes in Semi-Arid America. Univ. Ariz. Press. Tucson, Ariz.

SOKOLOV, S. S. 1955. Trans. by E. R. Hope from Priroda. 7: 96–98. Defence. Scient. Inform. Service, DRB, Canada.

STEINHAUSER, F. 1960. Geof. e Met. Genova, 111: 112.

ERRATA

On the second page of Table 1, columns A through E are "Tree rings" and columns F and G are "Europe." In column F, the sequence from 1510–1570 should read +6, −1, +6, +5, +2, −4, and −2. In the key to column F of Table 1, 60°N should be 55°N.

32

Reprinted from New York Acad. Sci. Ann. **95**:605–622 (1961)

TREE RINGS AND CLIMATIC CHRONOLOGY

Derek J. Schove

St. David's College, Beckenham, Kent, England

Introduction

Tree rings in marginal climates provide indications of rainfall and temperature in individual seasons. In the United States, the practical needs of the water engineer in semiarid regions have justified the emphasis on local rainfall but, from the climatological viewpoint, it is more satisfactory to consider the pressure and pressure pattern over a wide area. Moreover, once the data are sufficiently extensive, past climatic fluctuations can be studied from a global viewpoint. Tree-ring series are accordingly being collected from different parts of the world, and preliminary results (subject to a more exhaustive computer analysis) of the new approach are available for four regions: the subarctic tree line in Scandinavia and Alaska; the semiarid regions of the temperate United States; drought-sensitive timbers of Northwestern Europe; and drought-sensitive timbers in low latitudes.

The Subarctic

A synthesis of North Scandinavian tree-ring series since 1460 A.D. has been published (Schove, 1954a). The constituent series showed the usual local variations, but all were affected by one common factor—the summer temperature—so that the combined index reflects the probable warmth of nearly 500 summers. Summer temperature in turn reflects wind and pressure so precisely that the probable pressure pattern of individual summers can be determined directly from the tree-ring index. The typical barometric distribution associated with narrow rings is thus shown by maps to be: low pressure over Finland and high over western Europe; this situation may be described as monsoonal, for it brings northerly winds with much cloud and rain to North Scandinavia. Similar maps for Scottish trees will be published shortly (Schove and Frewer, 1961).

Thirty-year moving means of this index reveal maxima and minima that may be significant. An f(T)-max. *ca. 1573** suggests the period of warm summers (Schove, 1949a) that ended suddenly about 1590, and (cf. Flohn, 1957, p. 204) the narrow rings of 1614 to 1650 (f(T)-min. *ca. 1628 or 1630*), 1708 to 1751 (f(T)-min. *ca. 1719*), and 1835 to 1870 (f(T)-min. *ca. 1849 to 1851*) as corresponding to the 3 phases of the Little Ice Age (cf. Table 2;* and C. J. Heusser, this monograph). However, cool summers in this area often synchronize with mild winters in Europe. Indeed tree growth was also low in the period of monsoonal summers from 1902 to 1928 (f(T)-min. *ca. 1913*) when compared with the preceding period of rapid growth (f(T)-max. *ca. 1886*).

Trends in the early part of the original series may, however, be spurious, and here they have been removed by the use of what I term "Relative Anomalies" (Schove, 1953e): each decadal value is an anomaly in relation to the enclosing 30-year mean. Thus in Table 1* the value for 1511 to 1520 represents the departure of the tree-ring width from the 1501 to 1530 normal.

* Schove, D. J. Solar Cycles and the Spectrum of Time Since 200 B.C. This monograph.

A similar index is being prepared for Alaska, using tree-ring series between West Alaska and the Mackenzie River, where a cool summer temperature impedes growth. The index given in the above-mentioned Table 1* was a preliminary one of this kind, based on information kindly supplied by Giddings and Oswalt (unpublished; see Schove, 1953c, pp. 107–109, 177, and 284) in their articles in *Tree Ring Bulletin*, 1943, 1948, and 1950.

Trends in the original series are inconsistent with those of Karlstrom's series (elsewhere in these pages) and, therefore, they too have been eliminated by the use of relative anomalies in TABLE 1. Nevertheless it is encouraging to find that one f(T)-min., *ca. 1830*, is very near the f(T)-min. evident in Karlstrom's series; the latter series, in so far as it is based on trees at selected sites above timber line, is a model that should be copied elsewhere, for such series should be given extra weight when included in the final index.

Relative anomalies of the Alaskan index suggest that summer warmth in Alaska coincides with summer warmth and dryness in Europe and, probably, also with warmth in the tropics, as there appears to be an inverse correlation with sunspot numbers and with relative decadal anomalies of the pressure parameter.

Semiarid North America

The United States tree-ring data (kindly supplied by Schulman and Smiley) are very accurately standardized, but the search for cycles has been unsuccessful. This is not surprising. Indeed regular cycles—such as one of 11 years— are not the type of oscillation that would be expected on meteorological grounds. In planning cycle analysis it is helpful to know what to look for first and, as far as tree rings are concerned, it is necessary to combine heterogenous series (1) that cover a very wide area, and (2) that are nevertheless affected by a common climatic factor.

There are several types of oscillation that can be studied in United States tree-ring data: (1) annual values associated with the "southern oscillation" of 2 to 3 years; (2) 5-year means associated with an apparent North American oscillation; (3) 5-year means with the pressure-parameter oscillation, including effects ascribed to the 11-year sunspot cycle; (4) 10-year means associated with, for example, fluctuations in the strength of successive sunspot cycles (cf. auroral numbers in Table 1*); (5) 30-year means reflecting major pressure oscillations (for example, 60 to 70 years); (6) major temperature oscillations (for example, 80 or 90 years); and (7) major solar variations (for example, the 200-year cycle).

The southern oscillation. Barometric pressures for the Indian Ocean are available since 1796 (Schove and Berlage, in preparation), and the data collected by Schulman (1956) show correlations that may at first seem surprising. Rainfall in many Pacific regions thus fluctuates positively with these annual pressure values. Tucson, Ariz., and Santa Fe, N.Mex., tend to follow this "Pacific" pattern. The United States is in a transition zone and the opposite or "Indian" pattern is followed by the rainfall of nearby Colorado and upper Colorado and by the upper Missouri River runoff in Montana. A

* Schove, D. J. Solar Cycles and the Spectrum of Time Since 200 B.C. This monograph.

similar contrast in tree-ring reactions can be noticed between Pacific sites to the south (for example, east-central California, west-central Mexico, Mexico, and Big Bend, Tex.) and Indian sites east of the Rockies (Banff, Alta., Canada, and the Missouri River basin).   In some cases there is a phase difference of as much as 12 months.   All these relationships are nevertheless weak and unreliable, and North American tree rings have not so far been included in our attempts to study the southern oscillation in past centuries.

A negative correlation of certain United States tree-ring data with the Scandinavian summer temperature index (Schove, 1954a) has been announced by Cross (1961), who is measuring the effects of snow pack in different years on the shape of tree rings at Mount Rainier, Wash.

Correlations of annual tree-ring values with N-S pressure differences or the north Pacific oscillation in different seasons should prove most illuminating but, as far as I am aware, this has not yet been attempted.   Likewise, qualitative features (cf. Schove and Lowther, 1957, p. 89) should prove significant.

North American oscillation.   There is an oscillation in 5-year moving means of the climatic elements of the United States that might seem to be some North American oscillation.   During the period 1901 to 1930 its wave length coincided with that of the sunspot cycle, which therefore appeared to be a convincing cause of the fluctuations shown by charts of moving 5-year means (for example, as in Tannehill, 1947).   The most obvious effect of this oscillation is the negative correlation between air (and sea) temperatures on the Pacific coast, on the one hand, and the air (and sea) temperatures for the remainder of the United States on the other.   Pressure is higher on the Pacific coast when it is colder in the southwest, but this pressure anomaly is opposed to that in the tropics of the Old World.

After 1920 the sea-temperature information published by Rodewald (1956) leads to the following differential values for the Atlantic minus the Pacific coasts:

1920 to 1924 ca.	−1
1925 to 1929 ca.	−11
1930 to 1934 ca.	−1
1935 to 1939 ca.	+1
1940 to 1944 ca.	−8
1945 to 1949 ca.	+8
1950 to 1954 ca.	+13

(Temperature is in degrees centigrade multiplied by 10, expressed as departures from 1925 to 1954 normals.)   Presumably these values will be found to match well with temperatures in the United States (cf. Mitchell, this monograph) as a whole, and with pressure off the Pacific coast.   The main peaks are the 5-year periods centered at 1881, 1894, 1910, 1922, 1937, and 1952(?) with the intervening minima at 1886, 1905, 1916, 1927, and 1941.

TABLE 1, based on curves in Tannehill (1947), illustrates the patterns at a time when the cycle was about 10 years.   This oscillation arises from differences in pressure pattern often associated with the sunspot cycle initially through what is termed a world pressure parameter.   There is therefore no separate North American oscillation, but 5-year means of responsive American tree rings may help us to determine pre-1796 values of this important index discussed below.

There is another cycle that affects Greenland, Canada, and the northeastern part of the United States, and this seems to be an oscillation in the upper air trough in the westerlies over the Davis Strait. Five-year moving means indicate pressure maxima (1878, 1887, 1893 . . . 1908, 1917, 1925, 1932/1933, 1941, 1948, and 1956 in W. Greenland) that reflect the parameter changes discussed below and that also indicate a cycle that changed from 7 to 8 years about 1900 (when the general pressure fell).

The world pressure parameter. World pressure maps of changes by 5-year periods have been constructed from barometric information received as a result of international appeals (cf. WMO Bulletin, January, 1959, p. 40). Changes can be grouped into two opposing types, according to the sign of the pressure change in west Greenland. A change of opposite sign but smaller

TABLE 1

CENTRAL DATES OF MOVING FIVE-YEAR MEANS OF CLIMATIC ELEMENTS IN
THE UNITED STATES

Sunspot maxima (actual year)		1883	1894	1907	1917
Pacific coast pressure	Max.	1881/1882	1893	1910	1920/1923
Pacific coast temperature	Min.	1881	1894	1910	1920/1921
United States (especially the East) temperature	Max.	1880	1898	ca. 1910	1920/1923
Pacific coast pressure	Min.	1886	1905	1915	1926
Pacific coast temperature	Max.	1888	1904/1905	1915/1916	1926
United States (especially the East) temperature	Min.	1886	1904	1917/1918	1928
Sunspot minima (actual year)		1889	1901	1913	1923

magnitude occurs simultaneously in the Indian Ocean where mean pressures can be accurately determined for over a century. Indeed, at present, it is convenient to use the reversed mean of several areas in the neighborhood of the Indian Ocean as the basis of what is termed the "first approximation to the pressure parameter". Successive approximations of increasing complexity will be made, but this is already a close approximation, and values are given in TABLE 2.

The similarity with the southern oscillation is thus obvious. However in the ordinary southern oscillation, as Berlage (this volume) has made clear, there is little correlation with the pressure in Greenland that, if anything, in individual years tends to agree with the Indian Ocean rather than the Pacific. In the case of these 5-year changes it now appears as if the southern oscillation is geared together with the North Atlantic oscillation. Moreover this oscillation is rather more extensive: the term Indian Ocean has now expanded to include the Cape of Good Hope and southeast Australia!

The "signature" of the pressure parameter during the 9 lustra of the period 1871 to 1915 is given by the mnemonic code sequence P, O, N, O, P, O, N, P,

N where P and N refer to relative peaks—positive and negative respectively—and O to those 5-year periods such as 1876/1880 that were transitional. This suggests an oscillation of decreasing wave length, and does not always agree with the sunspot cycle. This pattern is followed for: (1) reversed pressure

TABLE 2

THE PRESSURE PARAMETER AND FIVE-YEAR MEANS OF SIGNIFICANT CLIMATIC INDICES

Lustrum*	Reversed Indian Ocean pressure mb. × 100†	Signature‡	Sign change from previous lustrum§	Tropical temperature CX 100 ‖	Sunspot turning points ¶
1823	−?			(+)	N
1828	+?		+	(−)	P
1833	−?		−	(+10)	N
1838	|?		+	(−20)	P
1843	(−)		−	(+10)	N
1848	(+)		+	(−10)	P
1853	(+4)		−	(+10)	N
1858	+9		0	(−0)	P
1863	+19		+	(−20)	O
1868	−9		−	(+10)	N
1873	+9	P	+	−2	P
1878	−15	O	−	+11	N
1883	−21	N	−	−14	P
1888	−12	O	+	−9	N
1893	+9	P	+	−15	P
1898	+1	O	−	+15	O
1903	−11	N	−	−1	N
1908	+7	P	+	−11	P
1913	−13	N	−	+9	N
1918	+11	P	+	−12	P
1923	+13	or P	0	−1	N
1928	−4	N	−	+15	P
1933	+7	P	+	+9	N
1938	+5	or P	(0)	+15	P
1943			(−)	+21	N
1948			(+)	+21	P
1953			−	+	N
1958				+	P

* Each lustrum is denoted by its middle year, for example, 1843 = 1841 to 1845.

† First approximation to pressure parameter (relative to 1881 to 1940 normals); the relative maximum in 1921 to 1925 is not representative of better approximations.

‡ Mnemonic signature of relative maxima (P) and minima (N).

§ For calculation of agreement coefficients. Values after 1935 are subject to correction when, for example, Greenland pressures become available.

‖ Date from Callendar (1961) relative to 1901 to 1930 normals. Estimates before 1870 are partly based on additional information kindly supplied by Callendar.

¶ Relative maxima and minima by lustra.

over the Indian Ocean and neighboring areas; (2) rainfall for regions of Indian type, and therefore the Nile flood area; (3) temperature in the tropics generally, with the notable exception of 1881/1885 (Krakatoa?); (4) pressure in southwest Greenland (except for 1871/1875), in temperate South America, and probably the Pacific; (5) ice in Davis Strait less that in Iceland and Newfoundland, except for the periods 1881/1900; (6) rainfall reversed in the south of the United States between the Rockies and the Mississippi River (also many

South American regions for most of the periods); and (7) sunspots, but not between 1876 and 1890 (cf. also divergence after 1920).

With such an extensive list of correlations, it will be a simple matter of arithmetic to combine tree-ring series from different parts of the world to evaluate the values of this parameter back at least to the years 1 to 5 A.D.! The principle of selection could be the agreement coefficient (cf. Schove and Lowther, 1957) between successive 5-year changes.

Many vague effects hitherto ascribed to the 11-year sunspot cycle show a much better relationship with this world pressure parameter, which presumably reflects more exactly than sunspot numbers the effective solar radiation. It is unlikely that tree-ring widths by themselves could be used to correct any errors in my dates of historical maxima (cf. Schove, 1955b), but it will be interesting to watch for divergencies such as have occurred since 1920 between the sunspot and parameter maxima.

Five-year means of tree-ring changes in North America again fall into two groups, but some series where the annual data were of Pacific-type are now of Indian-type, that is, tree-growth over 5 years is greater when Indian Ocean pressure is lower. Nevertheless the marginal location of the region has again made it advisable to omit both series in investigating the pressure parameter itself. This marginal location also explains why the sunspot cycle rarely shows up in United States trees.

The sunspot cycle does appear in some United States trees if my auroral minima (Schove, 1955b) are used instead of the rigid 11.4-year sunspot cycle conceived by Douglass. The central Pueblo series described by Douglass (1936, Vol. 111: 106, 107) reflects the cycle in much the same way as a rainfall series of Pacific-type, but with about 2 years delay between sunspot minima and tree-ring maxima. Presumably the apparent sunspot effect is again indirect, and the differences between United States trees of the two types might indicate not so much the sunspot cycle but the pressure parameter fluctuations as reflected in a North American oscillation.

Ten-year means and auroral numbers. The sunspot cycle is almost eliminated if overlapping 10- or 11-year means are used. It is surprising therefore to find that sunspot numbers are now more and not less important. At the time when the auroral numbers of Table 1* were prepared, it did not seem likely to me that meteorological correlations would be significant.

The parameter, not the auroral numbers, should again be regarded as the immediate factor necessary to explain climatic anomalies and, although Faegri's rule is not true in the contrast between 5- and 10-year changes, the maps of the changes fall roughly into the same two phases of the fundamental "standing wave" of southern-oscillation type. A first approximation to this parameter is provided by the tree-ring indices of Table 1,* column E (cf. also column D).

The effects of these changes of pressure pattern have sometimes been noted and ascribed to other causes. Thus from 1880 to 1945 there was an accidental 22-year cycle that appears as an attractive scapegoat. However, a number of similar curves from different parts of the world assembled by Yamamoto (1957), although ascribed by other climatologists to coincidence, do fit well the patterns anticipated.

* Schove, D. J. Solar Cycles and the Spectrum of Time Since 200 B.C. This monograph.

The signature of the 10-year pattern is, in the instrumental period, best shown in decades of the form 06-15, for then the mnemonic for successive decades from ca. 1870 to ca. 1930 reads PNPNOPN, although the period 1866 to 1875 is not a period of minimum pressure in the Indian Ocean (cf. TABLE 2) itself. That sequence is nevertheless characteristic of both climatic anomalies and sunspot changes, and it provides an objective criterion when tree-ring data are used to determine the decadal pressure parameter in past centuries.

Once again a high parameter normally means wetness at Indian-type stations and dryness at Pacific-type stations. This time, however, most good tree-ring series, from North Platte, Nebr., to Arkansas and across the Rockies to Colorado and even southern California (but not, for example, Tioga Pass) are—in the period of the signature—of Indian type. A column (D) has been included in TABLE 1* to represent the mean growth over this region but, as all the trees concerned are within the United States, anomalies peculiar to the United States also play their part. This explains why the disagreement percentage by lustra with an Argentine series of Pacific-type trees (40 per cent) was less instead of more than 50 per cent.

It will be possible to obtain a good index of the parameter from tree-ring series in the neighborhood of the Indian Ocean, but the values of Table 1,* column E, are included to illustrate the method rather than provide the results, as they are based on the only three sets of standardized series available at the time of writing.

The North American series provide meanwhile a useful approximation to the parameter, for the use of running 10-year means reveals the same sequence of t-max. and t-min. as found thermometrically and as mentioned above. Moreover these warm-dry and cold-moist waves are now seen to be the phases of pressure parameter waves associated with decadal sunspot minima and maxima respectively. The dryness of the dry phase extends from the old-world tropics to North America as far north as Montreal, P.Q., Canada.

Low sunspot numbers about 1810 thus fit with a warm wave evident in the parameter and in the Alaskan trees and a "long dry" period (noted below) in the United States.

Sunspots are still relatively low about 1830 when the next low-parameter pattern appears. Again trees in Alaska and in the Rockies reflect a warm-dry wave dated instrumentally to ca. 1829/1830. The rapid rise to an s-max. ca. 1835/1840 is, however, associated with the usual high parameter pattern and cold wave. In northern Italy the t-min. is dated again rather early (ca. 1833), but in the United States the instrumental date of ca. 1839 is perhaps a little late in relation to the characteristic Japanese famines of 1833 to 1839.

The next warm and cold waves occur in rapid succession in the United States about 1865 and 1870 respectively. These dates do not fit the parameter approximation given in TABLE 2, but the sudden revival of sunspot activity from s-min. to s-max. in the 1860s resembled that of the 1830s and seems to have given similar effects (for example, in Japan, as noted by Arakawa, 1955 and 1960, p. 112).

The warm wave of 1876 to 1881 followed the sudden collapse of sunspot activity and is this time reflected in a very considerable decrease in the pressure

* Schove, D. J. Solar Cycles and the Spectrum of Time Since 200 B.C. This monograph.

parameter associated with droughts in several continents. The cold wave followed about 1888/1890, ahead of the parameter and the s-max.; it was possibly partly induced by the eruption of Krakatoa, in the East Indies, and it extended to Central Europe.

About 1900, there was the most striking s-min. of recent times, again with widespread subtropical droughts. There was, at the same time, a well-marked minimum in the various effects of the parameter change. The pressure pattern change was described by me (Schove, in Kraus, 1956, p. 358) before I appreciated the part played by sunspots. The warm, dry phase occurred about 1898, both in the tropics and in the United States, but it was not until ca. 1920 that a further high value occurred of either the s-max. or of the pressure parameter.

A p-min. about 1930 fits a minimum in the parameter and a further warm wave, although this is less clearly defined. The t-max. ca. 1934 is nevertheless common to the eastern United States and to Central Europe.

Although subsequent waves also are indefinite (cf. transatlantic t-min. ca. 1938, t-max. ca. 1948) the inevitable fall from the present high sunspot numbers will presumably be associated with a sudden fall in the parameter, and a further warm, dry wave extending from the old-world tropics to the United States may yet be associated with drought famines, as in 1877/1878.

The relations between the various indices of Table 1* will be discussed in a forthcoming study of the history of European climate.

Thirty-year trends, the major pressure oscillation ca. 1890 to 1920. Homogeneous pressure series for different parts of the world have been prepared since my FIGURE 3* was first published and it is becoming possible to determine the dates of the P-max. and P-min. in low latitudes. It is now clear that south steering is not important south of 40° north but that, instead, the dates assembled indicate a major pressure oscillation, a seesaw between the Pacific and the Indian Oceans that resembles the southern oscillation. It is interesting to find that "superoscillation" has a different wave length from the major temperature oscillation dominant in the west Greenland center of action.

The pattern is thus:

P-min. Pacific Ocean ca. 1890 to 1891   P-max. Pacific Ocean ca. 1920
P-max. Indian Ocean ca. 1890 to 1891   P-min. Indian Ocean ca. 1923

The Pacific P-min. *ca. 1890 to 1891* included the United States Southwest and even Argentina. Simultaneously the Indian P-max. extended from southeastern Australia to Syria, and reinforced the south-steering high that was migrating from central Norway *ca. 1889* to Austria *ca. 1894*. Standard deviations of most climatological elements were reduced even in Europe at this time. In the west, the Indian area probably extended across Africa to Brazil and the southeastern states of the United States, but uncertain barometric records make it difficult to be specific about this. The nodal zone from the Rockies to perhaps southwestern Brazil would seem to have been a narrow belt with winds that were appreciably anomalous in relation to 20th-century normals, and rainfall distributions must also have been peculiar.

A P-max. *ca. 1919/1922* affected much of the Pacific extending from Welling-

* Schove, D. J. Solar Cycles and the Spectrum of Time Since 200 B.C. This monograph.

ton, N.Z. and, once again, to the United States Southwest and across South America to the Argentine.

Almost simultaneously, the P-min. of *ca. 1923/1925* affected the same area as the p-max. of *ca. 1890/1891*. Thus it extended from the southeastern states of the United States (*ca. 1923/1925*) to parts of northwestern Europe on the one hand, and from the subtropical belt to Portugal and Syria on the other. India, Java, and even Hong Kong are also included. In the Southern Hemisphere, it extended from southeastern Australia west-north-west and, probably, across Africa to Northeastern Brazil.

The P-max. and P-min. often correspond to an R-min. and R-max., but there are exceptions, often within short distances of one another. Pacific-type reactions occur in the northwestern United States (cf. *Monthly Weather Review* **63**, pp. 19–23) but, in the Rockies, Indian reactions are more common. Tree-ring maxima and minima faithfully reflect these differing responses and, in the fully standardized series of Schulman (1956) the R-max. and the f(R)-max. often agree to the exact year.

An inverse correlation exists between 30-year curves of tree growth in Pacific areas such as Banff and Jasper, Alta., Canada, on the one hand and the southern and eastern sites (for example, southern California, Rio Grande, and Colorado) on the other. Let us suppose that differences between the two areas reflect major pressure oscillations of the same geographical pattern as that of the half-wave *ca. 1890* to *ca. 1920*. We can then determine the dates of barometric maxima and minima in regions of Pacific-type as follows:

ca. 1530	P-min. Important	(SW in Europe)
ca. 1560	P-max. Important	(NE in Europe)
ca. 1590	P-min. Important	
ca. 1635/1645	P-max. Important	
ca. 1670	Minor p-min.	
ca. 1685	P-max.	
ca. 1710	Important p-min.	
ca. 1735	P-max.	
ca. 1760	Irregular p-min. Minor	(SW in northwestern Europe)
ca. 1790	P-max.	
ca. 1815	P-min.	
ca. 1840	P-max.	(NE in northwestern Europe)
ca. 1885	P-min. (Model)	(NE in northwestern Europe)
ca. 1920	P-max. (Model)	(SW in northwestern Europe)

These results are evidently significant in United States climatic history but, until an index can be found that correlates inversely with an Indian Ocean tree-ring index, we must suppose that a complex function of more than one parameter is involved. In particular, we must note: (1) there is no inverse correlation with Table 1* column E (Java minus Pacific); (2) there is no obvious correlation with sunspot numbers; there was an S-min. ca. 1890, but it lasted until *ca. 1910* and there was no S-max. *ca. 1920* to *1925*; (3) wind variations in Europe (indicated from Figures 2 and 6* by the bracketed figures) are independent; (4) there is no regular cycle, such as a Brückner cycle of 35 years; and (5) another parameter implicated in the above dates is the major temperature oscillation that will be discussed below.

Thirty-year means, the major temperature oscillation. The temperature

* Schove, D. J.   Solar Cycles and the Spectrum of Time Since 200 B.C.   This monograph.

changes, for example in Greenland, reflect another climatic oscillation with a wave length—in the past 2 centuries—of the order of 80 years, thus comparable (although out of phase) with the long sunspot cycle.

The climatic amelioration that set in suddenly in the subarctic about 1895 is associated with climatic changes elsewhere, such as are indicated by Kraus (in these pages), who notes the rainfall decrease on the eastern side of continents. A similar drying (Schulman, 1956, Figure 14) in tree-ring areas of the Pacific slope in western states can also be noted in TABLE 3.

The signature of a single wave is insufficient for us to use, for example, for lakes in Oregon or tree rings in British Columbia, as inverted thermometers, despite the excellent correlation (cf. Shutler *et al.*, this monograph) of pluvials and glacials in the past. Nevertheless if, for the sake of argument, we use the same series of trees to date the phases of the Little Ice Age, we would have: Cold I *ca. 1605*, Warm *ca. 1630*, Cold II *ca. 1675, ca. 1715* and *ca. 1750*, Warm *ca. 1780*, and Cold III *ca. 1820*.

TABLE 3
THE MAJOR TEMPERATURE OSCILLATION

	Ice		Trees		Rain
	Davis Strait	Iceland	Pacific slope		Eastern United States
	Speerschneider	Koch, Schell, Lamb	Schulman		Lysgaard
f(T)-max.	ca. 1854 or 1841	ca. 1850	ca. 1845	R-min.	ca. 1850
f(T)-min.	ca. 1872	ca. 1878	ca. 1875	R-max.	ca. 1875
f(T)-max.	ca. 1930	ca. 1933	ca. 1930	R-min.	ca. 1920

Comparisons between the two half centuries 1851 to 1900 and 1901 to 1950, such as are made by Schell (in these pages) for the year as a whole, are of special interest in this connection. Preliminary results (received in response to my appeal) for the 12 separate months suggest that, although the westerlies have become stronger in all the winter months (November to April) in Europe, the trend has been toward an increased northerliness in summer months (June to October), as if monsoon effects were more significant and solar radiation more effective in the period 1901 to 1950.

Long cycles since ca. 25 B.C.: the 200-year solar cycle. Tree-ring data in the Southwest enable many wet and dry phases to be identified before 1100. Well-marked f(R)-max. can be dated *ca. 25 B.C., ca. 194 A.D., ca. 545 to 580, ca. 900*, and *ca. 1020*, for example, and well-marked f(R)-min. *A.D. 27, ca. 250, ca. 770* and *ca. 970 to 1000*. Lack of evidence prevents us from determining the dates of S-max. and S-min. in some cases, but there is again no obvious correspondence and indeed no obvious periodicity. The 200-year cycle in sunspot activity (Schove, 1955*b*, pp. 143, 144; Dewey, 1960) is possibly matched in a drought cycle noted by Schulman between 1200 and 1500, but the latter oscillation may well be accidental; we must await standardized series from the

4000-year-old trees now being investigated before such long cycles can be determined from tree-ring analysis.

Drought-Sensitive Timbers of Northwestern Europe Since 100 B.C.

Oak trees in northwestern Europe are not as long-lived as United States trees but, at most sites even in a normal English summer, they do not get sufficient water for maximum growth, so that the rings are drought-sensitive. Archaeological timber from the Roman and Anglo-Saxon periods can be dated approximately (Schove, 1959, pp. 288–290). Later medieval timber is securely dated, and the narrow rings are often explained by droughts recorded in the chronicles. In the New World, the only comparable evidence is in the Aztec records of Mexico. However, old-world chronicles are not always reliable, and the tree-ring evidence in one instance led (Schove and Lowther, 1957) to the discovery that a date (1231) in a medieval printed chronicle was incorrect; in another it threw doubt on the 7th-century drought of St. Wilfrid, a drought used by C. E. P. Brooks as evidence for his belief that the 7th century was the driest phase of the Dark Ages. Meanwhile, meteorological extracts from unpublished manorial rolls have now been published and found (Titow, 1960, p. 365) to confirm the tree-ring dating of droughts in the 13th century.

The English oak trees show cycles that are probably physiological and are certainly not of solar origin. They also show intriguing climatic fluctuations. In the Roman period, for instance, drought-sensitive trees in southeastern England show minima: termed f(R)-min. (cf. above) *ca. 100 B.C., ca. 100 A.D., ca. 160, ca. 205, ca. 255*, and f(R)-max. *ca. 40 B.C., A.D. 135, 185 and 225* (Lowther's measurements and approximate dating on archaeological grounds). Tree-growth over 30-year periods probably depends on, for example, temperature as well as annual rainfall but, with the aid of the "British Tree-Ring Project," it is hoped to prepare a standardized series and to discover what the other factors are. In the meantime we must be cautious about similarities between minima such as 1246 to 1290 that are common (cf. FIGURE 1) to English oak and to trees in the United States semiarid zone.

Low Latitudes

Drought-sensitive trees in low latitudes are especially useful, but few standardized series are available. The pioneer work of Berlage (this volume) in Java is only now being followed up in southwestern Asia and Africa. Individual series show little indication of the sunspot cycle, but they do show very clear influence of tropical pressure patterns. The Java series, compared with pressure maps (unpublished) can be shown to reflect the southern oscillation in general and pressure in Australia in particular. By summation of different series—allowing for phase lags where necessary—it is hoped to obtain an index of Indian Ocean pressure (and thus the southern oscillation) extending back many centuries. Comparison with the dates of sunspot minimum (Schove, 1955b) will then make it possible to investigate further the rules indicated by Berlage (in these pages) and Visser (1959). The decadal values in Table 1* in the column termed "Java minus Pacific," although unrepresentative at present, show how an index could be constructed to reflect the pressure parameter and the major pressure oscillation.

* Schove, D. J. Solar Cycles and the Spectrum of Time Since 200 B.C. This monograph.

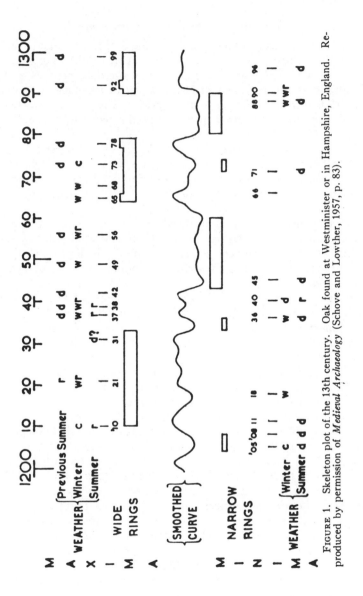

FIGURE 1. Skeleton plot of the 13th century. Oak found at Westminster or in Hampshire, England. Reproduced by permission of *Medieval Archaeology* (Schove and Lowther, 1957, p. 83).

In the concluding section, a survey will be made of the last 185 years' weather in the United States in the light of the principles described in my two papers in this monograph.

Climatic Patterns in the United States Since 1776

The year 1776 happens to mark an early stage of a long dry period in the semiarid parts of North America. It is also a convenient date to begin our review because, after the War of Independence, the documentary and instrumental evidence from other parts of North America is sufficient to determine characteristic features of the various decades. The results of this study, deliberately simplified, are now considered in the light of the south-steering hypothesis.

The 1770s are indeed generally dry, not only in New England but also, by the second half of the decade, in the semiarid belt as a whole. The drought of 1777/1778, according to Hawley, almost exterminated the Hopi Indians. A tropical warm wave is dated in Central Europe 1776/1778 and no doubt extended to the United States about this date.

A minor wet phase (r-max.) commenced in the east Rockies of Canada ca. 1785, moved south to the 40th parallel ca. 1790, and to the 30th parallel ca. 1795. The 1790s were a typical easterly decade with dryness in the northwestern (southwestern Canada) parts and wetness in the southeastern parts of the tree-ring area.

The long dry period proper thus began in southwestern Canada, reached the upper Colorado by 1798, the Gila valley in 1800, the Rio Grande by 1804, and southern California by 1805. In the 1810s, wetness affected southwestern Canada and adjacent Pacific states, but the change from dry to wet over most of the area is dated fairly precisely as 1825.

The wet phase, from 1825 to 1840, which has been associated with Little Ice Age III coincides with a cold period—both in the United States and the European continent—in the 1830s, and the wetness of this phase is unique. The whole tree-ring area was affected from Canada to Mexico more completely than in the earlier wet phases (I and II): 1601 to 1620, and the 1740s. At San Francisco, documentary evidence had independently suggested to Lynch that the period 1832 to 1840 must have been wet and, on the other side of the Rockies, references to the levels of lakes in North Dakota about 1830 and to Lake Huron in the 1830s suggest that levels were very high. Tree-ring evidence points to the greatest abnormality in the upper parts of the Arkansas and Rio Grande valleys. In the former region, for instance, a well-marked r-max. must have occurred *ca. 1830*. An r-max. is clearly indicated by the rain-gauge record at Marietta, Ohio, and by the Mississippi runoff record at Vicksburg, Miss., in each case at ca. 1827. Tree-ring evidence in the Rio Grande suggests an r-max. ca. 1836, so that once again south-steering seems to have occurred.

In low latitudes of the Old World, where reliable information is now available, this period can (see above) be subdivided into two phases: a warm-dry one, then a cold-wet one. These phases also extend to the Atlantic states of the United States, and near New York state can be dated (t-max. ca. 1826, r-min. ca. 1830; t-min. ca. 1835/1839, r-max. ca. 1839/1845).

The mid-century dry phase.   The "interstadial" between Little Ice Age IIIa and IIIb is marked by a short phase of remarkable warmth (a T-max. *ca. 1850*) in the Iceland-Davis Strait region of the Arctic, and a similar warm phase in the southern hemisphere is perhaps reflected in Heusser's glacier chart (in these pages). We find, as we should expect, a dry phase in the tree-ring zone of North America.

The mid-century dry phase began in southwestern Canada abruptly about 1840 and affected the whole tree-ring area (except the extreme southeast) in the course of the 1840s. The 1840s mark the culmination of the dry phase in the Pacific states and the pioneers that trekked to the west did so in a very

TABLE 4*

PRESSURE AND PRESSURE GRADIENT IN THE TRIANGLE TORONTO–NEW YORK
–ST. JOHNS RELATIVE TO 1841 TO 1845 MEANS

Pressure Ins. × 100		Type	Winds (New York state)	
			W	S
1836	+6	High; west type in north	−10	+1
1837	−3	Low, but dry SW	3	−5
1838	+1		6	−2
1839	+3	E in north	−1	−1
1840	+2½		2	+2
1841	−1	NE	−5	−4
1842	1	SW	−2	2
1843	0	W	4	0
1844	1		−2	0
1845	−2		6	4
1846	1		−4	−1
1847	3	S	−3	3
1848	1		1	1
1849	6	High E	−12	−4
1850	0		5	−4

* Wind data from Hough, calculated as percentages, using method as in Schove (1953*f*, 1961/1962 and unpublished). Notice that the p-min. ca. 1841 coincides with a max. in the north and west wind components, suggesting a south steering p-min. off the Atlantic coast.

difficult decade. The wagon tracks left in eastern Oregon in 1849 were subsequently covered by water; the water remained until the lake dried up in a later warm period in the 1920s when the tracks were recognized and photographed. In the Mississippi valley, rainfall and runoff records indicate that the r-min. occurred as early as ca. 1840 but, in California documentary evidence led Lynch (1931) to select the period 1840 to 1847 as the driest, and to note that a contemporary diarist (Belden) remembered the dry phase as 1841 to 1851 to 1852. In the 1840s it would seem that easterly winds were more frequent than usual in the United States, but not in Canada.

Barometer and wind records are available for this period for New England and neighboring parts of North America, and they have been used in TABLE 4.

The words "great American Desert" that appeared on some early maps across the western range, covering the western parts of Nebraska, Kansas, and Texas, are consistent with travellers' tales of the mid-century dry phase that continued on the eastern slopes of the Rockies into the 1850s. The 1850s were indeed probably westerly in type for the rainfall pattern (from ca. 1851/1865) was like that of the 1800s and the 1930s: in each of these decades the Pacific states of the west were now wet.

The moist phase 1865 to 1885. The new moist phase might be termed Little Ice Age III*b*: it coincides with ice increases after 1863 in the Davis Strait and with the T-min. in the Arctic (see above) *ca. 1875.* The wetness spread, about 1865, east of the Rockies and then affected the United States as a whole during the Civil War decades and throughout the 1870s. In the tree-ring zone an f(R)-max. is noted in southern California and the Rio Grande *ca. 1875 to 1882;*

TABLE 5

DECADAL PATTERNS IN THE UNITED STATES
(RELATIVE TO 1911/1940 NORMALS)

Decade	Pressure mid-United States	Pressure gradient	South minus north	Rainfall		Temperature		Thirty-year type
				All United States	West Central	North Central	SE	
1881–1890	−10	(−)	(E)	(>100)	(>100)	−17	−1	
1891–1900	−10	+1	(W)	98	100	−9	−5	
1901–1910	0	−3	(E)	101	105	−6	−6	Easterly
1911–1920	+10	−7	(E)	102	103	−9	−5	P-max.
1921–1930	+5	+1	(W)	100	101	−1	+2	
1931–1940	−2	+9	(W)	97	93	+9	+4	Westerly
1941–1950	−5	(−)	(E)	(>100)	(>100)	(+)	(+)	Frontal B (?)

Pressure from *World Weather Records* (1959 only) in ×100 relative to 1921/1950 and pressure gradient data from graphs given by Kincer (1940).

Rainfall data in percentages derived from data in Tannehill (1947, Tables II and IV, pp. 240 and 244).

Temperature data from *World Weather Records* kindly supplied by Callendar (°F × 10).

North Central = Marquette and Bismarck (corrected). Other stations show a similar tendency, provided "Bismarck" record published in *World Weather Records* was corrected for "city" influence at a rate of 2° F per century.

South-east = Nashville, Galveston, and Key West.

Pressure and bracketed information added since the 1950 table (in *Quarterly Journal of the Royal Meteorological Society*) was compiled.

however, the dryness spread back northward during the 1870s and 1880s reaching southwestern Canada about 1890. In low latitudes, there are three subdivisions of this phase: warm, cold, and then warm.

The dry westerly phase 1885 to 1905. The dry phase was at its height in the 1890s—the driest decade in the semiarid zone since the 1650s or the 1580s—and the dryness then extended from central Canada to Mexico. A tropical warm phase coincided with an anticyclonic phase from the north: the P-max. (Figure 3*). The f(R)-min. in southern California, the Rio Grande, and the Colorado *ca. 1888/1889,* is significantly close to the period of the P-min. in the Pacific. The decadal patterns can now be interpreted statistically according to TABLE 5.

The transition of the 1900s. The 1900s resembled the 1860s. Wet conditions that had begun earlier (1895) in southwestern Canada affected the United

* Schove, D. J. Solar Cycles and the Spectrum of Time Since 200 B.C. This monograph.

States generally, with an r-max. ca. 1904 and ca. 1910. Wetness spread south in the middle of the decade. On the Pacific coast this was the wettest decade, but the decade ca. 1910 marks the principal r-max. in the far west and, as westerly winds decreased in frequency, the rest of the tree-ring zone gradually became wet. Indeed, pressure maps suggested an easterly phase *ca. 1905* (cf. Schove, 1950a, p. 161).

The wet and easterly 1910s. The wetness of the 1910s stands out in the tree-ring area as a moist phase almost comparable with the 1830s. Rainfall was much above average on the southern and eastern slopes of the Rockies. Dry conditions had nevertheless reached the Jasper area of Alberta, Canada, by 1912, and the dry phase reached the western states during the decade, the Pacific coast (sheltered from the east) being drier in this decade than in the period 1886 1895. The south steering of the wet phase about 1915 is well illustrated in a map (drawn by L. F. Page in *Monthly Weather Review*, 1937, p. 16) revealing that the period 1916/1935 was to be drier in the northern half of the United States and wetter in the southern half than the preceding period 1896 to 1915. The R-max., indicated by rainfall records and tree rings alike occurred in the northeastern part of the tree-ring area *ca. 1905/1911*, but in' most of the remainder of the area *ca. 1917/1920*; in the latter areas, the date thus almost coincides with the Pacific P-max. of the major pressure oscillation.

The transition of the 1920s. Dryness in the northwestern part and wetness in the southeastern part of the semiarid zone recalls the 1840s, the 1790s, and the 1510s. The drought (associated with an r-min. on the British Columbia coast ca. 1927) was especially severe in Washington and Oregon, and tree-ring evidence indicates the ending of the wet phase by 1929 in the upper Missouri, by 1930 in the upper Colorado, and by 1933 in the Gila River basin of southern Arizona. The easterly phase was being replaced by a westerly one and, again, this change about 1930 is reflected in the rainfall map (Landsberg, 1960, p. 1524) showing the trend between the periods 1906 to 1930 and 1931 to 1955. This time it is the southern and the leeward slopes of the Rockies that become drier.

The dry westerly 1930s. The abnormal westerly phase of the 1930s was associated with dryness generally in the lee of the Rockies and with the notorious dust-bowl droughts of the plains. The great Salt Lake in Utah now shrank to a record low level and, in the course of the decade, dryness set in in Mexico. The r-min. from Toronto, Ont., Canada, to the Rockies is dated ca. 1934 and corresponds precisely to a t-max. extending from Norway to Ontario. The R-min., indicated either by rainfall records or tree rings, is dated *ca. 1919/1925* on the western side of the Canadian Rockies but, farther south, is dated ca. 1929/1935 and corresponds closely to the T-max. of the major temperature oscillation.

The wet 1940s. Information for the 1940s was not available to me when, using south steering as a guide, I wrote (Schove, 1950, p. 161): "An easterly excess had already commenced in North Canada by 1930 and spread to South Canada during the next decade. This is presumably associated with a frontal-B stage in the U. S. A. in the forties with a reversal of trends and prospects of higher rainfall in that country."

These suggestions have now been confirmed by the maps of Rodewald (1952) of the temperature and rainfall changes between 1941 to 1950 and 1931 to 1940. The return to the rainfall pattern of the 1910s (cf. Wagner, 1929) is linked with the change to a more easterly type. Barometric pressure was generally lower.

The drought zone continued to steer south, affecting Arizona from 1934 to 1953 and, in this region, rainfall was 20 per cent less than in the 1930s. Possibly the recent long drought in Texas in the 1950s should be regarded as its final phase in the United States. However, tree-ring and rainfall changes in the 1940s indicated a new drying phase approaching from the north, and past experience reveals that south steering cannot be relied upon for long-range forecasts south of 40° N. Tree rings, in short, prove invaluable in precise climatic chronology, although they cannot yet be used to predict the future!

Conclusions

In my paper titled "Solar Cycles and the Spectrum of Time Since 200 B.C." (this monograph), auroral and European weather history was represented in diagrams and decadal indices. In the present paper, tree-ring indices established separately are used to measure the climatic oscillations affecting the weather history respectively (1) of the subarctic, (2) of North America, and (3) of low latitudes in general. As the various indices are successively refined, it is hoped to obtain a coherent picture of global pressure and temperature fluctuations over the last 2000 years. The preceding application of this wider approach to North America since 1776 shows how the several types of oscillation discussed explain the changes of climatic pattern in the instrumental period. Transatlantic temperature waves are noted that pose problems to be solved by Anglo-American cooperation. The Spectrum of Time technique implies collaboration from various disciplines and from various parts of the world, and I should be most grateful for further information and indices that can be utilized in a project that at one time seemed almost impossible.

Acknowledgments

Most of the other contributors to this volume have also helped in supplying information for Spectrum of Time studies. Acknowledgments have been given in the appropriate articles but, as far as United States evidence is concerned, I should like to thank especially, in addition to those whose names are mentioned in the text: E. Antevs, G. W. Brier, E. R. Dewey, R. W. Fairbridge, W. S. Glock, J. M. Havens, T. N. V. Karlstrom, F. P. Keen, H. H. Lamb, H. Landsberg, I. I. Schell, Waldo E. Smith, M. K. Thomas, H. C. Willett, and L. W. Wing.

References*

ARAKAWA, H. 1960. Selected papers on climatic change. Meteorol. Research Inst. Tokyo. (See his article in Weather, 1957, 12.)

CALLENDER, G. S. 1961. Temperature fluctuations and trends over the earth. Quart. J. Roy. Meteorol. Soc. **87**: 1–12.

CROSS, C. I. 1961. Abstract only in U.G.I. Stockholm Abstracts.

DEWEY, E. R. The 200 year cycle in the length of the sunspot cycle. J. Cycle Research. **9**(2).

* This list omits important articles by other cited contributors whose bibliographies should be consulted separately.

DOUGLASS, A. E. 1936. Climatic cycles and tree growth. Carnegie Inst. Washington Publ. **289**: 111.

KRAUS, E. B. 1956. Discussion on his papers in Quart. J. Roy. Meterol. Soc. **82**: 358.

LANDSBERG, H. E. 1960. Note on the recent climatic fluctuation in the United States. J. Geophys. Research. **65**: 1519–1525.

LYNCH, H. B. 1931. Rainfall and stream runoff in southern California since 1789. Metropolitan Water District of So. Calif., Los Angeles.

RODEWALD, M. 1952. Geografiska Annaler. Stockholm, Sweden.

RODEWALD, M. 1956. Deutsche Hydrog. Zeit. Hamburg. **9**(Heft. 4): 182–186.

SCHOVE, D. J. 1949*d*. European raininess and European temperatures, A.D. 1500–1950. Quart. J. Roy. Meteorol. Soc. **75**: 175–179; 181.

SCHOVE, D. J. 1953*e*. MSc. Thesis. London, England.

SCHOVE, D. J. 1953*f*. Proces-Verbaux des Seances de l'Assocation de Meterologie Memoires et Discussions. Paper 10: 187–193.

SCHOVE, D. J. 1954*a*. Summer temperatures and tree rings in north Scandinavia, A.D. 1461–1950. Geogr. Ann. Stockholm, Sweden. **36**(H 1–2): 40–80.

SCHOVE, D. J. 1955*b*. The sunspot cycle 649 B.C.–A.D. 2000. J. Brit. Astron. Assoc. **66**(2): 59–61.

SCHOVE, D. J. & A. W. G. LOWTHER. 1957. Tree rings and medieval archaeology. Medieval Archaeol., **1**: 78–95.

SCHOVE, D. J. 1959. Cross dating of Anglo-Saxon timbers at Old Windsor and Southampton. Medieval Archaeol. **3**: 288–290.

SCHOVE, D. J. & A. W. G. LOWTHER. 1957. Tree rings and medieval archaeology. Medieval Archaeol. **1**: 78-95.

SCHOVE, D. J. 1961/1962. Annual winds in north-west Europe since A.D. 1625–1960. Geografiska Annaler, Stockholm, Sweden.

SCHOVE, D. J. & A. FREWER. 1961. Tree rings in the Cairngorms, 1900–1956. Scottish Forestry, **15**(2): 63–71.

SCHULMAN, E. 1956. Dendroclimatic Changes in Semi-Arid America. Univ. Ariz. Press. Tucson, Ariz.

TANNEHILL, I. R. 1947. Drought. Princeton Univ. Press. Princeton, N. J.

TITOW, J. 1960. Economic History Review. Cambridge, England.

VISSER, S. W. 1959. (Kon. Ned. Met. Inst.) **75**.

WAGNER, T. 1929. Geografiska Annaler. Stockholm. **11**: 71.

YAMAMOTO, T. 1957. Long-term variations of the jet stream (in Japanese, trans. E. R. Hope. Defense Inform. Service, DRB, Canada.) Kagahu, **27**: 630–631.

ERRATA

Page 612, paragraph 6, the P-min and P-max dates should be italicized since they represent 30-year periods dated by their fifteenth year. Page 613, line 4, "p-max" should be "P-max," and in the list "p-min" should be "P-min." Page 615, the Roman dates of Lowther and the Anglo Saxon dates of Schove are incorrect. Page 616, the thirteenth-century curve was based on only a few specimens. Page 620, line 9 from the bottom of the page, 1929/1935 should be italicized since it relates to 30-year periods 1915/1944 and 1921/1950.

33

Reprinted from *Weather* **24**:390–395 (1969)

THE BIENNIAL OSCILLATION, TREE RINGS AND SUNSPOTS

By D. J. SCHOVE

IN many parts of Europe the summer of 1968 was cool and wet, whereas the summer of 1969 has been warm and dry. This two-year cycle has been evident in English summers ever since 1910, and readers of the interesting articles published in *Weather* in the wet August of 1968 (Wright 1968, Davis 1968) may now be expecting another bad summer in the even year 1970. Certainly, Davis (1967) has been justified meteorologically in suggesting that we should have had double holidays in the odd years and remain at work during the even ones! We can imagine a meteorologist saying to his fiancée " Let's postpone our marriage until the odd year 1971 when we're likely to get a fine honeymoon ". This may sound facetious, but there once was a Director of the Meteorological Office who worked out the date of his honeymoon from weather cycles, and he chose the first fortnight in September. That is usually safe enough, but it proved to be the wettest fortnight of the year. And indeed, the moral of this article, if it has to have one, is that this alternation of dry and wet summers is not likely to last.

The biennial oscillation is one of a group of cycles in the atmosphere that are somehow connected with the sunspot cycle. The periods of natural rebound in the atmosphere are too short to fit the 11-year cycle, so that the solar effect is broken up into four to five sub-cycles. A child swinging freely reaches a higher " amplitude " if its father gives an additional push every fourth or fifth swing, thus providing reinforcement at the critical moment. Computer analysis in the usual way can neither prove nor disprove such an hypothesis, but during the past half-century the two-year cycle has been one-fifth of the solar cycle. Indeed, I previously suggested that this idea of the sun playing the part of the father – if I may mix my metaphors – could be investigated by " a separate analysis of the periods 1870–1905 and 1907–57, when the solar cycles were respectively 12 and 10 years " (Schove 1964, p. 463). The diagram of wine-harvests (Wright 1968) does show that in 1907–57 the period was nearer 2·0 than the normal 2·2 years, and the same diagram suggests that in previous times when the solar cycle was only 10 years, as in 1820–50, the same shortening of period occurred.

Such coincidences are curious, but in order to investigate the effects of sunspot cycles we need to go to long climatological series. The longest series giving indications of raininess are tree-rings. Two pilot studies of missing and narrow tree-rings (Schove 1969, 1970) reveal something of the links between these cycles and the sun.

MISSING RINGS

Missing-ring years in the American South-West are those in which the rings in certain trees fail to occur at all – almost invariably a sign of severe and persistent drought. Such years are indicated, on the graphs of Schulman (1956), by small circles, and they have been listed and weighted (1 to 10) here according to the number of separate pages on which these same years appeared (weights bracketed):

A.D. 279; 302; 452, 455: 522, 526 (2), 563 (3), 567, 585, 590 (2); 611, 640, 645 (2), 660: 738 (2), 741, 742: 1005, '41, 46, 67: 1150, '69; 1243, 51 58 (2), 83, 95, 99: 1304, '07, 61: 1437 (2), 1446, '49 (2), '55 (3), 60 (2), 65, 81 (2), 96 (3); 1500, 06 (2), 07 (2), 10, 16, 22, 31, 32 (3), 42, 80, 82, 84 (3), 85, 90, 91, 98: 1601, 04, 06, 12, 22, 26, 32 (4), 54, 70, 76, 85 (2), 86 (4), 1707, 29, 35, 42, 44, 56, 82 (2): 1801, 10, 13 (2), 20, 28, 45, 47, 57, 61 (4), 64 (2), 70, 79 (2), 82, 93, 97 (2), 99 (7); 1902, 04 (10 cases), 25, (2), 31 (2) and 34 (6). This sample is not homogeneous with time for several reasons. Missing rings are characteristic of very old trees, and the period 1830–1935 is over-represented. For the present investigation, nevertheless, this is the very period in which the dates of sunspot maxima are precisely known, and the bias is thus an advantage.

Each year listed was compared with the year of the *nearest* sunspot maximum. The year 1934 is three years before the sunspot maximum of 1937 and thus belongs to the (X − 3) class; it is considered ' odd' in the sense of this paper. The dates of maxima before 1700 were obtained from Schove (1955), using such minor revisions (*cf*. Schove 1967) as had been made prior to the commencement of this missing-ring study. There is a difficulty here, as sunspot cycles are not always 11 years long. In the present case we are not very interested in classes five years away from maximum, so there is no need to total the number of years in each class and use that as denominator. Instead, we let each year belong to one class only, and add the class (X + 6) to the class (X − 5), and the class (X − 6) to the class (X + 5) and so on, so that the 11 classes, at least since 1700, were effectively equal as far as curves (a) and (b) of Fig. 1 are concerned.

The distribution of missing rings relative to the year of sunspot maximum is given in Fig. 1 curve (b). This may look random, but such rings are much more frequent in the odd years, (X − 3), (X − 1), (X + 1), (X + 3), than the even (X − 4), (X − 2), X or year of maximum, (X + 2), the ratio being 53:22 – an oddness ratio of 2·4. These totals differ from chance at the 1% level.

NARROW RINGS

In the second pilot study, narrow rings for the period 1800–1940 were taken to be those in the lowest decile (or 10%) of the thickness range. The threshold so defined was extended also to preceding years back to 1500, and the distribution through this period is shown in Fig. 1 (curve a). The oddness ratio varies according to the precision with which the dates of sunspot maxima can be specified. In the sixteenth century, maximum-years (X) had been amended (through the evidence of aurorae) to: 1504, 19, 28, 39, 47, 57, 70, 81 and 93, and the oddness ratio was 35:18 or 1·9. Since 1800, however, the oddness ratio has been 46:12 or nearly 4, significant at the 1% level.

Narrow-ring years that occurred in more than one of the tree-series studied were as follows: 1208, 36, 58, 63, 78, 79; 1306, 24, 61, 63; 1403, 07, 13, 25, 37, 38, 44, 45, 46, 50, 55, 60, 65, 71, 75, 1492, 95, 96, 97; 1500, 05, 06, 16, 22, 28, 38, 42, 52, 73, 79, 80, 84, 85, 90, 91; 1600, 25, 26, 32, 37, 45, 54, 68, 70, 75, 84, 85, 86; 1705, 07, 09, 16, 29, 33, 35, 37, 1752, 65, 73, 78, 80, 82, 89; 1809, 13, 22, 42, 45, 47, 51, 61, 63, 64, 71, 80, 82, 88, 89, 96, 99; 1900, 02, 04, 25, 34, 36 and 1943.

Tree-ring series from Mexico and Java were also studied, but this time percentages of long-term mean ring-widths were used. The pattern in Mexico (based on Schulman 1956, Table 81) was very similar to that in the U.S.A., with relative maxima (i.e. raininess) in the four even years and relative minima in the four odd years. The pattern in Java (based on De Boer 1951, col. 5) showed opposite tendencies, well-exemplified in curve (c) of Fig. 1. This curve is based on moderate cycles only (sunspot maxima between 80 and 100), but the means for all types of sunspot cycles indicated 104% for odd and 97% for even years in Java (since 1700), and exactly the opposite for west central Mexico (since 1640). In curve (d) of Fig. 1 the differences, in per cent, are plotted for each of the years of the solar cycle.

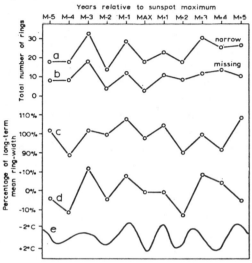

Fig. 1.   The 2·2 year oscillation in tree rings relative to sunspot maxima (a) narrow rings (lowest decile), AD 1500 onwards; (b) missing rings, AD 279 onwards; (c) tree-rings from Java (moderate solar cycles only); (d) Java (all cycles since 1700) *minus* Mexico (all cycles since 1640); (e) Aden temperature (60 mb) reversed

The interpretation of these results involves the concept of an upper high (Schove 1963), with its western boundary (or discontinuity) over Ecuador and its eastern boundary near Ocean Island (170°E). Sunspots, which can cause an eruption of air from this Indian Ocean upper high (cf. Schove 1964, p. 464 and Fig. 7), evidently cause it to wobble like a jelly. When the pressure is high in the eastern Indian Ocean it is dry in Java; when it is high in the eastern Pacific Ocean it is dry in Mexico. In the American South-West, a dry winter occurs at the same time, but a prolonged drought into summer (for the reason, see Schove 1963, Fig. 6), leading to a missing or a narrow ring, can be shown to correspond to years when the Indian Ocean pressure rises (using the data in Schove and Berlage 1965).

Let us suppose that the 'aerial fountains' caused by sunspots arise alternately at the western and eastern discontinuities of this upper high affecting the associated Western and Eastern Crescents (see Schove 1963, Fig. 3). They could in this way give rise to alternating west and east phases in the lower stratosphere.

A recent analysis of the biennial oscillation (Belmont and Dartt 1968) concludes that at levels between 500 and 100 mb " the wave appears to originate in subtropical latitudes in the vicinity of 80°W". This seems consistent with a connection with the discontinuity at 200 mb over Ecuador. The biennial oscillation that affects English summers is related to the remarkable 2·2 year cycle in the equatorial stratosphere, and the lowest curve, (e), of Fig. 1, which will be explained below, does indeed look very similar to the tree-ring curves.

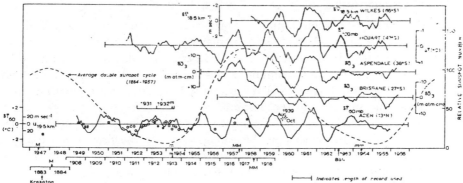

Fig. 2. Solar activity and breaks in the quasi-biennial circulation

EXPLANATION OF SYMBOLS FOR 19·5 KM ZONAL WINDS IN THE 1908–39 EPOCH

● 1883, Equatorial east-west orbit of Krakatoa dust veil front, Aug.–Sept. 1883.—E. D. Archibald (*Rep. Krakatoa Committee*, Roy. Soc., London).

△ 1908, Shirati, Lake Victoria (1½°S), E. Africa Exp. A. Berson. (*Acad. Sci. Rep.*, Berlin 1910).

□ 1908, Dar-es-Salam (7½°S), E. Africa Exp. A. Berson. (*Acad. Sci. Rep.*, Berlin 1910).

○ 1910–13, 15–16, 1918 Batavia (Jakarta, 6°S), W. Van Bemmelen (*Roy. Magn. & Met. Obs.*, Verh. 1 & 6, 1920).

⊙ 1917, Bandoeng (7°S), W. Van Bemmelen (*Roy. Magn. & Met. Obs.*, Verh, 1 & 6, 1920).

✱ 1932–33, (" Polar Year ") Mogadisho (2°S), M. Bossolasco (*Geof. Pu. Appl.* 14, 139).

+ 1939, Traverses 2° to 5°S, Finnish Acad. Sc. Atl. Exp., A. Vuorela (*Ann. Ac. Sc. Fenn.* 1, 79, 1950).

Reproduced from Berson and Kulkarni (1968) by kind permission

THE EQUATORIAL OSCILLATION

" There is only one regular cycle in meteorology longer than the seasonal year", I wrote in 1964 (p. 457), adding: "this is the remarkable alternation of west and east winds in the equatorial stratosphere, with a period of 2·2 years ". Since 1964, however, even this cycle has begun to falter. We have direct or indirect evidence for easterly phases at 60 mb (19½ km) in 1883 . . . , 1909.7, 11.8, 14.1, 16.2, 18.2. . . . 1923, 25, about 1928.2, . . ., 34, 36 . . about 39.2 and 41.2, 43 , . . about 46.2 and 48.2, 50.1, 52.4, 54.8, 56.9, 59.2, 60.8 to 61.2 and late 62. (Alterations to the previous list in Schove 1964 are based on spring ozone data in Angell and Korshover (1967, Fig. 5), and on the Aden temperature evidence provided by Berson and Kulkarni in our Fig. 2). To show the amplitude of the biennial aspect of the oscillation, wind (*v*), temperature (*T*) and ozone data in Fig. 2 are plotted as 12-month moving means minus 24-month moving

" normals ". This is an effective ' band-pass filter '. Historic data since the year of Krakatoa have been incorporated in the diagram in relation to the double sunspot cycle of 22 years. The biennial oscillation hardly shows at sunspot minimum, and its amplitude increased in 1954, three years before sunspot maximum, (X − 3), and decreased in 1953 and 1963, (X + 6) years. The Aden temperature curve from 1949 to 1965 in this diagram has been used for curve (e) of Fig. 1. The sign has been reversed so that cold easterly phases are made to correspond with dry years in the U.S.A. and Mexico, and with wet years in Java.

ENGLISH SUMMERS

The biennial oscillation, both in tree-rings and in the tropical stratosphere, must be a terrestrial sub-cycle of the solar cycle of 11 years, with a wavelength of one-fifth. In the tree-rings studied here, the sub-cycle is symmetrical with respect to the sunspot maximum. This sub-cycle resembles the ticking of an old and rusty grandfather clock that repeatedly has to be restarted with a push. In our case the sunspots do the pushing, being effective about one year after sunspot minimum (cf. Fig. 2) or three years before sunspot maximum (Fig. 1).

After all this, the reader must be asking, what are the prospects – if any – for summer in the even year 1970? Let us consider the sunspots first. In the present century, maximum years have usually ended in a 7 (e.g., 1907, 17, . . . 37, 47, 57). Can we argue that that is why our odd years have been better than the even? Our best honey-summers (Davis 1968, Fig. 2) between 1935, an (X − 2) year, and 1961, an (X + 4) year, although mainly odd-numbered years, have been even with respect to sunspot maximum, as if English rainfall behaves like that of Java and in opposition to that in the New World. Even if this were more than a half-truth (which it isn't, as I hope to show in a sequel to this article), the one prediction that seems certain (c.f. Schove 1955) is that we are now in for a succession of longer sunspot cycles.

There have certainly been periods, like the present, when odd-numbered summers have been better (1205–41, 1463–77, 1581–1615, 1690–1703, 1718–60), and others when they have been worse (1101–15, 1248–74, 1336–84, 1414–30, 1484–1506, 1529–56, 1617–36, 1820–52). However, I cannot find in the central dates either the theoretical 26-year period (Wright 1968), or any obvious link with sunspots.

Meanwhile, if any meteorologist is thinking of getting married, my advice is: don't take any account of weather cycles when you fix your honeymoon-date.

REFERENCES

ANGELL, J. K. and   1967   Biennial variation in springtime temperature and total
KORSHOVER, J.   ozone in extratropical latitudes. *Mon. Wea. Rev.*, 95, pp.
                           757–762
                   1968   Additional evidence for quasi-biennial variations in
                           tropospheric parameters. *Ibid.*, 96, pp. 778–784
BELMONT, A. D. and 1968   Variation with longitude of the quasi-biennial oscillation.
DARTT, D. G.   *Ibid.*, 96, pp. 767–777

BERSON, F. A. and
  KULKARNI, R. N.   1968   Sunspot cycle and the quasi-biennial oscillation. *Nature*, **217**, pp. 1133–4

BOER, H. J. DE   1951   Tree-ring measurements and weather fluctuations in Java from AD 1514. *Proc. Kon. Ned. Akad. Wetens.*, Amsterdam, B **54**, pp. 194–209

DAVIS, N. E.   1967   The summers of north-west Europe. *Met. Mag.*, **96**, pp. 178–187

                    1968   An optimum summer weather index. *Weather*, **23**, pp. 305–317

SCHOVE, D. J.   1955   The sunspot cycle, 649 BC to AD 2000. *Journ. Geophys. Res.*, **60**, pp. 127–146

                    1961   Solar cycles and the spectrum of time. *Annals of the New York Acad. Sci.*, **95**, pp. 107–123 and 605–622

                    1963   Models of the southern oscillation in the 300–100 mb layer and the basis of seasonal forecasting. *Geof. Pura e Applicata*, **55**, pp. 249–261

                    1964   Solar cycles and equatorial climates. *Geol. Rundschau*, **54** (1), pp. 448–477

                    1967   Sunspot cycles. In: Ed. R. W. Fairbridge, *Encyclopaedia of atmospheric sciences and astrogeology*, New York (Reinhold), pp. 963–968

                    1969   Proc. Solar Terrestrial Conference, Brussels (Presses Academiques Européennes)

                    1970   (in progress)

SCHOVE, D. J. and
  BERLAGE, H. P.   1965   Pressure anomalies in the Indian Ocean, 1796–1960. *Pure and Applied Geophysics*, **61**, pp. 86–98

SCHULMAN, E.   1956   *Dendroclimatic changes in semiarid America.* Univ. of Arizona Press.

WRIGHT, P. B.   1968   Wine harvests in Luxembourg and the biennial oscillation in European summers. *Weather*, **23** (8), pp. 300–304

[*Editor's Note:* In 1976 a drought in an even year marked the end of the oddness rule that had lasted in northwestern Europe for about 70 years.

A later paper by Schove is "The Biennial and Triennial Cycles: Stratospheric Reversals," *Jour. Interdisciplinary Cycle Research* **3**:349–354, and see also pp. 361–363, 409–411 (1972).

An unpublished computer study made by the Meteorological Office, Bracknell, England, reveals that there is a biennial oscillation of pressure between the northern and southern hemispheres near sunspot maxima. Hot, dry summers in the Alps were mainly even years at c. 1670–1730 but, as in England, they were odd years at c. 1910–1975 (X-ray data). The biennial oscillation since 1960 at the 50 mb level has shown easterly peaks at 1963.4, 66.0, 69.0, 71.0, 72.9, 75.4, 77.7, 79.9, and 1982.]

34

Reprinted from *Weather* **26**:201–209 (1971)

BIENNIAL OSCILLATIONS AND SOLAR CYCLES, AD 1490-1970

By D. J. SCHOVE

St David's College, Beckenham, Kent

" IN many parts of Europe the summer of 1968 was cool and wet, whereas the summer of 1969 (was) warm and dry. This two-year cycle has been evident in English summers ever since 1910 and readers . . . may now be expecting another bad summer in the even year 1970 ". Thus ran the opening remarks of a previous article on ' The biennial oscillation, tree-rings and sunspots ' (Schove 1969a).

It would indeed be ' odd ', some people said, if we had two good summers in succession, and they invoked the ' law of averages ' to confirm their suspicions that it was best to take advantage of the de-restricted travel allowances and go abroad in 1970. The ' law of averages ' has no logical existence, although the biennial oscillation can be shown to exist in the period 1907-1967. The mean monthly pattern of odd-minus-even calendar years, through the period 1899 to 1964 (1939-46 omitted) was studied over the northern hemisphere by Murray and Moffitt (1969), and their maps published in *Weather*.

The point of the previous article was to suggest that Nature did not react differently to odd and even years in the AD calendar merely because they were odd or even – that would be *too* remarkable. It was suggested that Nature did react differently to odd and even years relative to the year in which sunspot maximum occurred. That is almost equally remarkable, but it is usually true. The last sunspot maximum was 1968.9, so that 1970 was even with regard to the solar maximum and should have behaved like an odd year in the sense of Murray and Moffitt (1969). Indeed, despite the heavy rain in August, the summer of 1970 was generally good. In our previous article we concluded that the moral was: " this alternation of dry and wet summers is not likely to last ". We shall show in this article that when sunspot activity is reduced we do not get a biennial oscillation at all; instead we get a triennial cycle, and this may well be the moral of the present sequel. The pattern of fluctuations for strong and weak solar cycles is quite distinct.

EQUATORIAL INFLATION MODEL

The writer uses a simple model to illustrate *how* the solar effects work, although he is well aware that far more complicated processes are involved; his model (Schove 1963) must not, therefore, be taken too literally. We know

(Heastie and Stephenson 1960) that an upper high lies over the equator in the Indian Ocean region, and we suppose that eruptions of air from this occur near strong solar maxima. These eruptions can account only for feeble and temporary oscillations such as those of 11- and 22-years. We must adapt the model to interpret the more reliable biennial oscillations.

This simple pattern of the upper high is illustrated in Fig. 1. If pressure is above normal throughout the troposphere over the Indian Ocean – and this is usually the case when surface pressure (c.f. Schove and Berlage 1965) is also above normal – wind convergence in depth occurs towards a point near Ocean Island, and wind divergence or diffluence occurs on the opposite side of the upper high near Ecuador. Normally east winds along the equator may be supposed to carry the extra air back into the upper high, perhaps strengthening it in the process.

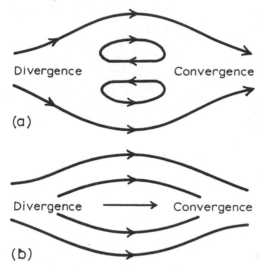

Divergence   Convergence

(a)

Divergence   ⟶   Convergence

(b)

Fig. 1. The principle of equatorial inflation.
(a) Easterly phase: normal upper air flow. Divergence region is often near Ecuador (80°W) and convergence near Ocean Island (170°E) (Schove 1963). Surface pressure anomalies, where positive, are similar.
(b) Westerly phase: in upper troposphere or, presumably, in stratosphere. Equatorial westerly winds are due to increased divergence and convergence resulting from changes in flow at lower levels.

Normal 200-mb contours of the upper high show two separate cells north and south of the equator with easterlies intervening (Fig. 1 (a)); when the surface pressure-anomalies are positive they also show the same geographical pattern. Lifting of the 200-mb level in the tropics (c.f. Schove 1967 b) tends to be sudden (1951.3, 1953.0, throughout 1957) whereas falls are slow (1952, mid 1953 to 1955, 1958–60). During periods of sudden rise, westerly components, at least in East Africa, occasionally increase (as in January 1958) as if inflation of the upper high at the equator itself is being produced from below by wind convergence. The pattern of Fig. 1 (a) gives way to the pattern of Fig. 1 (b). This 'principle of equatorial inflation' (Schove 1963, p. 254) seems to account for the persistence of pressure anomalies so characteristic of equatorial latitudes.

We can assume that the same principle applies to the stratosphere, that easterly winds in its lower layers, e.g. at 50 mb, lead to convergence and raise the contours above until a single, elongated high (Fig. 1 (b)) is produced at, say, 20 mb; at this level the high is less durable. Equatorial easterlies return, first at higher then at lower levels. At 30 mb (about 24 km), westerly phases have occurred in a regular 2·2 year cycle with maxima *c.* 1947.2, *c.* 1949.2, *c.* 1951.2, and approximately 1953.6, 1955.9, 1958.3, 1960.3, 1964.5 (see Dartt and Belmont 1970, Figs. 5 and 3).

The period of 2·2 years may be a natural cycle in the stratosphere, but the solar cycle of 11 years has more effect on the upper tropospheric high, at least up to the 200-mb level where equatorial inflation occurs on a different scale. Tropospheric pressure, in the central Indian Ocean beneath the upper high, rises slowly during the period of sunspot minimum, but, about three years before sunspot maximum (the $(X - 3)$ years of our previous article), both oscillations seem to be involved. Tree-rings give evidence (Schove 1969a) in the even years (relative to sunspot maximum) of decreased rainfall in Java and increased rainfall in Mexico and the American South-west, i.e. near the upper convergence and divergence regions in Fig. 1. Nile flood data indicate that, in this case, Ethiopian rainfall behaves like that of Mexico.

EFFECTS AT HIGHER LATITUDES

Anomalies of surface pressure associated with specific years relative to the solar maximum were mapped by Scherhag (1950, Figs. 7 to 12). During the period 1897 to 1940 the pressures were slightly less low in the Greenland trough area in years that were, in that sense, even.

Pressure patterns away from the tropics are complicated by various feed-back effects. Nevertheless, the years of sunspot maximum have usually been odd from 1907 to 1957, so that years that were *odd* calendrically were *even* with respect to sunspot maximum. We may test our conclusions, therefore, from Scherhag's maps against the maps of Murray and Moffitt (1969, pp. 384–385). From March onwards the significantly high pressures tend to occur between 45° and 70°N, and the significantly low pressures between 20° and 45°N. Over the Indian Ocean, pressure was low in calendrically-odd years from 1912 to 1948 although, as we shall explain, other longer tropospheric cycles dominate surface pressure changes in that area and the rule was reversed in the 1950s. The persistence of an oddness rule throughout the sixty-year period 1910 to 1969 is not a global phenomenon, and in Britain must be partly accidental, although some even longer spells can be detected from varve-* analysis in the Late Glacial (c.f. Schove 1971).

AMPLITUDE OF SOLAR CYCLES

Solar cycles vary considerably in their amplitude. The sunspot ' number ' for 1957 was 190·2 and the official maximum at 1957.9 was 201·3. The recent maximum in 1968.9 was 111. This last sunspot cycle was still ' strong ', although the slowness of the decline since 1968 suggests that we are entering a period of longer cycles. In previous articles (Schove 1955, 1967a), sunspot cycles since AD 300 were classified as very strong (SS), strong (S), moderate (M), weak

---

Varves are thin, alternating layers of coarse and fine sediment laid down in glacial lakes as a result of seasonal variations in deposition.

(W), and very weak (WW), with an occasional intermediate (MS or WM) type. 'Very strong' cycles since 1749 are those with maxima of 140 or more; 'strong' are those with maxima 110 or more, and so on. The classifications specified in the 1955 article were used, as in the previous article, to calculate 'oddness ratios' and differences between 'even' and 'odd' years (relative to solar maximum). An 'oddness ratio' was there defined as the ratio of occurrences (e.g. of missing or narrow tree-rings, indicative in semi-arid America of drought) in the odd years $(X - 3)$, $(X - 1)$, $(X + 1)$, $(X + 3)$ compared with those in the even years $(X - 4)$, $(X - 2)$, (X), $(X + 2)$, taking X as the year of the maximum.

TABLE 1.   The oddness rule and intensity of solar activity: the difference between tree-rings in odd and even years relative to the sunspot maximum $(X - 4)$ to $(X + 3)$, expressed as ratios and as differences of per cent growth. For sources see Schove (1969a). The results are significant, although for SS cycles in Mexico only 32 years were available.

Sunspot activity	Oddness ratios USA		Percentage growth difference Even minus odd	
	Missing rings	Narrow rings	Mexico mean	Java mean
SS (Very strong)	1	0·8	1	− 7
S (Strong)	14	3·3	7	− 5
MS + M	2	1·3	10	− 4
WM, W, WW	2	2·1	10	− 1

MEXICO AND INDONESIA

The oddness ratios for the United States tree-rings, and the percentage differences for tree-rings in Mexico and Java, are shown in Table 1. Near the 'divergence' and 'convergence' points of Fig. 1 – near Ecuador and Ocean Island – it would seem that varying solar amplitudes do not upset the rule. In Mexico, even years appear to be drier for all types of cycle; in the USA this effect broke down with very strong cycles, when the ratio was near unity – that expected by chance. In Java, even years were wetter for both strong and weak solar cycles. We do notice, however, that the biennial oscillation associated with the sunspot cycles is, in the USA and Java, most pronounced when the sunspot activity is strong. Indeed in many parts of the world there is a threshold above which the biennial oscillation appears. The thresholds for the several curves of Fig. 2 are not identical, partly because of the varying size of the samples, but the contrast with Fig. 4 makes the existence of thresholds obvious.

The biennial oscillation was pronounced in both German tree-rings and Franco-Swiss vintage data in the 1530's (Schove 1964, p. 463) when the sunspot cycle was strong (Fig. 3). The first of a series of strong maxima occurred in 1527–8 (= 1528.0) and another in c. 1538. The even summers 1532, 34, 36, 38 and 40 were all dry, but 1529, 1533 and 1539 were wet over north-west Europe.

Late vintages in northern France occur when cool, cloudy air spreads in, either from the north-north-east or south-west, and it is interesting to find that in even years vintages are late, although in England it is drier than usual. This probably means that in 'strong' cycles the coolness is associated with the north-north-east type, when summers are dry in most parts of the British Isles.

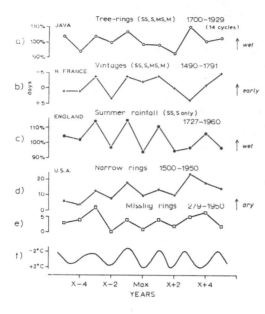

Fig. 2. The 2·2-year oscillation and stronger solar maxima. (a) Java. Tree-rings, 1700–1929. Indicative of raininess. (b) North France. Wine-harvest dates, 1490–1791. Indicative of warmth (Data from Ladurie 1966, 1967) (Negative). (c) England. Summer rain. 1727–1960, %. (Data kindly provided by Meteorological Office). (d) USA. Narrow rings. Drought indications. AD 1500–1950. (e) USA. Missing rings. Drought indications. AD. 279–1950. (f) Aden temperature (60 mb), reversed. 1949–65. Reflects the biennial oscillation in the stratosphere. Sources in Schove (1969)

WEAK CYCLES AND LONGER OSCILLATIONS

When solar cycles are weak, at least in Europe, there is no biennial oscillation. Instead, vague oscillations of about three years appear. These triennial oscillations, like the biennial ones, are still geared to the sunspot cycle. There is a tendency to symmetry with respect to sunspot maximum, the peaks in the curves of Fig. 4 occurring especially at $(X - 3\frac{1}{2})$, X and $(X + 3)$. The threshold for England was a sunspot number at maximum of 110, but 80 was used for the other curves.

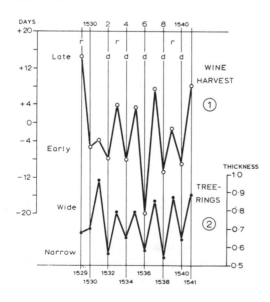

Fig. 3. The biennial sawtooth signature, 1530–1541. d, r = dry, or wet, in the months V to VIII (north-west Europe). The upper curve shows wine-harvests; they were early in the dry, even years in France and Switzerland. The lower curve shows tree-rings of oak at Odenwald; they were wide in the wetter, odd years. (Based on Ladurie 1967, Fig. 234 and p. 233)

Low pressure over the Indian Ocean corresponds with wet years in Java and dry years in Chile. Over the central Indian Ocean even the three-year oscillation tends to be swamped by an 11-year oscillation that has a main low at sunspot maximum and three peaks at $(X - 2\frac{1}{2})$, $(X + 2)$ and $(X + 4\frac{1}{2})$; this pattern occurs with only slight modifications for 'strong', 'moderate' and 'weak' cycles.

ENGLISH SUMMERS AND SOLAR AMPLITUDE

The English summer rainfall curve in Fig. 4 shows an unexpected correlation with the sunspot curve. In the year of solar maximum the rainfall was 120% of normal and four years later only 83%. The eleven individual percentages are themselves each a mean for eleven separate solar cycles, and the Chi-squared test rates the distribution as significant at the 6% level. As the year of sunspot maximum can usually be predicted in advance, this result may be useful in planning our English summer holidays – but the next maximum-year is not due until near 1980!

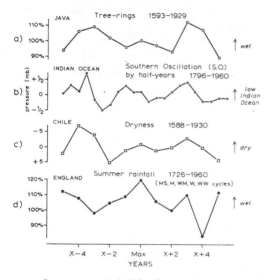

Fig. 4. Triennial oscillations and the weaker sunspot maxima, WM + W + WW (a) Java. Tree-rings, 1593–1929. Indicative of raininess. (b) Indian Ocean countries. Reversed pressure by half-years, 1796–1960 (Schove and Berlage 1965). (c) Chile. Dryness, 1588–1930 (Data from Taulis 1934). (d) England. Summer rainfall. Percentage of 1916–50 normals. 1726–1960 (MS + M + WM + W + WW)

Summer rainfall in the maximum sunspot year is very different in 'strong' cycles (Fig. 2), when it was the driest, and in weaker cycles (Fig. 4), when it was the wettest. This contrast suggested that there might also be a systematic trend with solar amplitude of the temperatures of English summers. Summer temperatures for Central England were kindly sent by Professor Manley (based on Manley 1958, 1963) extending from 1671 to 1968, and compared with the solar data (Schove 1967a). These data were used for Fig. 5, which shows very clearly the result expected. Summer temperatures in a series from Philadelphia (Landsberg et al., 1968) show a similar curve.

The full Table 2 shows how summer temperature varied in the course of solar cycles of specified amplitudes. Bold and italic type are used to emphasize apparently non-random features, according to sign, even though some of these

TABLE 2. Summer temperature in central England in relation to maximum and amplitude of solar cycle (Departure from 15°C in tenths-degree C)

Amplitude of solar cycle	Number of cycles	−5	−4	−3	−2	−1	X	+1	+2	+3	=Odd =Even +4	+5	Means Even	Means Odd	Difference (even minus odd)
SS	5	+6	+4	+6	**+7**	0	**+15**	+7	**+11**	+7	+1	+6	**+9**	+5	+4
S	6	+4	+3	+1	**+8**	+3	+4	+4	−1	+4	+5	+4	+3	+3	+1
MS	2	−2	−1	**+10**	+3	+5	−5	+1	−7	−5	+3	−3	−3	+3	−5
M	2	−3	+1	+7	+6	**+13**	−3	+9	+5	+6	*−1*	**+11**	+2	+9	−7
WM	2	−1	+1	**+17**	+4	+1	−2	−5	+9	+1	+3	**+16**	+3	+3	*−1*
W	3	+1	*−10*	+4	+3	+3	*−13*	0	+4	+2	**+14**	0	−4	+2	−6
WW	5	+2	+3	−1	+3	+6	−2	+2	+6	+6	−1	−2	+3	+3	−1

Apparently non-random features are emphasised by the use of **bold** and *italic* type. Compare column X and Fig. 5.

would disappear in a three-tier contingency table. Even years are cooler unless the solar cycles are ' strong '. The two-year cycle occurs, but with reversed phase, when the solar cycles are ' moderate '.

Data have been collected for computer studies to determine the pressure patterns involved in the biennial and triennial cycles of this article. We may then be able to determine how much reality lies behind the models of Fig. 1. In conclusion, we mention two applications of these results that are described in more detail elsewhere.

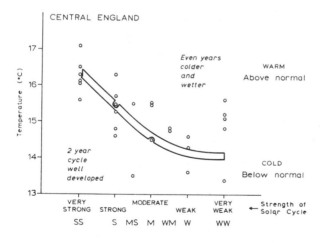

Fig. 5. Summer temperature in the years of sunspot maximum (X) and the amplitude of solar cycles 1674, 1685–1968

FAMINE AND REVOLT

We have found above that both the wettest and driest summers have occurred at sunspot maximum, so it would therefore seem reasonable to expect a correlation of famines and revolts with the solar cycle. There are again characteristically distinct patterns for different solar amplitudes and for different parts of Europe, but for Europe as a whole, famines and scarcities since 1500 were nearly twice as frequent in each of the years $(X - 2)$ to $(X + 1)$ as in the years $(X - 4)$. In north-west Europe, the solar effects were most pronounced in weak solar cycles, such as the 1690's and 1790's, scarcities increasing steadily from 5% in the years $(X - 5)$ and $(X - 4)$ to 13% in the years X and $(X + 1)$. Revolts in the same region increased from 4% in the years $(X - 5)$ and $(X - 4)$ to 18% in the years $(X + 1)$. The list of revolts used was taken from Sorokin (1937), who had no preconceptions about sunspots.

VARVE-TELECONNECTIONS

If we assume that swings between biennial and triennial oscillations during the late glacial epoch corresponded to swings between 'strong' and 'weak' solar cycles in the same manner as they do today, we can use varves as a means of determining solar changes during the Ice Age. With this aim in mind, such tendencies were among various non-random elements selected from varve-series for a computer program. The results of the study are not yet available, but in preparing them it proved possible to use some of these non-random features for cross-dating, first of all the Finnish with the Swedish varves, then the north Canadian with the south Canadian varves, and finally the American with the European. Fuller details of the methods and the results are given in Schove 1969b, 1971a, b.

SUMMERS OF THE SEVENTIES

Our study of sunspots and weather has enabled us to interpret the past, but we would like to use it to determine the future as well. Let us consider the alternatives for 1971, which will be:

(a) an odd year calendrically and, if it is behaving according to type, we can consult the maps of Murray and Moffitt (1969). Higher pressure than normal should develop in May over southern Scandinavia and spread SW beyond Ireland by August.

(b) an odd year with respect to the recent sunspot maximum and, if the biennial oscillation had adjusted itself to this (it went wrong in 1805 and 1718), we should expect exactly the opposite, with wetness in Britain, and SW winds in northern France.

(c) an $(X + 3)$ year, so its characteristics depend on the amplitude of the current cycle. Table 2 suggests a normal summer of $15.4°C \pm \frac{1}{2}°C$ for a 'strong' solar cycle.

The recent rule, that maximum sunspot years are odd and end in 7 (Schove 1967), will certainly not apply to the next few cycles, and if we are going back to weaker activity and 12-year cycles, as was the case in 1910, perhaps we are due for a change to triennial cycles. Triennial cycles were discussed by Brooks and Glasspoole (1928, p. 179), who noted how 1888, 91, 94, 1900, 03, 06 and 09 had all been wet years, but it was only about 1910, just when meteorologists had begun to use the cycle for long-range forecasting, that the pattern changed to the even years – 1910, 12, 16, 18, 20 and 22 –

starting a run that has continued, in the summers, intermittently to the present day.

Perhaps we should now start to investigate the triennial oscillations, which in the period 1880–1910 appear to have begun in the ' Eastern Crescent ' near the ' convergence ' point of Fig. 1 (c.f. Schove 1964, Fig. 6), probably in the autumns of the northern hemisphere. Perhaps it is time we gave the computers their next meal!

REFERENCES

BROOKS, C. E. P. and GLASSPOOLE, J.	1928	*British floods and droughts*. London (E. Benn)
DARTT, D. G. and BELMONT, A. D.	1970	A global analysis of the variability of the quasi-biennial oscillation. *Quart. J. R. Met. Soc.*, **96**, pp. 186–194
HEASTIE, H. and STEPHENSON, P. M.	1960	Upper winds over the world. Parts I and II. Met. Off., *Geophys. Mem.* No. 103
LADURIE, E.	1967	*Histoire du climat depuis l'an mil*. Paris (Flammarion)
LANDSBERG, H. E. *et al.*	1968	University of Maryland, *Tech. Note* BN–571, Table VI
MANLEY, G.	1958	On the frequency of snowfall in metropolitan England. *Quart. J. R. Met. Soc.*, **84**, pp. 70–72
	1963	Seventeenth-century London temperatures: some further experiments. *Weather*, **18**, pp. 98–105
MURRAY, R. and MOFFITT, B. J.	1969	Monthly patterns of the quasi-biennial oscillation. *Ibid.*, **24**, pp. 382–389
SCHERHAG, R.	1950	Betrachtungen z. allgemeinen atmosphärischen Zirkulation. *Deut. Hydrog. Zeit.*, **3**, pp. 108–111
SCHOVE, D. J.	1955	The sunspot cycle, 649 BC to AD 2000. *J. Geophys. Res.*, **60**, pp. 127–146
	1963	Models of the southern oscillation in the 300–100-mb layer. *Geof. Pura Appl.*, **55**, pp. 249–261
	1964	Solar cycles and equatorial climates. *Geol. Rund.*, **54**, pp. 448–477
	1967a	Sunspot cycles. In: *Encyclopaedia of atmospheric sciences and astrology*. Ed.: R. W. Fairbridge. New York (Reinhold), pp. 963–968
	1967b	Anomalies at the 200-mb level in the tropics, 1950–1962. *Quart. J. R. Met. Soc.*, **93**, pp. 138–9
	1969a	The biennial oscillation, tree-rings and sunspots. *Weather*, **24**, pp. 390–396
	1969b	Reviews of three books on Swedish geochronology. *Geog. J.*, **135**, pp. 594–596
	1971a	A varve-teleconnection project. INQUA, Paris (1969)
	1971b	*Geograf. Ann.* (in press)
SCHOVE, D. J. and BERLAGE, H. P.	1965	Pressure anomalies in the Indian Ocean, 1796–1960. *Pure Appl. Geophys.*, **61**, pp. 86–98
SOROKIN, P. A.	1937	*Social and cultural dynamics*, **3**. New York (American Book Co.). Appendix pp. 578 ff
TAULIS, E.	1934	De la distribution des pluies au Chili. La périodicité des pluies depuis quatre cents ans. *Mat. études calam.*, **9**, pp. 1–20

ERRATUM

The Manley (1958) reference is not the correct one. Instead see the data reprinted in Manley, G., 1974, Central England Temperatures: Monthly Means 1659 to 1973, *Royal Meteorol. Soc. Quart. Jour.* **100**:389–405.

35

Reprinted from pages 38–40, 45–48, and 50–52 of *Internat. Afr. Inst. (London)*
Environ. Spec. Rep. 6, D. Dalby, R. J. Harrison-Church, and F. Bezzaz, eds., 1977

AFRICAN DROUGHTS AND THE SPECTRUM OF TIME

D. J. Schove

THE meteorological factors of droughts past and present were briefly considered in the Abstract 'African Drought and Weather History' (in Dalby and Harrison Church 1973: 29–30). The historical information received for the Sahel and Ethiopia is here considered in more detail, especially in relation to the Nile Flood; the Nile levels are highly correlated with summer rainfall in Ethiopia, the Sahel and the Indian Ocean countries generally.

The northern limit of the summer rains of the south Sahara is a sharp one, and today the northern limit of settlement is generally the 15th Parallel although in the Niger bend the towns of Timbuktu and Gao still exist at $17°N$. The northward decrease of rainfall is illustrated in the Upper Nile region. At El Obeid ($13°N$) in Kordofan rainfall averages 13 inches, at Khartoum ($15°N$) it averages 6 or 7 inches, whilst at Merowe ($18°N$) it is only about 1 inch. Three different belts of population can accordingly be distinguished. To the south of $13°N$ — near rivers or oases — settled villagers flourish. In the dry grasslands of the middle belt (c.$13°N$) cattle-breeding nomads (Baḳḳāra) are found, wheras camel-breeding nomads (Abbāla) wander in the dry steppe lands near $15°N$. Near $18°N$ the desert is now almost uninhabited. A very slight movement of this belt makes all the difference between plenty and starvation to people of the Sahel.

Pluvials and Interpluvials
Climatic oscillations over the past 40,000 years are discussed in a series of papers edited by Rognon (1976) and some pluvials and interpluvial periods of the post-glacial have been dated by radiocarbon (Schove 1969, 1975), and expressed in charts (e.g. Shaw 1976) and maps (e.g. Street and Grove 1976). At the time of the Saharan Neolithic the wetter periods were generally the hotter periods (fig. 1) and these pluvials are thus classified as *Hot*. In recent centuries Maley (1976) has shown that on the other hand Lake Chad levels were higher during the colder periods and the wet period of the Little Ice Age, c.1590/1850 is therefore classified as a *Cold* Subpluvial, the recent dry period occurring in a century which has so far been outstandingly warm.

To separate the various factors involved, the writer distinguishes different types of pluvial as follows:

(D) *Deglaciation Pluvials* as in the USA in the Late Glacial.
(I) *Indian Ocean Pluvials* notably c.10,000/5000 B.P. where B.P. means radiocarbon years before the present.
(H) *Hot Pluvials*, as in the Saharan Neolithic of fig. 2, notably c.5350 B.P. which corresponds to c.4850 B.C. in absolute tree-ring chronology.
(W) *Winter Pluvials* notably near 8000 and 4400 B.P. which affect some Jan/Apr regimes especially in the New World and to some extent the northern Sahara.
(C) *Cold Pluvials* as in the Little Ice Age c.A.D. 1700 (cf. fig. 3).

After the Ice Age there was thus an Indian Ocean pluvial, and 'in Africa, for about four centuries, the equatorial rain belts swung further north in August and further south in February. Lake Chad grew into Lake Megachad and Nile Floods 12 m above the present normal occurred' (Schove 1969, p. 227). This short phase culminated c.9250 B.P. and probably affected India and Australia as well. As we shall explain below the recent drought is

[*Editor's Note:* A spectral analysis of the Nile Flood data will be published soon by R. W. Fairbridge and S. Hameed (personal communication, May 1983), who find cycles of 77 and 18.4 years, the latter in the Winter (Equatorial) Nile only.]

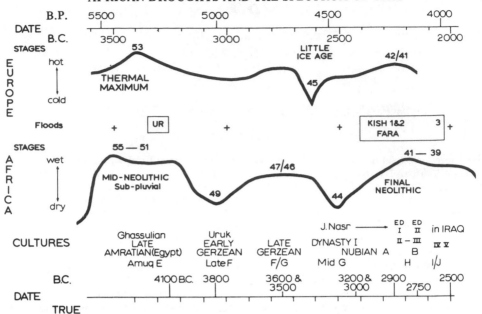

Fig. 4.1. 'Hot' pluvials of the Saharan Neolithic c.4500/2800 B.C. The figures above the curve are radio-carbon centuries before the present. Stade 45 thus means 4550 B.P. which is known to be about 3370 B.C. (Reproduced from Schove 1969, Plate II, *Palaeoecology of Africa* 4 by kind permission of E. M. Van Zinderen Bakker)

partly the converse of this pattern as barometric pressure in the Indian Ocean was high during the decade 1964–1973.

Nile Flood Records
In Ancient Egypt periods of famine due to failure of the summer rains in Ethiopia (and probably the Sahel) are recorded with low Nile levels. They often led to a Time of Troubles between the Dynasties, and are being studied by Barbara Bell (1975). Tree-ring and other evidence will enable these periods (c.2180 and c.1765/1630 B.C. with a period of better floods c.1840/1765 B.C.) to be precisely dated. At present the conventional date for the beginning of the Middle Kingdom is 1991 B.C.

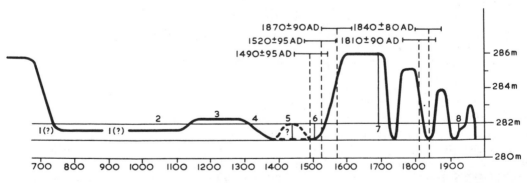

Fig. 4.2. Lake Chad levels since about A.D. 700. (After the schematic curve by J. Maley, 1976, *Palaeo-ecology of Africa* 9, ed. E. M. Van Zinderen Bakker.)

284

A fairly continuous curve has been established for the Graeco-Roman period thanks to the help of Madame Bonneau, whose works on the Nile provided the basic data. Prolonged periods of very low levels A.D. 53/63, 151/170 and 254/265 may have been partly responsible for economic crises in the Roman world, and other low level periods are 253/242 B.C., 192/187 B.C., 139/130 B.C., 50/42 B.C., A.D. 12/19 and 111/124. Some of these failures of the Ethiopian summer rain would seem to be comparable in duration and severity with that of 1965/73.

From A.D. 622 the original Nile records were usually reliable and there are several fifteenth century lists, inaccurately used in Toussoun's well-known catalogue (1925). Help has been obtained from the late Professor Popper, from Dr. Beshir and other contributors to the Spectrum of Time project who have sent translations from earlier Arabic sources such as al-Hakam.

The earliest known Ethiopian drought is that in the *Mashafa Senkesar* known to have occurred within the period 831/849 (Pankhurst 1972/3). The Emperor is said to have banished the leader of the Ethiopian church, stating that 'all our men are dying of the plague (i.e. evidently famine-pestilence rather than true plague), all our beasts and cattle have perished, and God hath restrained the heavens so that they cannot rain upon our land . . .' This presumably synchronises with the period of weak Nile Floods reported c.832/842 in Egypt.

Further runs of weak Niles occurred c.939/953 and 963/968 (recorded by Yahya, the continuator of Eutychius) and in the eleventh century in 1006/1013 and in 1054/1065.

The next occasion when our Ethiopian source refers to drought is within the period 1131/1145, evidently corresponding to the period of low Niles reported in Egypt c.1141/44. In Ethiopia 'famine and plague (famine-pestilence) broke out in the land, and the rain would not fall on the fields, and great tribulation came upon the people.'

The terrible famine about 1200 in Egypt is described by 'Abd al-Latif from personal experience. In the summer of 1200 cannibalism became common. In the thirteenth century we have the first references to famine so far collected from Sudanese annals, and these refer to 1252, 1258 and 1272, although (judging from the Nile Flood data) the last famine may have been due to too much rather than too little rain. Weak Niles from c.1295 to c.1340 coincide with a well-known famine period in Europe.

At the beginning of the fifteenth century, a further series of weak Niles from c.1400 to 1405 prompted the still useful treatise on Egyptian famines by Maqrīzī.

In a forthcoming study, R. S. Herring uses the summer Nile Flood data to date medieval drought periods as 758/787, 827/848, 939/950, 1009/1017, 1199/1232, 1281/1335 and 1400/09 (his Chart A-1), and these periods are in excellent agreement with the documentary evidence.

The corrected fifteenth century Nile records are very accurate and will be referred to below. However, for the following (Ottoman) period, 1520/1700, good Nile statistics are proving difficult to find. Some archives of Turkish administration have been studied by Stanford Shaw, and are remarkably detailed. 'While the (Turkish) records of Egypt have been known to the scholarly community since 1930, small use has been made of them' is a comment made by J. E. Mandeville (1966: 312). It has been suggested to me that the missing Nile data may still be tied up in sacks in Istanbul. Any information leading to their rediscovery will be greatly appreciated!

Translations and the 'Spectrum of Time'
As part of the Spectrum of Time project, chronicles and annals in Arabic, Portuguese, Ethiopian and other languages are being combed for astronomical, meteorological and medical items (cf. Schove 1974), and the information so collected is being checked for chronology and plotted on maps for each year. There are still many sources that have not been investigated, and current studies by Dr. H. J. Fisher (in his drought seminar at the School of Oriental and African Studies), Miss S. Nicolson (1976) and Miss Y. R. Merritt (thesis in progress) indicate the wealth of new material coming to light. Some of this is cited briefly below.

[*Editor's Note:* Figures 4.3, 4.4, 4.5, and 4.7, and material relating to African climatic history have been omitted.]

The Southern Oscillation

Similar patterns were found when the rainfall for the Sahel was considered separately. This is because the dryness is linked closely with what is known as the Southern Oscillation. Half-yearly indices back to 1796 were tabulated by Schove and Berlage (1965) whereas seasonal data 1851 to 1974 are available in Wright (1975).

The S.O. is essentially a pressure see-saw between the Pacific and the Indian Ocean, high Pacific pressures being regarded as positive. Most parts of Africa have more rain (in most seasons) when the S.O. is positive and such places are said to be of Indian type as far as the seasons in question are concerned. The year 1949 was an exception to the usual rule.

The S.O. is not yet explained. My own models were based on the hypothesis of a feed-back mechanism in equatorial winds near the top of the troposphere, and I thus assumed that the high pressure in the 'Eastern Crescent' between Australia and Bangladesh in the northern autumn would be a sign of drought in the following year in Africa (Schove 1963). However, air—sea interaction was neglected in my models, and many people believe this is more important (see Dalby and Harrison Church 1973: 29) than upper winds (an authoritative study by Professor Newell is expected shortly).

Documentary evidence in low latitude regions suggests that high pressure years in the Indian Ocean region (negative S.O.) can be dated e.g. 1200, 1403, 1450, c.1540, c.1543, c.1555, 1562/3, 1594/97, 1601/05, 1613/14, 1630, 1634, 1637, 1640, 1644/47, 1650/51, 53, 59/60, 64/65, 68/69, c.1674, 85/86, 92/95, c.1702, 1710/11, 1720, 23, 25, 29, c.1734, 37, 44/46, 54, 59, 62, 69, 72, 76, 82/84, 90/91, 96, 1803 and so on, including especially 1876/7, 1913/14, 1939/41, 1965/6 and 1972. Many parts of Africa are of 'Indian' type and had droughts in these years.

Any station where the reaction is of the opposite or 'Pacific' type would be expected to have droughts in e.g. 1611, 1640, 1654/57, 76/7, 1701/02, 04/10, 14, 22, 27, 38/42, 50/53, 56/58, 71, 74, 77, 79/80, 86/88, 1797, 1800/02, 1808/13, 18/19, 48/49, 61/63, 1870, 92, 1910, 16/17, 23, 38, 54/56, 62/63, 71 and 73/74. Tree-ring evidence thus suggests that parts of Morocco are of Pacific type.

Rainfall can be a clue to show even the trend of the S.O. Thus rainfall at Luanda (9°S 13°E), where it falls mostly around March, is indicative of the S.O. index of the half-year October to the *following* March, as a comparison with our indices (Schove and Berlage 1965) revealed.

Other Pressure Patterns

Some other pressure anomaly patterns may prove useful in explaining droughts peculiar to parts of Africa. Over Africa as a whole pressure may be above normal (as in 1897, 99, 1912, 1953, 1958 . . .) or below normal (as in 1944, 1950, 1954/56 . . .). Sometimes it is high in the north and low in the south and this (see fig. 5) occurred in 1913, 1949, 1957 and 1959; or it may be the other way round, as in 1940/42, 1947 and 1960.

In northern Africa some years are characterised by frequent westerly (as in 1885, 90, 92, 1901, 1909 and 35/36) or southwesterly (as in 1895, 1901, 15, 30, 41/42 and 47) winds. In other years easterly (1882, 1932 and 1945) or northeasterly (1903, 13, 38, 43 and 48/9) winds were more frequent than usual. These patterns are relevant more to winter-rain regimes than to the Sahel.

The annals of the nomads make it possible to compare the wind patterns and the movements of the nomads. In NW Africa they moved further from their mean positions (say, 25°N 10°W) in southwesterly years, when heavy rain would fall in unusual situations. The nomads may not have known much about the winds, but news of unexpected grass (as in 24°N 6°W) seems to have travelled faster than the wind!

European Weather and Sahel Drought

Summer season pressure anomalies since c.1740 in northern Europe have been tabulated (Schove 1957), and the corresponding (unpublished) maps include data for southern Europe as well. A secondary cause of drought in the Sahel is evidently high pressure in the region Azores—Scandinavia, and maps drawn by Namias (1974) confirm this for individual stations. These maps indicate that high pressure in Russia and low pressure in North Africa synchronises with summer drought.

A map drawn by Jenkinson (1975) reveals the overall northern hemisphere pattern, and a further map due to Jenkinson shows that on the other hand, in the Spring season (April—May) immediately preceding a drought, pressure tends to be *below* normal near the Faroes.

The summer pressure anomaly maps indicate that the NW European situation could have favoured dryness in the Sahel in the following years: 1716, 19, 23, 32, 36, 42, 47 and 48; 1762, 65, 73, 79, 82, 86, 94, 1800, 01, 03, 08, 11 to 14 inclusive, 18, 21, 25 to 27, 35, 37, 42, 47, 57 to 59, 64 and 65, 69 and 70, 75 and 76, 84 and especially 85, 96, 99, 1901, 06, 08, 11, especially 1913, 14, 21, 25, 32, 37, 40, but especially 1949 and 1976. Westerly weather in Ireland in the April—May season of such years as 1755, 1779, 1784 and 1803 could have favoured a drought in the summer according to Jenkinson's correlation.

Five-Year Changes

The rules affecting climatic fluctuations vary with the time-period in question. The writer made for the late Professor Scherhag a special study of the pressure difference between adjacent five-year periods from 1846/50 minus 1841/45 using data to 1960. The maps have not yet found a publisher, but it was clear that much of the variance could be explained by a principal component or Eigenvector that was quite distinct from any pattern found in seasonal forecasting. It has been briefly described in Schove 1961a and 1964, and consists essentially of an exchange of pressure between the Upper High and the various Upper Troughs. The Upper High over the Indian Ocean represents a bulge or 'hill' of the atmosphere and the Upper Troughs over the Davis Strait and elsewhere represent 'valleys'. A simplified map was published (Schove 1964: 458) and a typical map of what we may call the First Eigenvector is presented here (fig. 6).

The variable involved was termed the Pressure Parameter (Schove 1961a: 608) and the pressure in Indian Ocean countries (a more extensive region than for the Southern Oscillation) was used as a first approximation, the anomaly reversed being termed the P.P.

A study was also made of the First Eigenvector for the summer season; the pattern this time was different but again the geographical pattern corresponded to the contours of the Upper Air so that for instance in northern Norway the pressure-trend although higher in winter is lower in late summer when the parameter trend is positive (cf. fig. 6).

The Meteorological Office (M.O.13), Bracknell, has kindly plotted computer maps of the four seasons in recent years, again as five-year differences, and it is hoped that Eigenvectors can be determined in due course for each season. The chart for the summer season in the drought period compared with the preceding five years shows the trend for pressure to rise in those parts of the atmosphere which normally bulge (e.g. the Indian Ocean again) and to fall in the regions of the summer troughs (cf. Lamb 1972: fig. 32, p. 76).

A drought in the Sahel can be due to the European pressure pattern as in 1949, but the series of droughts necessary for a widespread famine is invariably associated with high pressure in the Indian Ocean countries. The similarity between five-year moving means of rainfall and reversed pressure in the Indian Ocean is very clear when charts are plotted.

Peaks of the P.P., likely to have been associated with times of plenty in the Sahel, were listed in Dalby and Harrison Church (1973: 30) as follows: c.1826/9, 40, 49, 61, 72, 92/4, 1908, 17/18, 22, 37, 54 and 62. A further peak may be occurring now.

On the next page of that report appeared Jenkinson's five-year running means for the Sahel (Central position, say, at 13°N 5°E) with maxima centred at 1907, 20/22, 31/37 and 54/59. When the Sudan and Indian summer rains were combined Winstanley (1973) found peaks centred at 1908, 1918/22, 33, 43, 52/57, c.1963/4 (relative), usually similar to the low pressure phases in the Indian Ocean.

287

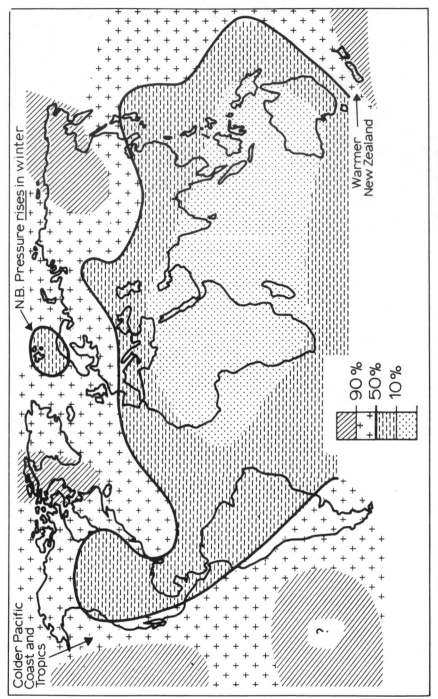

Fig. 4.6. The sign-agreement of pressure trends between adjacent five-year periods, 1841/1960, with the P.P., where the P.P. is the reversed pressure anomaly of Indian Ocean Countries (see Schove and Berlage 1965)

288

Minima of the P.P., that is maxima of Indian Ocean pressure, were listed (Dalby and Harrison Church 1973: 30) as c.1823/4, 33/36, 44, 54, 66/67, 79, 86, 1903, 13, 20, 29/32, 42 and c.1966. Again Winstanley's dry phases in the Sudan and western India are somewhat similar: 1900/2, 13, 20 (relative) and 26 (also relatively dry), but also c.1939, 1947 and 1970. Jenkinson's driest periods c.1912, 1942 and 1970 are associated.

Tree-ring and other evidence makes it possible to extend the list of turning points of the P.P. back through the centuries. The fine records kept in Mameluke Egypt, although we have no information from Australia and very little from the Sahel, reveal that P.P. maxima (five years of plenty) are probably dated: 1383, 90, 97, 1409, 15, 23, 30, 41, 55, 65, c.1477, c.1489, 1501 and 1514. Minima, corresponding often to five-year hardship phases with droughts in Africa, are likewise dated 1387/9, 93, 1403, 19, 26/7, 36 (relatively), especially 1450/1, 60, c.1470, c.1484, 96, 1506/08 and 1519. This pattern is being used to interpret the chronology of Indian tree-rings sent by Professor Agrawal; X-rays of the tree-rings have been kindly prepared by Mr. E. L. Hughes.

The winter/spring rainfall north of the Sahara often has a Pacific-type response to the P.P., for Winstanley noted that peaks occurred there c.1907, 1913, c.1926/8 (relative), c.1938/41, c.1955 and c.1970 and that minima occurred c.1900, 21, 32, 49/51 (relative) and c.1962.

Details of the pressure anomalies of the drought period 1968–72 has been kindly provided by Namias from his 1974 paper and this shows high pressure over NW Europe and low from Egypt to the West Indies, favouring intensified NE winds and dryness in the Sahel (see fig. 7).

A chart of the summer pressure *difference* between 1969/73 and 1964/68 was one of many kindly prepared for me by the Meteorological Office (M.O.13), Bracknell. Its details are too complex for reproduction here but comparison with the normal pattern of troughs and ridges confirms that it indicates a falling pressure parameter, that is, the pressure fell where the troughs usually are in that season (e.g. Davis Strait, N. Pacific) and rose where the atmosphere bulges into ridges (e.g. Alaska, SE Greenland, E and NE Urals).

Decadal Changes

Decadal changes have been ploted by Goudie (1972) and obey a further set of rules. The winter rain regions of the north and south (e.g. Alexandria, and Cape Town) show a general decline from c.1898 to c.1938 associated with the poleward movement of the climatic belts emphasised by Lamb (this volume chapter 3).

In the Sudan and in SE Africa (the southern summer monsoon area) the reversed Indian Ocean pressure is still the dominant factor, most places (except e.g. Freetown) being again of Indian rather than Pacific type. Ten-year means of the Indian Ocean pressure were thus low c.1893, c.1919/24 and c.1952 but high c.1884, c.1900/03/09, c.1929, c. 1944 and presumably c.1967.

In the Sahel, Jenkinson (in Dalby and Harrison Church 1973) found that the wettest decades were c.1933 and c.1955 and the driest were c.1914, c.1944 and c.1967. These dates approximate those of the Senegal and Niger river-flow curves (Goudie 1972: fig. 7, p. 27) except that the middle maximum is there dated c.1928/32 – a period of *high* Indian Ocean pressure.

The summer rains of SSE Africa show tendencies the reverse of those at Cape Town with principal dry phases c.1899, c.1924/27 and c.1946 and principal wet phases c.1882 (and c.92/3), c.1911 and c.1931.

In North Africa winter rains seem to have extended further south c.1710, c.1780, c.1800, c.1815, c.1955 and less far c.1825 and c.1877. Tunisian stations (Goudie 1972: 34) showed maxima c.1918, 35/39 and 58 and minima c.1928, c.1944/48 and at or after 1968. In the Egyptian delta, however, winter rains were high c.1898, low c.1936 and high c.1947.

Thirty-year Changes

A superoscillation between the Western and Eastern hemispheres had turning points about 1890 and 1920; Cape Town rainfall (maximum c.1890) being of Pacific and Port Elizabeth (maximum c.1920) of Indian type on this time-scale (Schove 1961b: 261). At Malta, for instance, the driest thirty-year period (P. K. Mitchell personal communication) is centred at 1890 and the wettest at 1919 and is thus typically 'Indian' in this respect.

The period 1900/40 is an outstandingly mild one with low pressure in the Arctic. Mild phases appear to be associated with lowered levels of Lake Chad, as fig. 4 indicates.

Rainfall of Indian dry zone stations shows the same effect.

Principal Components (Eigenvectors)

The principal components of the pressure pattern in the Tropics (30°N–30°S) were kindly calculated and mapped by the Meteorological Office from the manuscript maps of Professor H. H. Lamb and A. I. Johnson (1966) for the months of January and July through the years 1854 to 1969. The first component or Eigenvector map (unpublished) shows high pressure from Mauritius to Ecuador with low pressure in the Pacific and the S. Indian Ocean and this pattern was dominant in the period 1920/45. This corresponds to an 'old' negative S.O. (see Schove 1963). The second Eigenvector has high pressure in the N. Sahara and from East New Guinea to SW Africa and corresponds especially to a 'new' negative S.O. and, in any July, this type, with low pressure in S. Brazil and the Pacific, may foreshadow an August drought in the Sahel.

Sunspots and Droughts

Many writers (cf. Wood, this volume chapter 7) have shown that droughts in Indian Ocean countries are more frequent when sunspots are low or weakening. Indian Ocean pressure is lowered (and rainfall increased) just before a strong solar maximum (Schove 1964: fig. 7, p. 465), especially in the year preceding, but in weak cycles pressure is above normal even at maximum. The pressure changes found in 1964 for the period since 1796 were compared with the Nile Flood data before that date, and the patterns of change illustrated in fig. 8 are as we should expect.

The abnormally high sunspot numbers of the late 1950s were associated with low pressure in 1955/6, well before maximum, and the weaker cycle of the late 1960s was associated with a much smaller pressure-fall from mid-1965 to late 1966. In this sense the trend to weaker sunspot cycles can be held as a further factor partially responsible for the recent drought. In the Dakar rainfall series, 1887/1973, Landsberg (1975) found a ten-year cycle that corresponds to the solar cycle, which was shorter than usual through much of this period.

The declining sunspot numbers over the last two decades appear to be a partial cause of the recent drought, as many earlier periods of prolonged Nile failures, as in the eleventh and fifteenth centuries (e.g. 1400/09), occurred on a similar downward swing of the solar curve.

Other Cycles

Another cycle found in the variance spectra by Landsberg was the three-year cycle. The two-year cycle is inhibited in low latitude regions because of the persistence of type, and this same persistence of type explains why the many 'seven-year famines' in the sources are not entirely fictitious.

Conclusions

The drought decade of 1964/73 was the worst such decade since at least the 1830s but other prolonged dry periods have been detected from Nile Flood evidence in earlier centuries since Graeco-Roman times; before that various very dry interpluvials can now be dated by radiocarbon.

The causes of these droughts may include:

(a) High pressure under the atmospheric 'bulges' over the Indian Ocean, etc., over a five-year period. (Low Pressure Parameter)
(b) Rising pressure in the extreme north of Africa and falling pressure near South Africa.
(c) High pressure in summer in the Azores–Scandinavia region.

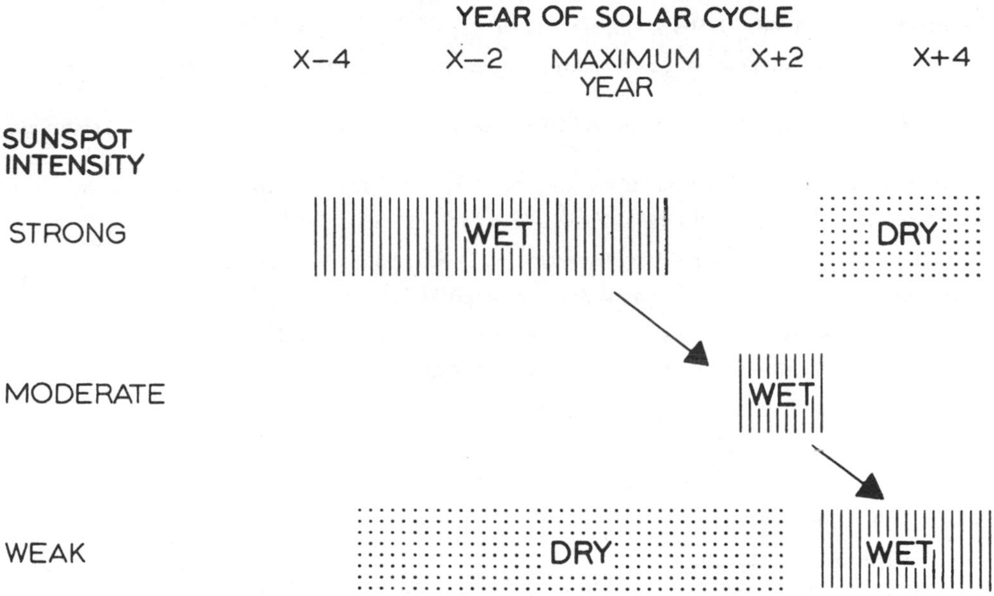

YEAR OF SOLAR CYCLE

X−4   X−2   MAXIMUM   X+2   X+4
YEAR

SUNSPOT
INTENSITY

STRONG ║║║║║║║║ WET ║║║║║║║   ∷∷∷ DRY ∷∷∷

MODERATE   WET

WEAK   ∷∷∷ DRY ∷∷∷   ║║║ WET ║║║

Fig. 4.8. Ethiopian rainfall (Nile Flood) and Sunspot Cycles of differing intensity-classes, A.D. 622/1522

(d) A positive second Eigenvector with tendencies in the summer months for high pressure in the E and SW Indian Ocean and low in the Pacific.
(e) Declining sunspot numbers.

In so far as any of these trends can be extrapolated into the future we can foresee whether and when further droughts may be expected. Sunspot numbers can be predicted to about 1985 with a fair probability of success and from these predictions estimates of Indian Ocean pressure anomalies in the two half-years can also be predicted.

REFERENCES

Bell, Barbara   1975   *Amer. Jnl. Archaeol.* 79: 223–69.

Cissoko, S-M.   1968   *Bull. IFAN* 30B(3): 806–21.

Dalby, D. and Harrison Church, R. J. (eds.)   1973   *Drought in Africa.* London: School of Oriental and African Studies.

Dubief, Jean   1947   *Travaux de l'Inst. de Recherches Sahariennes* e.

Goudie, A. S.   1972   Research Paper 4, School of Geography, Oxford.

Herring, R. S.   forthcoming   'Hydrology and chronology,' in J. B. Webster (ed.) *Chronology in African History.*

Jenkinson, A. F.   1975   In *Proceedings of the World Meteorological Organisation/IAMAP Symposium on Long Term Climatic Fluctuations, 1975.*

Lamb, H. H.   1972   *Climate: Past, Present and Future.* London: Methuen.

────── and Johnson, A. I.   1966   Geophysical Memoir 110, London Met. Office.

Landsberg, H. E.   1975   *Arch. f. Klimatologie* (Wien).

Maley, J.   1976   *Palaeoecology of Africa* (South Africa) 9.

Mandeville, J. E.   1966   *J. Amer. Oriental Soc.* 86.

Namias, J.   1974   In *Proceedings of the International Tropical Meteorological Meeting, Nairobi.* American Meteorological Society.

Newell, R. A.   forthcoming   *The General Circulation of the Tropical Atmosphere.* Vol. 3. Cambridge, Mass.: MIT.

Nicolson, S. E.   1976   *A Climatic Chronology for Africa.* Ph.D. thesis, University of Wisconsin.

Pankhurst, R.   1972/73   *Ethiopian Medical Journal* 11: 233–6.

Rognon, P. et al.   1976   *Rev. Géogr. Phys. et Géol. Dyn.* (2ème sér.) 18(2–3): 147–282.

Schove, D. J.   1954   *Geografiska Annaler* (Stockholm) 36: 40–80.

——— 1961a   *Annals of the New York Acad. Sc.* 95(1): 107–23, 605–22.

——— 1961b   *Geofisica Pura e App.* (Milan) 49: 255–63.

——— 1963   *Geofisica Pura e App.* 55: 249–61.

——— 1964   *Geologischen Rundschau* 54: 448–77.

——— 1968   *Jnl. Brit. Astr. Assn.* 78: 91–8.

——— 1969   *Bull. IFAN* 31A: 224–9.

——— 1971   *Weather* 26: 201–9.

——— 1974   In *Proceedings of the XXIII Congress of the History of Medicine, London, 1972.* 1265–72.

——— 1975   Discussion of paper by A. T. Grove, *Geographical Jnl.* 141: 195–6.

——— and Berlage, H. P.   1965   *Pure and Applied Geophysics* (Basle)-61: 219–31.

Shaw, P. K.   1976   *World Archaeology.*

Sikes, S. K.   1972   *Lake Chad.* London: Methuen.

Street, F. A. and Grove, A. T.   1976   *Nature* 261: 385–90.

Toussoun, Prince Omar   1925   *Mémoire sur l'Histoire du Nil.* Cairo.

Webster, J. B. (ed.)   forthcoming   *Chronology of African History.*

Winstanley, D.   1973   *Nature* 245: 189–94.

Wright, P. B.   1975   *An Index of the Southern Oscillation.* Climatic Research Unit, University of East Anglia, CRU RP4.

[*Editor's Note:* Details of the effects of different classes of cycles on the S.O. and on Indian Ocean (reversed) pressure are given in *Sun and Climate* (1980, Centre National d'Etudes Spatiales, Toulouse, pp. 89-90), and make it possible to give a fair prediction of the droughts of 1983 as follows:

There will again be a risk of severe droughts in India and Ethiopia in 1983 or 1984. In the meantime low pressure is characteristic of years $X - \frac{1}{2}$ to $X + 2\frac{1}{2}$ and this suggests that in SSE Africa and Australia monsoon rains should be mainly good until 1982 or 1983. Until then, winters in W. Canada may be cool and in the Gulf states of the USA warm, as is usual with a positive Southern Oscillation.

The 1961 prediction (p. 257) that dryness "from the old-world tropics to the United States may yet be associated with drought famines as in 1877/1878" has been regarded as an anticipation of the Africa drought of 1965/1972, but 1982/1983 has indeed been similar to 1877/1878 and 1782/1783.]

36

Solar–terrestrial influences on weather and climate

George L. Siscoe

Department of Atmospheric Sciences, University of California, Los Angeles, California 90024

During the past century over 1,000 articles have been published claiming or refuting a correlation between some aspect of solar activity and some feature of terrestrial weather or climate. Nevertheless, the sense of progress that should attend such an outpouring of 'results' has been absent for most of this period. The problem all along has been to separate a suspected Sun–weather signal from the characteristically noisy background of both systems. The present decade may be witnessing the first evidence of progress in this field. Three independent investigations have revealed what seem to be well resolved Sun–weather signals, although it is still too early to have unreserved confidence in all cases. The three correlations are between terrestrial climate and Maunder Minimum–type solar activity variations, a regional drought cycle and the 22-yr solar magnetic cycle, and winter hemisphere atmospheric circulation and passages by the Earth of solar sector boundaries in the solar wind. The apparent emergence of clear Sun–weather signals stimulated numerous searches for underlying physical causal links.

ACTIVE interest in the field of Sun–weather relationships is probably higher at the present time than it has ever been in the past. Since 1972 there have been six major national or international conferences on the subject. The literature is correspondingly very large, and readers interested in going beyond what will be emphasised in this short review should consult the proceedings that have (or will have) emerged from some of these conferences[1-3], and also the extensive references contained in a recent review by Pittock[4].

Solar variability

The subject of Sun–weather relationships is built out of correlations, real or apparent, between changes in solar activity and changes in terrestrial weather and climate. No matter how variable might be the terrestrial atmosphere, if the Sun did not vary, there would be no such subject. It is therefore useful to begin with a brief account of solar variations. These fall naturally into distinct time divisions, the most familiar of which is the 11-yr sunspot cycle. There is also a 22-yr oscillation in the magnetic polarity of the leading member of sunspot pairs, the so-called Hale magnetic cycle. A longer variation of very roughly 80 yr, referred to as the Wolf–Gleissberg cycle, is seen in sunspot cycle amplitudes, as measured for example by the annual mean sunspot number.

On an even longer timescale, there is an important, recently rediscovered kind of solar variation which is typified by the now-famous Maunder Minimum, an interval in the history of solar activity extending approximately from AD 1645 to 1715 (ref. 5). The Maunder Minimum is remarkable for the near-complete absence of spots on the face of the Sun, and a concurrent absence from terrestrial skies of the aurora, a good local indicator of solar activity. It is now generally believed that ^{14}C anomalies detected in tree rings of known, specific ages mark times in the past when other weak solar activity episodes have occurred[6,7]. Episodes of anomalously strong solar activity are indicated as well. The intervals between anomalies vary, but are perhaps characteristically around 400 yr. There is also a hint of a slower variation with a period around 2,500 yr associated with clustering of anomalies of the same sign.

On timescales less than the 11-yr period of the sunspot cycle, there are annual and semi-annual variations in geomagnetic and auroral activity resulting from orbital variations in the efficiency with which solar activity energises the Earth's magnetosphere. On still shorter timescales, there is the 27-d period of the solar rotation, and the variable, but approximately 8-d spacings between passages by the Earth of solar sector boundaries. In addition, at irregular intervals, solar flares produce geomagnetic storms which persist typically for 1–2 d.

Terrestrial atmospheric responses to all of the solar activity variation types listed above have been claimed or suggested, with the exception of the annual and semi-annual variations which would be impossible to detect against the background of the yearly succession of the seasons. A panel composed of participants at the symposium held recently at Ohio State University came to the historically unprecedented conclusion that three of the claimed correlations must now be granted the status of reasonable to virtual certainty. These three, correlations with Maunder Minimum-type episodes, the Hale magnetic cycle, and sector boundary passages, will be emphasised in the limited space available here. Other possible correlations, such as with the sunspot cycle and with the solar rotation, however, retain the controversial status that has traditionally characterised claims of Sun–weather correlations.

Maunder Minimum-type episodes

Atmospheric responses to the kind of changes in the Sun for which the Maunder Minimum is archetypal are of the clearest sort, global changes in climate[5-7]. The Maunder Minimum itself together with the preceding Spörer Minimum (AD 1420–1570) coincide with the two coldest excursions within the Little Ice Age, a period encompassing approximately the fifteenth to seventeenth centuries, when the surface temperatures in Europe and America were depressed by 0.5–1°C. In the case of the Maunder Minimum, the climatic event is linked directly to anaomalous solar behaviour through an ample record of telescopic observations of the Sun covering the period in question. Earlier climatic excursions, such as the Spörer Minimum and the Mediaeval Climatic Optimum, a globally warm period during the twelfth century, are linked to solar changes through proxy data. Foremost among these is the tree-ring record of ^{14}C anomalies, which are believed to result when a change in the level of solar activity varies the flux of ^{14}C-producing galactic cosmic rays that can penetrate the Solar System to Earth's orbit. High solar activity reduces the galactic cosmic ray flux at Earth, which depresses the rate of ^{14}C production. The opposite condition characterises periods of attenuated solar activity. When the curve of tree ring ^{14}C anomalies is

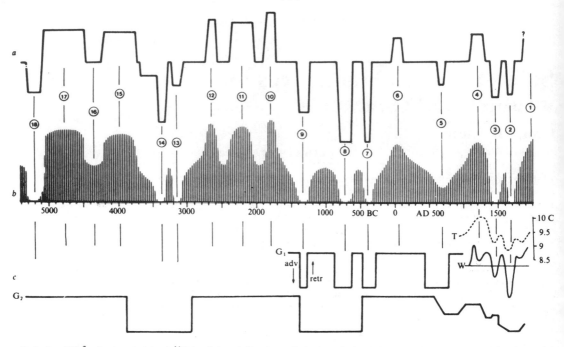

Fig. 1 From Eddy[7]. *a*, Persistent deviations in ^{14}C plotted schematically and normalised to feature 2 (Maunder Minimum); downward excursions refer to increased relative ^{14}C and imply decreased solar activity. *b*, Interpretation of curve *a* as a long-term envelope (of possible sunspot cycle) that minimises in features 2, 3, and so on, and maximises at 4, 6, and so on. *c*, Four estimates of past climate. Step-curve G_1: times of advance and retreat of Alpine glaciers, after Le Roy Ladurie[16]. Curve G_2: same for worldwide glacier fluctuations, from Denton and Karlen[17]. Curve T: estimate of mean annual temperature in England (scale at right), after Lamb[18]. Curve W: winter-severity index for Paris–London area, from Lamb[19], downward is colder.

compared with the curve of climatic variations, in the words of J. A. Eddy, who has convincingly promoted and applied the field of long-term solar changes, "the correspondence, feature for feature, is almost the fit of a key in a lock" (see Fig. 1)[7].

Before the ^{14}C studies, the same long-term Sun–climate correlation had been recognised by J. R. Bray[8,9], using the solar activity index constructed by J. D. Schove[10] from historical observations of aurorae and naked-eye sunspots. A more recent and more extensive compilation of historical aurorae which combines occidental and oriental observations also shows the existence of major solar activity features corresponding to the major climatic excursions[11]. Although the presently known auroral record becomes too fragmentary before the first few centuries of the Christian Era to identify solar activity episodes of the Maunder type, it may serve to validate for that purpose the use of the much longer record of tree-ring ^{14}C anomalies.

Hale magnetic cycle

The correlation that has been found linking atmospheric behaviour and the (approximately) 22-yr solar magnetic cycle involves a modulation of regional climate. The Great American Desert, which is centred roughly in the southwestern US and northern Mexico, has long been known to expand and contract its area of influence with a timescale of about 20 yr producing cycles of drought in the region west of the Mississippi River. The 'dust bowl' of the 1930s and the droughts of the 1950s and 1970s are the most recent of its intensifications and expansions. Many people have suspected and presented evidence to show that the drought cycle is related to the Hale double sunspot cycle[12-14]. Until recently the record length was too short and the drought cycle was not quantified, so that statistical tests of the proposed correlations were difficult to perform.

Mitchell and Stockton have now succeeded in quantifying the temporal climatic behaviour of the desert with a single parameter, yearly index extending back to at least AD 1700 constructed out of tree-ring indices from 40 regionally distributed, climatically sensitive sites[15]. (Actually three families with four sets of such indices were generated to establish internal consistency of the procedure.)

Power spectral analyses of the obtained data sets, so-called drought area indices (DAI), revealed well defined concentrations of power at periods near 22 yr, at formal significance levels ranging from 95% to more than 99%. The analyses were based on sets of 262 data points from the time interval 1700–1962, which included approximately 12 complete Hale cycles. Phase locking between the succession of Hale magnetic cycles and the cycles of drought area was demonstrated to exist at formal significance levels exceeding 99%. Thirdly, the amplitudes of the drought cycles vary from one to the next in a way that bears a strong resemblance to the envelope of the amplitudes of the Hale sunspot cycle. The linear correlation between the two data sets is statistically significant at between the 95 and 99% levels. Weak sunspot cycle amplitudes tend to correspond to minimum drought conditions, but the authors are quick to point out that the apparent correlation is not a reliable basis for operational climate prediction.

Although Pittock[4] has urged caution in accepting at face value the inferred Sun–weather correlation because "complex questions are involved in assessing the true statistical significance of the result", the Hale solar cycle–regional climate connection must be considered at least a serious possibility in view of its having survived the three levels of statistical tests. Full confidence in the result will most likely be achieved only if it is independently confirmed by other researchers who, like Mitchell and Stockton, have a critical attitude toward testing its

reality. Such confirmation has been secured by the third of the
cited correlations, which is described next.

[*Editor's Note:* Material has been omitted at this point.]

1. Bandeen, W. R. & Maran, S. P. (eds) *Possible Relationships between Solar Activity and Meteorological Phenomena* (NASA SP-366, Washington, DC, 1975).
2. Johnson, F. S. (ed.) *Collection of Extended Summaries of Contributions Presented at Joint IAGA/IAMAP Assembly,* Seattle, Washington, Joint Symposium C, 22 Aug.–3 Sept., 1977 (NCAR, Boulder, Colorado, 1978).
3. *Proceedings of Ohio State University Symposium/ Workshop on Solar Terrestrial Influences on Weather and Climate,* July 24–28, 1978 (in the press).
4. Pittock, A. B: *Rev. Geophys. Space Phys.* **16,** 400 (1978).
5. Eddy, J. A. *Science* **192,** 1189 (1976).
6. Eddy, J. A. in *Physics of Solar Planetary Environments* 958 (Am. Geophys. Union, Washington, DC, 1976).
7. Eddy, J. A. *Climatic Change* **1,** 173 (1977).
8. Bray, J. R. *Science* **168,** 571 (1970).
9. Bray, J. R. *Science* **171,** 1242 (1971).
10. Schove, D. J. *J. geophys. Res.* **60,** 127 (1955).
11. Keimatsu, M. *Ann. Sci. Kanazawa Univ.* **13,** 1 (1976).
12. Borchert, J. R. *Ann. Ass. Am. Geogr.* **61,** 1 (1971).
13. Marshall, J. R. thesis, Univ. Kansas (1972).
14. Thompson, L. M. *J. Soil Water Conserv.* **28,** 87 (1973).
15. Mitchell, J. M., Jr, Stockton, C. W. & Meko, D. M. in *Proceedings of Ohio State University Symposium/ Workshop on Solar-Terrestrial Influences on Weather and Climate,* July 24–28, 1978 (in the press).
16. Le Roy Ladurie, E. *Histoire du Climat depuis l'an mil* (Flammarion, Paris, translated by B. Bray, Doubleday, New York, 1971).
17. Denton, G. H. & Karlen, W. *Quatern. Res.* **3,** 155 (1973).
18. Lamb, H. H. *Climate: Present, Past and Future* **1** (Methuen, London, 1972).
19. Gates, W. L. & Mintz, Y. *Understanding Climate Change,* Appendix A (Natn. Acad. Sci., U.S.A., 1975).
20. Shapiro, R. *J. Meteor.* **13,** 335 (1956).
21. Woodbridge, D., Macdonald, N. J. & Pohrte, T. W. *J. geophys. Res.* **64,** 331 (1959).
22. Macdonald, N. J. & Roberts, W. O. *J. geophys. Res.* **65,** 529 (1960).
23. Macdonald, N. J. & Roberts, W. O. *J. Meteor.* **18,** 116 (1961).
24. Roberts, W. O. & Olson, R. H. *J. atmos. Sci.* **30,** 135 (1973).
25. Roberts, W. O. & Olson, R. H. *Rev. Geophys. Space Phys.* **11,** 731 (1973).
26. Wilcox, J. M. & Ness, N. F. *J. geophys. Res.* **70,** 5793 (1965).
27. Wilcox, J. M., Scherrer, P. H., Svalgaard, L., Roberts, W. O. & Olson, R. H. *Science* **180,** 185 (1973).
28. Wilcox, J. M. *et al. J. atmos. Sci.* **31,** 581 (1974).
29. Wilcox, J. M., Svalgaard, L. & Scherrer, P. H. *Nature* **255,** 539 (1975).
30. Wilcox, J. M. *J. atmos. terres. Phys.* **37,** 237 (1975).
31. Wilcox, J. M. *Science* **192,** 745 (1976).
32. Wilcox, J. M., Svalgaard, L. & Scherrer, Ph. H. *J. atmos. Sci.* **33,** 1113 (1976).

[*Editor's Note:* Maps showing regions in the northern hemisphere affected by the 11-year and 22-year cycles will be published soon by M. Kelly in a volume edited by B. M. McCormac (for earlier volume, see McCormac and Seliga, 1979, on our p. 236). A review paper on new research has been published in *Royal Meteorol. Soc. Quart. Jour.* by Pittock, 1983.]

Part VII

VARVE AND GEOLOGICAL CYCLES

Editor's Comments
on Papers 37 Through 40

37   ANDERSON
Excerpts from *Solar-Terrestrial Climatic Patterns in Varved Sediments*

38   SCHOVE
Excerpts from *Solar Cycles and Equatorial Climates*

39   SCHOVE
Tree-Ring and Varve Scales Combined, c. 13500 B.C. to A.D. 1977

40   SCHOVE
Varve-Chronologies and Their Teleconnections, 14000-750 B.C.

A geophysical method of determining the dates and amplitudes of solar cycles is urgently needed. At present the only good clue to solar variability before 300 B.C. is that provided by the "wiggles" in the radiocarbon curve and, as for the 11-year cycle, we have no accepted proof of its early existence at all. In this section we consider the evidence for cycles in varves, ice-cores, and tree-rings.

In the precomputer age, the 11-year cycle was still easily "found" in both tree-rings and varves, but at the New York Symposium of 1961 (Fairbridge, 1961) power-spectra results became available. The contrast between *before* and *after* is reflected in Anderson's Figs. 1 and 5 in Paper 37.

Most varve series studied are too short for cycle analysis, but Anderson's (1961, 1963) 22-year cycle, in both halves of the 470 Devonian (Ireton) series and in Jurassic varves, looks significant. However, Anderson is now working on much longer series from the Upper Permian Castile formation and, whereas he finds salinity cycles of 2700 years (cf. Anderson in Schove and Fairbridge, 1983), he finds no evidence of either an 11-year or a 22-year cycle. There is, nevertheless, a clear 200-year period (personal communication) and a complex pattern of oscillations in the 70-year to 400-year range. Some of these may be solar in origin.

Upper Permian series in Brazil from the Itararé formation (placed as Carboniferous in Table 1 of Schove et al., 1958) have been investigated by Maximum Entropy Spectral Analysis (Marcia, 1977) and cycles of 7.6 and 10.0 were apparent in short series. On the other hand, in the northern hemisphere the Upper Permian varves of the Upper Kama deposit (cf. Richter-Bernburg, 1964, for Zechstein deposits) showed no evidence of solar cycles (Fig. 3 in Monin and Vulis, 1971) and the weak cycles that might be inferred from their curve have values of 5.9 and 15½ (and possibly 2.8, 3.7, 5.1, 8¼, and 10½) years.

In the Quaternary the results so far are equally inconsistent. Cycles of 22, 35, and 85 years were claimed in Early Pleistocene varves of northwestern Texas. In the Late Glacial, 748 Estonian varves (about 12,000 b.p.) and 780 Finnish varves (c. 11,000 b.p.) seemed to have cycles of 2.24, 2.56, 3.60, 4.3, 6.0 (see Paper 38), 7.0, about 10.8, and 32 years (Siren and Hari, 1971). Glacial varves from the Early Holocene were investigated by Agterberg and Banerjee (1969), who found two weak cycles of 10 and 19 years in a series of 537 varves measured separately for summer and winter layers and, in addition, a 14-year cycle in the summer silt only. The maxima of the 10-year cycle was alternately higher and lower, suggestive of the modern pattern of odd and even solar cycles (see Fig. 3 in our Introduction).

The Later Holocene varves from Lake Van in Armenia are of particular interest and are being studied by Kempe and Degens. In preliminary tests (Kempe, 1977, p. 174) a cycle of 70 years was found in the period 1396 to 577 B.C. and in short series a sunspot cycle (Kempe and Degens, 1979, Figs. 7 and 8, pp. 317, 318) of about 10 years appears. As in tree-ring data (see Parts V and VI) there is no consistency in these various periodicities.

The hope of finding persistent cycles common to geological and modern periods led me to present the table of recent cycles (Paper 38) in my paper presented at a geological congress in Cologne. For the last few centuries these cycles have indeed been confirmed by MESA analysis and we can now specify them precisely as 2.17 (quasibiennial), 2.7, 3.03 (triennial), 3.33, 3.9, 4.98, and 7.0. The errors are generally less than .05 years; however, over the last seven centuries the *upper trough cycle* (see Paper 32) seems to have averaged 7.4 or 7.5 judging from the analysis of ice-cores in Greenland and tree-rings in Finland (Siren and Hari, 1971). Solar connections are suspected but pressure anomaly studies are needed (see Wagner, 1971; Ol', 1969) so that the reasons for these preferred values can be investigated. At present we have no satisfactory proofs that these precise periods existed before A.D. 1200.

Changes in the magnetic intensity, declination, and inclination of Late Glacial varves are believed by Mörner (1978a, 1978b) to reflect the 11-year and 22-year cycles. Conceivably the clay might respond more rapidly to the magnetic fluctuations that occur at sunspot maxima but at present geologists are awaiting confirmation with longer series. The precise dating of varves is meanwhile desirable and clues are provided in Papers 39 and 40.

The intensity of sunspot maxima may eventually be unscrambled from the 2-year and 3-year cycle patterns as the German curve (Fig. 5 in Paper 39) suggests. The American trees show *biennial cycle* peaks at about 35 years to 40 years after active solar periods (Table I in Paper 39; see also Harris, 1980; Schove and Fairbridge, 1983) thus often synchronizing with radiocarbon minima. Triennial cycles are associated with sunspot minima and with weak solar maxima. The curves for the Late Glacial (Figs. 3 and 4 in Paper 39) provide clues only in the 30-year to 200-year range, as other factors—notably nearness to the glacier front—are important and make it impossible to specify a significant vertical scale.

The *remainders* (in the sense of Paper 20) of the middle years of 5-year maxima and minima can be examined for features that persist from one century to another as this suggests a meteorological response to a solar cycle of 11.1 years ($9 \times 11.11 = 100$ years). Indeed, in Asia Minor the moist 5-year runs of each century of the period 1501 to 951 B.C. show a similar pattern; if we convert the dates to positive *Bristlecone years* (B.Y.) the period becomes 6500 to 7070 B.Y. since (where B.Y. + B.C. = 8001) the last two digits of the middle years tend to follow the 11-times table, that is, they have a remainder near zero. The drought years are about 5 years to 6 years later and, adding $5\frac{1}{2}$ years to the previous set and taking the mean of both sets, we get the *phase* of dryness, defined as *remainders* by half centuries as follows: 6500–6550 to 7000–7050 B.Y.—7, 6, 8, 6, 6, 3, 4, 8, 8, 4, 3, and 9. This phase-persistence suggests a *weather* cycle averaging 11 years, and similar studies of biennial cycles, oddness indices, and other parameters might perhaps confirm a solar connection. Years such as 6761 B.Y. (1240 B.C.), giving a *remainder* of 6, would be expected to be near a sunspot minimum. Papers 39 and 40 also show how the absolute chronology of long varve series can be checked, but it is now necessary to cross-date longer series of varves with tree-rings so that the length and amplitudes of cycles can be determined.

REFERENCES

Agterberg, F. P., and I. Banerjee, 1969, Stochastic Model for the Deposition of Varves in Glacial Lake Barlow-Ojibway, Ontario, Canada, *Canadian Jour. Earth Sci.* **6:**625–652. (Cycles c. 7500 B.C.)

Anderson, R.Y., 1961, Solar-Terrestrial Climatic Patterns in Varve Sediments, *New York Acad. Sci. Annals* **95:**424–439. (See Paper 37 and Schove and Fairbridge, eds., 1983.)

Anderson, R. Y., 1963, Harmonic Analysis of Varve Time Series, *Jour. Geophys. Research* **68:**877–893.

Fairbridge, R. W., ed., 1961, Solar Variations, Climatic Change and Related Geophysical Problems, *New York Acad. Sci. Annals, Vol. 95,* 740p.

Harris, P. R., 1980, *A Summer Temperature Index for Western USA A.D. 590–1960,* B.A. geography dissertation, Cambridge, England. (Summary in Schove and Fairbridge, eds., 1983.)

Kempe, S., 1977, Hydrographie, Warven-Chronologie und Organische Geochemie des Van Sees Ost-Turkei, *Inst. der Universitat Hamburg Geologisch-Palaontologisch Mitt.* **47:**125–228. (See summary in Schove and Fairbridge, eds., 1983.)

Kempe, S., and E. T. Degens, 1979, Varves in the Black Sea and in Lake Van (Turkey), in *Moraines and Varves,* C. Schlüchter, ed., Balkema, Rotterdam, Netherlands, pp. 309–318. (Review in *Bull. Inst. Archaeol. No. 18,* 1981.)

Marcia, E., 1977, *Estudo da Variacao Paleosecular de Campo Magnetico da Terra,* Dissertacao de Mestrado, Instituto Astronomico e Geofisico, Sao Paulo. (See summary in Schove and Fairbridge, eds., 1983.)

Monin, A. S., and I. L. Vulis, 1971, On the Spectra of Long-Period Oscillations of Geophysical Parameters, *Tellus* **23:**337–345.

Mörner, N-A., 1978a, *Paleomagnetism and Varved Clays (Abstract),* Israeli Association of Science, 10th Congress, Jerusalem.

Mörner, N-A., 1978b, Annual and Interannual Magnetic Variations in Varved Clay, *Jour. Interdisciplinary Cycle Research* **9:**229–241.

Ol', A. I., 1969, Manifestation of the 22-year Solar Activity Cycle in the Earth's Climate, *Arkt. Antarkt. Nauc Issled Inst. Leningrad* **289:**116–131.

Richter-Bernburg, G., 1964, Solar Cycles and Other Climatic Periods in Varvitic Evaporites, in *Problems in Palaeoclimatology,* A. E. M. Nairn, ed., Wiley, New York, 517p. (Uncritical pioneer work.)

Schove, D. J., and R. W. Fairbridge, eds., 1983, *Ice-cores, Varves, and Tree-rings,* Balkema, London.

Schove, D. J., A. E. M. Nairn, and N. D. Opdyke, 1958, The Climatic Geography of the Permian, *Geog. Annaler* **40:**216–231.

Siren, G., and P. Hari, 1971, Coinciding Periodicity in Recent Tree-rings and Glacial Clay Sediments, *Rep. Kevo-Subarctic Stat. Helsinki* **8:**155–157.

Wagner, A. J., 1971, Long-period Variations in Seasonal Sea-Level Pressure Over the Northern Hemisphere, *Monthly Weather Rev.* **99:**49–66.

37

Reprinted from pages 424–426, 427, 429–435, and 437–439 of New York Acad. Sci.
Ann. **95**:424–439 (1961)

SOLAR-TERRESTRIAL CLIMATIC PATTERNS
IN VARVED SEDIMENTS*

Roger Y. Anderson

University of New Mexico, Albuquerque, N. Mex.

INTRODUCTION

Joseph Barrell (1917) in speaking of a thinly laminated anhydrite said: "Such a series of beds is worthy of very careful analytical study as a detailed meteorological record of the past." Barrell was referring to the principle that the thickness of an annual sedimentary layer is a function of climate. It is theoretically possible to translate climatic changes into sediment layers of varying thickness and, if there is a solar effect on climate, it should be recorded in varves. This influence might be demonstrated in several ways, and this paper presents the evidence compiled to date.

Seasonal Interpretation

It is fair to state at the outset that there has not always been agreement that thinly laminated sedimentary deposits represent seasonal deposition. The most impressive evidence has come from studies of recent sediments where varves are in the process of formation. At least 35 examples have been reported from Europe and the Near East where most of the work has been done: 5 from North America and 3 from the Far East. The generalizations derived from the study of these deposits show that there is one outstanding feature common to nearly all of them. This is the deposition of some sort of an organic layer each year (Bradley, 1937, p. 32). This may be a well-defined layer of aquatic algal ooze or may be represented by only a gradual transition from a light to dark color. Inorganic glacial varves are an exception, but virtually all of the recent clastic varves and all but one (?) of the recent evaporite sequences including halite have this feature. The organic layer represents a yearly interval of high growth and mortality in the aquatic population; once the bottom water of a lake, lagoon, fjord, or bay becomes stagnant and unventilated, it takes only the addition of a slight amount of clastic material or chemical precipitate to form a recognizable varve. The most important complication of the previous generalization is caused by changes in the growth and mortality of aquatic organisms. There may be several such intervals in one year, although this factor does not seem too important, and the supernumerary layer can often be recognized as such. More confusing are diatom-organic laminae where diatom flowering and deposition are coincident with organic deposition. In this case, several periods of flowering may result in a multiplicity of layers each year (Deevey, 1953, p. 291).

If the organic-layer criterion is applied to ancient varves it is found that most of the reported examples do represent seasonal deposition. The greatest exceptions are iron-stain laminations, where the Liesegang mechanism (Stetson, 1933; Sugawara, 1934) is capable of producing varvelike layers and alternating

* This work was supported in part by Research Grant NSF-G17144 from the Earth Sciences Program of the National Science Foundation, Washington, D.C.

sand, silt, and clay laminae without an organic fraction. Also the strong seasonal control on sedimentation exerted by the freeze and thaw of ice can form glacial varves without a well-defined organic layer. A list of varved deposits would be too long to be included here, but it is safe to say that every geologic period from each continent has many examples. These would include varves whose second component consisted of sand, silt, clay, calcite, dolomite, anhydrite, halite, diatoms, and calcitic shell material. In addition some iron, potash salt, bittern, and silica laminae might also be included.

Climatic Interpretation

Varve thickness can be an approximate index of precipitation and temperature. Other factors such as evaporation, storminess, and insolation also enter in but probably cannot be treated separately in ancient varves. Unfortunately, even the effects of temperature and precipitation are difficult to separate, but estimates can be made after a study of the varve type and the environmental setting of the basin. For example, thickness changes in the clastic layer of organic-silt laminae would most likely represent changes in runoff and, indirectly, precipitation. In special cases pollen analysis of microtome sections of individual laminae can reveal the season of greatest runoff and, perhaps, the type of precipitation (Welton, 1944; Anderson, 1961). The thickness of a glacial varve is thought to be an index of glacial melting and, indirectly, of temperature. The interpretation of calcite varves is difficult because thickness changes may be the result of the dilution of basin water by precipitation, or they may be caused by changes in CO_2 content arising from variations in water temperature, evaporation, or organic activity. In the case of evaporitic anhydrite laminae forming in a lagoonal basin with little interior drainage, the effects of temperature and evaporation would overshadow those of precipitation. Obviously the problems of interpretation are complex, and each basin must be considered separately in order to determine to what extent temperature, precipitation, or both are being measured by varve-thickness changes.

Comparison of recent varves with meteorologic observations will be necessary before the influence of climatic factors on varve deposition can be fully known. Little work has been done. Shostakovich (1931, 1936) presented data relating varve thickness in Pert Lake, Finland, and Saki Lake, Crimea, to local precipitation. He found that a twofold increase in seasonal precipitation brought about nearly the same increase in varve thickness. He also found a close relationship between the thickest varves and very wet years and between the thinnest varves and very dry years. In a different type of study, Granar (1956) related varve thickness in Nyland Fjord, Sweden, to discharge records for associated rivers. For some intervals the correlation he obtained was weak, but for others the correspondence was outstanding. Late-glacial and post-glacial varve correlations by DeGeer and Antevs in Europe and North America are another point of evidence. Their success is indirect testimony that the climatic changes recorded in varves are sufficiently strong and consistent to be traceable over a wide area.

One can state that there is a relationship between seasonal weather and varve thickness, and preliminary studies indicate that it is an important one.

R. Y. Anderson

EVIDENCE OF A SOLAR INFLUENCE ON TERRESTRIAL CLIMATE

Recognition of Solar Periods

One way of demonstrating a solar-climatic influence is to recognize the 11- or 22-year solar cycle* in varve data. Meteorologists seem divided on the question of a solar period in the data from the earth's lower atmosphere. This in itself suggests that if there is an effect it is slight, and might appear in the varve record only under favorable circumstances. So many investigators have found evidence for an 11-year period in meteorologic data that a solar influence would seem a certainty (Nupen and Kageorge, 1958, p. 5). Other meteorologists argue that the statistical techniques used in these studies were of doubtful validity and that no good pertinent evidence has ever been established (F. W. Ward, Jr., personal communication, 1959).

The number of reported occurrences of the 11- and 22-year period in varved sediments is also impressive (FIGURE 1). The same argument of inadequate statistical technique could also be used against the evidence. In general, analysis has been either graphic-visual, or has employed some form of harmonic analysis. Graphic plots of published varve series leads one to suspect that many of the reported occurrences of the 11-year solar period are probably erroneous. Some are merely averages between prominent maxima that are a great deal more variable than the 11-year sunspot maxima intervals (Bradley, 1929, FIGURE 15). In most cases this variable interval between prominent maxima seems to average about 5 years. However, some of the curves published by Korn (1938, FIGURES 14 and 15) show a remarkably uniform interval at about 11 years, and parts of other series show isolated groups of maxima about 11 years apart that could represent the solar period (Anderson and Kirkland, 1960, FIGURE 9, units near 500).

[*Editor's Note:* Material has been omitted at this point. Anderson's later results with very long series are summarized in *Ice-cores, Varves and Tree-rings,* D. J. Schove and R. W. Fairbridge, eds., Balkema, London, 1983.]

---

* In this paper a cycle is considered any kind of phenomenon that tends to return to an original position, status, or value. Cycles may recur in a predictable sequence but need not appear at a regular time. A period or rhythm is considered a cycle that recurs at a predictable time. For example the waxing and waning of sunspot frequency is a cycle. The alternation of large and small maxima is another cycle, but the recurrence of maxima at approximate 11- and 22-year intervals is a period or rhythm.

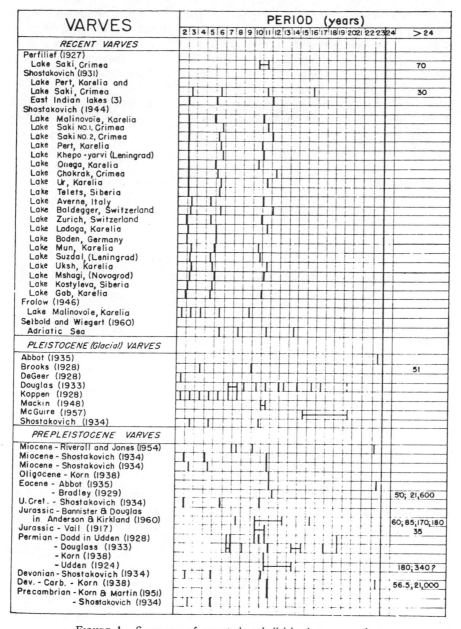

FIGURE 1. Summary of reported periodicities in varve series.

The most interesting series analyzed by Ward is a 470-U. sequence from a core of the Upper Devonian Ireton shale member of the Woodbend formation, Alberta, Canada. The laminations are made up of couplets of green-gray calcareous shale and brown organic-rich calcareous shale that average slightly under 2 mm. in thickness. The varves are marine, and were formed as a calcareous ooze with both a terrigenous and chemically precipitated component. Temperature exerted some control over the formation of the chemical precipitate. The controls for the clastic component might be oceanic circula-

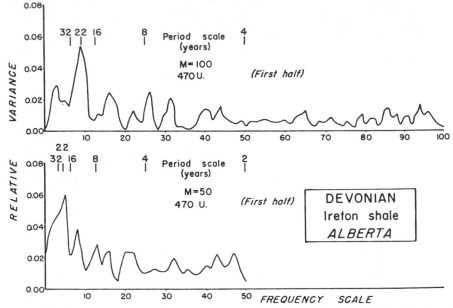

FIGURE 3. Power spectra of the Upper Devonian Ireton shale varves, marine Woodbend formation, Alberta, Canada (*first half*).

tion or storminess and precipitation. The power spectra from the Ireton shale show a prominent peak at the frequency 9 (lag = 100) corresponding to a period of 22 years (FIGURES 3 and 4). Ward (personal communication, 1960) believes that the peak is significant, but he would reserve judgment until a longer series can be analyzed from either end of the measured interval, or until an independent series is available. The series, however, is a long one in so far as meteorologic records are concerned. Spectra computed for each half show a peak at the same point, indicating that the period was present throughout the data and increasing the probability that the period has physical reality. The 22-year period is not obvious in a simple graphic plot of the varve-thickness changes but apparently is responsible for the group of prominent maxima in the smoothed curve of FIGURE 7. These maxima show a remarkably uniform recurrence at little more than 21 U. apart that closely resembles the spacing of the 22-year cycles in the modern sunspot record (compare with FIGURE 6).

Tentative results from harmonic analyses of several other series are available. The analyses were done by L. H. Koopmans of the Sandia Corporation, Albuquerque, N. M., using an IBM-704 computor and programs modified from those originally developed by Tukey. This brings the total number of spectra available to 12 as follows.

Analyzed by Ward (1957): Ireton shale, Upper Devonian, 470 U.; Todilto limestone, Upper Jurassic, 105 and 101 U.; and Puente formation, Miocene, 480 U.

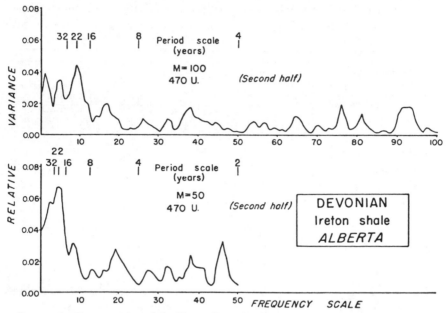

FIGURE 4. Power spectra of the Upper Devonian Ireton shale varves, marine Woodbend formation, Alberta, Canada (*second half*).

Analyzed by Koopmans (tentative): Todilto limestone, Upper Jurassic, 943 U.; Nyland Fjord, recent, 100 U.; Averne, recent, 157 U.; Chokrak, recent, 76 U.; Lake Onega, recent, 81 U.; Lake Saki, recent, 145 U.; Lake Teletz, recent, 50 U.; and Lake Ur, recent, 97 U.

Published sources referring to the above series are as follows: Todilto limestone, Anderson and Kirkland (1960); Puente formation, Riveroll and Jones (1954); Nyland Fjord, Granar (1956); Lakes Averne, Chokrak, Onega, Saki, Teletz, and Ur, Shostakovich (1934).

As might be expected, many of the peaks in the spectra of the above series are weak, and many are probably chance variations that do not reflect consistencies in the data. In order to facilitate interpretation, all the prominent peaks in the above spectra (except the short Todilto series) were plotted and the results summarized (FIGURE 5). The strongest peak in each series appears as a longer bar on the graph; peaks whose periods were less than about one fourth the length of the series were not plotted. Little is known of the technique for gathering some of the data. No attempt will be made here to evaluate climatic significance or data reliability and the 10 varve series must be assumed to be representative. If the peaks are randomly distributed there should be a gradual increase in the number of peaks toward the right of the chart.

Interpretation must be subjective, and some of the series are so short that the significance of the position of some of the peaks at the longer periods is low. The 11-year period is not present. Instead the frequencies near 12 to 14 years and 8 to 9 years seem better developed. This distribution could be due to chance, but if it persists as more series are analyzed one might expect a causal relationship. A Fourier analysis of recent varves from the Adriatic

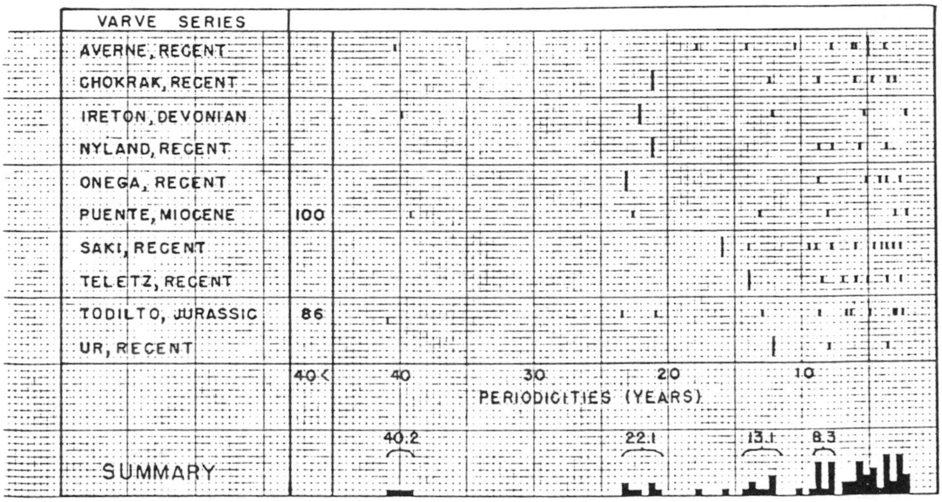

FIGURE 5. Summary of prominent peaks in spectra from 10 varve series.

area made by Seibold and Wiegert (1960) found tendencies near 6, 8, and 14 years throughout an 820-U. record that agree quite closely with the summary in FIGURE 5. They also found a weak 11-year period in their data. The complex of peaks between 2 and 6 years either represents "noise" or some factor of local climatic persistence that might account for the interpretation of periods between 2 and 6 years by Brooks (1928), Shostakovich (1931, 1934, 1944) (FIGURE 1), and Landsberg et al. (1959).

The appearance of several peaks near 22 years, the prominence of some of them, and the slight separation from adjacent peaks support the earlier supposition that the 22-year solar cycle has a slight influence on climate. One might speculate as to why a 22-year period and not the 11-year period should manifest itself in the varves. Since about 1850 the Zurich mean annual relative sunspot numbers show the 22-year solar cycle as higher peaks on alternate 11-year maxima. One estimate fixes the difference at about 22 per cent (Nicholson, 1929, p. 80). There is also a difference in the spacing of maxima

308

in alternate cycles. The 11-year influence, modified by terrestrial factors, might be so weakened or modified that the 22-year period, with a longer interval of climatic adjustment, could appear as the stronger of the 2 solar cycles. Alternatively the 22-year magnetic cycle of sunspots may exert a disproportionate control on weather and climate.

Meteorological series are too short for the study of the 22-year period over long intervals, and further work with varves and tree rings will probably yield the most conclusive results. In this respect it may be important that the examination of many long tree-ring series by Schulman (1945, 1956) led him to believe that the cycle near 22+ years was the strongest recurring phenomena in the general tree-ring index.

Comparison of Long-Term Trends

Trends between 50 and 500 years. With some evidence for a physical relationship between solar and climatic changes at the 22-year level it is desirable to compare trends that are somewhat longer. The 211-year record of sunspot data is the most reliable measurement available for comparison, and other solar phenomena can be related to it. In order to obtain a graphic idea of the observed changes greater than 30 years in length, the Zurich mean annual relative sunspot numbers were averaged over an arbitrary interval of 31 years and plotted (FIGURE 6). The curve is shorter than 211 years because 15 U. at each end were lost by the smoothing operation. The most obvious character of the curve is the significant change over a long period of time. The intervals between the 2 maxima are 68 years and more than 100 years. The interval between minima is about 86 years. The average number for the decade ca. 1775 is about 64, dropping rapidly to only 22 in the decade ca. 1810. A lengthening of the interval between 11-year maxima accompanied this strong long-term minimum. A less intense maximum and minimum occurred near 1850 and 1900 respectively, and the present maximum has already exceeded the one near 1775. Assuming that other solar phenomena are related to changes in sunspot activity, it is apparent that large oscillations on the order of 70 to more than 100 years are taking place within the observable record. The differences between long-term maxima and minima should be sufficient to be reflected in meteorologic observations and also in varved sequences if there is any effect at all on the 22-year level.

When the smoothed sunspot curve is compared to a similarly constructed curve from the Upper Devonian Ireton shale the result is impressive (FIGURE 7). The Ireton trends are of nearly the same duration as those of the observed sunspots. The 22-year period, already indicated in the power spectrum, is most obvious between units 200 and 340 and was strong enough to appear after the 31-year averaging. Furthermore the association of the prominent 22-year maxima in the Ireton sequence with the gradual buildup of an unusually strong maxima near unit 330 suggests that solar activity played some direct role in intensifying that trend. In order to obtain a graphic estimate of the amount of oscillation that might be due to random-thickness changes, thickness measurements were selected randomly from a pool of values and plotted, using the same smoothing formula. If nature had selected varve

309

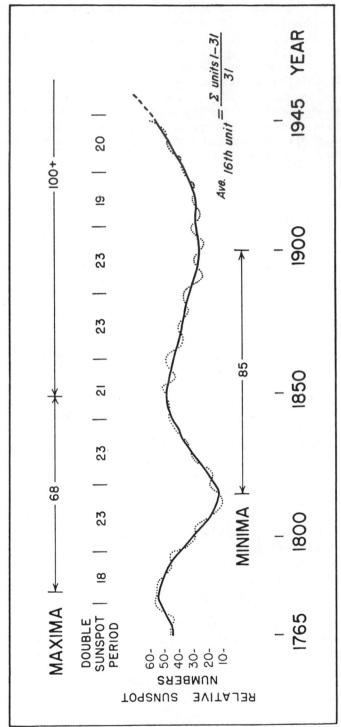

FIGURE 6. Smoothed curve of mean annual relative sunspot numbers (Zurich, Switzerland).

310

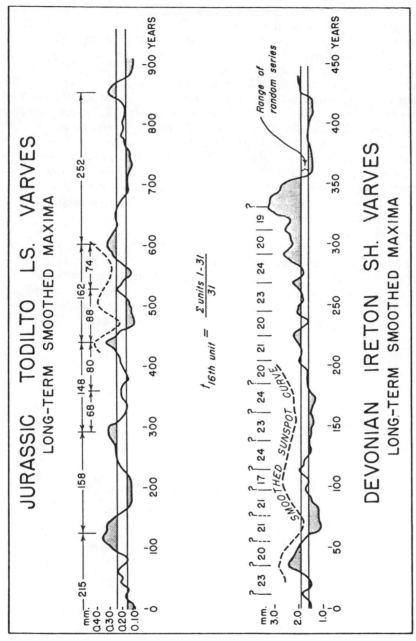

FIGURE 7. Smoothed curve of Upper Jurassic Todilto limestone varves, New Mexico, and the Upper Devonian Ireton shale varves, Alberta, Canada.

311

thickness individually, in a completely random manner, the natural and artificially selected curves should be similar. The random curve, however, is monotonous and shows none of the extreme variations or long-term trends found in the natural sequence. The extreme range of the randomly selected values is compared to the natural sequence in FIGURE 7.

A similar plot was made for 943 varves from the Jurassic Todilto limestone (FIGURE 7). The intervals between maxima that extend above the range of random values are between 148 and 252 years, and average 187 years. Smaller maxima appear between the larger ones and display a pattern of 68 to 88 years between maxima. These shorter trends resemble the smoothed sunspot curve even more closely than those in the Ireton sequence, and may be responsible for the strong 86-year peak in the harmonic analysis of the Todilto limestone by Koopmans (FIGURE 5). These quasi-periodic trends probably represent long-term persistence in temperature, precipitation and, perhaps, oceanic circulation, and their similarity to observed sunspot trends suggests a relationship.

Trends greater than 500 years. The 1592-year Todilto varve record (Anderson and Kirkland, 1960, figure 9) shows cycles on the order of 180 years through only about two thirds of the sequence. Both ends of the series have closer and weaker maxima. These differences may be the result of changes in the pattern on the order of 1000 years although a much longer sequence would be needed to see if the 86- or 180-year quasi-period would recur.

[*Editor's Note:* Material has been omitted at this point.]

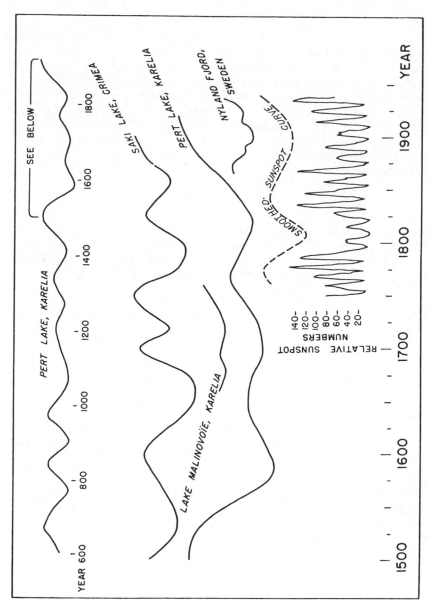

FIGURE 8. Comparison of long-term sunspot trends and varve thickness trends in recent lakes.

313

Conclusions

The 22-year solar period seems to be recorded in some varves, but the 11-year solar period may be modified by shorter-lived terrestrial factors so that it is unrecognizable or weak periodic tendencies of 12 to 14 and 7 to 9 years result. A complex of weak periodicities between 2 and 6 years may represent noise or local terrestrial climatic persistence. Longer varve trends with about 70 to 100 years between maxima may be related to similar long-term solar trends, and the varve record has revealed other quasi-periodic cycles of about double that length and of slightly greater magnitude. These observations comform to the generalization that both solar-climatic adjustment and total climatic effect increase with the length of oscillation.

Acknowledgments

I am grateful to F. W. Ward, Jr., of the Air Force Research Center, Cambridge, Mass., for analyses of the varve series shown in FIGURES 2, 3, and 4; to L. H. Koopmans of Sandia Corporation, Albuquerque, N. Mex., for the analyses represented in FIGURE 5; and to R. G. McCrossan of Imperial Oil Limited, Calgary, Alta., Canada, for the Ireton shale varve series.

References

ABBOT, C. G. 1935. Solar radiation and weather studies. Smithsonian Misc. Coll. 94(10): 89.

ANDERSON, R. Y. 1961. Evidence of seasonal lamination. Bull. Geol. Soc. Am. Abstr. In press.

ANDERSON, R. Y. & D. W. KIRKLAND. 1960. Origin, varves, and cycles of Jurassic Todilto formation, New Mexico. Bull. Am. Assoc. Petrol. Geol. 44(1): 37–52.

BARRELL, J. 1917. Rhythms and the measurements of geologic time. Bull. Geol. Soc. Am. 28: 745–904.

BLACKMAN, R. B. & J. W. TUKEY. 1958. The measurement of power spectra from the point of view of communications engineering. Bell System Tech. J. 37: 185–282, 485–569.

BRADLEY, W. H. 1929. The varves and climate of the Green River epoch. U. S. Geol. Survey Prof. Paper. 158: 87–110.

BRADLEY, W. H. 1937. Non-glacial varves with selected bibiography. Nat. Research Council Ann. Report. App. A. Rept. Committee on Geologic Time. : 32–42.

BROOKS, C. E. P. 1928. The problem of the varves. Quart. J. Roy. Meteorol. Soc. 54: 64–70.

DE GEER, G. 1928. Geochronology. Antiquity. 2: 308.

DEEVEY, E. S. 1953. Paleolimnology. In Climatic Change, Harlow Shapely, Ed. : 273–318. Harvard Univ. Press. Cambridge, Mass.

DOUGLASS, A. E. 1933. Tree growth and climatic cycles. Sci. Monthly. 37: 481–495.

FROLOW, V. 1946. Analyse de la série des varves du lac Malinovoïe. Acad. sci. Paris, compt. rend. 222: 669–672.

GILBERT, G. K. 1895. Sedimentary measurement of Cretaceous time. J. Geol. 3: 121–127.

GRANAR, L. 1956. Dating of recent fluvial sediments from the estuary of the Ångerman River (the period 1850–1950 A.D.). Geol. Fören. Stockholm, Förhandl. 78: 654–658.

KÖPPEN, V. 1928. Mehrjährige Temperatur schwankungen vor 8 bis 18 Jahrtausenden. Meteorol. Z. 45: 263–265.

KORN, H. 1938. Schichtung und absolute Zeit. Neues Jahrb. Mineral. Geol. Paläont., Stuttgart. 74A: 50–186.

KORN, H. & H. MARTIN. 1951. Cyclic sedimentation in varved sediments of the Nama system in South-West Africa. Trans. Geol. Soc. South Africa. 54: 65–67.

LANDSBERG, H. E., J. M. MITCHELL, JR. & H. I. CRUTCHER. 1959. Power spectrum analysis of climatological data for Woodstock College, Maryland. Monthly Weather Rev. 87: 283–298.

MACKIN, J. H. 1948. Possible sun-spot cycle in pre-Wisconsin varves in the Puget area, Washington [abs.]. Bull. Geol. Soc. Am. 59: 1376.

McGUIRE, R. H., JR. 1957. A study of some Lake Missoula varves [Mont.]. Compass. 34: 197–204.

NICHOLSON, S. B. 1929. Remarks on the sunspot cycle *in* Second Conference on Cycles. Carnegie Institute, Washington, D. C. : 79–80.

NUPEN, W. & M. KAGEORGE. 1958. Bibliography on solar-weather relationships. Amer. Met. Soc. Meteorological Abstracts and Bibliography, Malcolm Rigby, Editor. Washington, D. C. : 248.

PERFILIEF, B. W. 1927. Ten years of Soviet Science. Moscow. : 402, 403 (not seen, from Bradley, 1937 summary).

RIVEROLL, D. D. & B. C. JONES. 1954. Varves and foraminifera of a portion of the upper Puente formation (Upper Miocene), Puente, California. J. Paleontol. **28:** 121–131.

SHOSTAKOVICH, V. B. 1931. Die Bedeutung der Untersuchung der Bodenablagerungen der Seen für einige Fragen der Geophysik. Int. Ver. theor. angew. Limnol. Verh. Stuttgart. **5:** 307–317.

SHOSTAKOVICH, V. B. 1934. Mud deposits in lakes and the periodic fluctuation phenomena in nature. Zapiski State Hydrologic Institute (Zapiski Gos. Gidrolog. In-ta.). **13:** 95–138. In Russian.

SHOSTAKOVICH, V. B. 1936. Geschichtete Bodenablagerungen der Seen als Klima-Ann. Met. Z. **53:** 176–182.

SHOSTAKOVICH, V. B. 1944. An experiment on geochronological analysis of mud deposits of Malinovoie Lake in connection with the uplift of the shore of the White Sea. Bull. All-Union Geographical Society, U.S.S.R. (Georgraficheskoe obschestvo SSSR, Izvetsia). **76:** 203–206. In Russian.

SCHULMAN, E. 1945. Tree ring hydrology of the Colorado River Basin. Univ. Arizona Bull. 16, no. 4. Lab. of Tree Ring Research Bull. **2:** 35.

SCHULMAN, E. 1956. Dendroclimatic Changes in Semiarid America. Univ. Ariz. Press. Tucson, Ariz. 142p.

SEIBOLD, E. & R. WIEGERT. 1960. Untersuchungen des zeitlichen Ablaufs der Sedimentation im Malo Jezero (Mljet, Adria) auf Periodizitäten. Z. Geophys. **26:** 87–103.

STETSON, H. C. 1933. Scientific results of the "Nautilus" expedition, 1931; part 5, the bottom deposits. : 17–37. Cambridge, Mass.

STETSON, H. T. 1947. Sunspots in Action. Ronald Press. New York, N. Y.

SUGAWARA, K. 1934. Liesegang's stratification developed in the diatomaceous gyttja from Lake Haruna, and problems related to it. Bull. Chem. Soc. Japan. **9:** 402–409.

UDDEN, J. A. 1924. Laminated anhydrite in Texas. Bull. Geol. Soc. Am. **35:** 347–354.

UDDEN, J. A. 1928. Study of the laminated structure of certain drill cores obtained from Permian rocks of Texas. Carnegie Inst. Wash. Year Book. **27:** 363.

VAIL, O. E. 1917. Lithologic evidence of climatic pulsations. Science, (n.s.). **46:** 90–93.

WARD, F. W., JR. 1957. Power spectra of astrogeophysical and meteorological time series. Air Force Cambridge Research Center. Cambridge, Mass. 144p. Unpublished.

WELTON, M. 1944. Pollenanalytische, stratigraphische, und geochronologische Untersuchungen aus dem Faulenseemoos bei Spiez. Geobotanisches Institut Rübel. Veröffentlichungen. **21:** 201.

WILLET, H. C. 1953. Atmospheric and oceanic circulation as factors in glacial-interglacial changes of climate *In* Climatic Change, Harlow Shapley, Ed. 318 p. Harvard Univ. Press. Cambridge, Mass.

38

Reprinted from pages 448–449, 451–453, 456–462, 462–463, 464–466, 468, 470, and 474–477 of *Geol. Rundschau* **54**:448–477 (1964)

SOLAR CYCLES AND EQUATORIAL CLIMATES

By D. J. SCHOVE, *Beckenham*

With 12 figures

Summary

Solar cycles since 100 B. C. of 11.1, 22.2, 78 and 205 years are considered in the light of sunspot, auroral and radiocarbon evidence.

Equatorial climatic data is distinguished by a) persistence and b) cyclic tendencies. Certain wave-lengths are favoured, some by solar influences, others by atmospheric pulsations. The 11-year cycle is effective in the Tropics when the maximum sunspot number lies between 85 and 150.

Varve-cycles, from the Pre-cambrian to the Quaternary, are considered in the light of power and amplitude spectra. Certain climatological principles help us to locate the positions of past equators, and varves from such regions should prove most significant. The 22-year cycle in the Devonian may be of solar origin.

Longer wave-trains of lengths 1265 (Post-glacial), 40/80,000 (Würm, and perhaps Carboniferous) and even 20 million years (Mesozoic) are noted; these cycles may be accidental. There is a cycle of orogeny about 400 million years, but no characteristic climatic pattern of change has been demonstrated.

[*Editor's Note:* Introductory material including Table 1 and Figure 1 has been omitted.]

3. Aurorae since 500 B.C.

In the West, and especially in Germany and England, records of the aurora (Nordlicht) are more frequent. There are numerous portents and visions by night in the series Monumenta Germaniae Historica and, although no separate list of German aurorae before 1700 (cf. Gronau, Kassner for the period after 1700) is available, some of these have been extracted and included in a recent paper published in Prague by Link (1962). The pattern of the aurorae is (almost) the same as that of the sunspots.

Earlier examples could be cited (cf. Schove, 1950). That of December 786 led to "repentance" in Ireland and was variously described by the Franks as a "bow in the clouds" (Frag. Ann. Chesnii) and "terrible armies in heaven" (Ann. Lauresheim) or even a "rain of blood". However, if we limit ourselves to the years of the sunspot observations cited, we find that in 807, Feb. 26, "there appeared by night battle-arrays (acies) of wonderful size" (Ann. regn. Franc.) at the time of the lunar eclipse. This suggests that the sunspot may have crossed the central meridian of the sun about Feb. 25 and again about Mar. 25. In A. D. 840 aurorae were witnessed in Syria on Apr. 6 and Sep. 14, and in Japan about this time on several occasions. In Europe the auroral corona is described in Annal. Fuld 11 Rudolf about March 25, and the directions suggest that magnetic north may have been about 015° E. In China there are various records of dated displays in 840 seen by Chinese, Japanese and Indian monks, and one display was painted by an artist.

The fifteenth century is a weak auroral period but there are numerous references in the sources used by Weikinn (1958, pp. 277—476). Other promising manuscript sources are cited by Thorndike (1924 ff.) or Zinner (1941) in their various wellknown works. These would be most useful if extracted, and might necessitate some revision of fig. 1.

In sixteenth century Germany, visual representations of the aurorae become frequent. Thus a coloured broadsheet printed at Nürnberg portrays a religious vision of a battle seen in the sky (reproduced in Hess, 1911, Abb. 4). The date, 1554, July 24, coincides with the date of what is obviously an aurora seen in the Rhenish Palatinate and described (cf. Link, 1962, p. 60) by Frytsch in 1563. A display in 1570, Jan. 12 is represented by a woodcut by Michael Manger printed at Augsburg (reproduced in Ellison, 1955).

With the help of collaborators in the "Spectrum of Time" project, named in previous articles, sufficient evidence was obtained from both East and West to estimate the probable date of nearly all the solar cycles since A. D. 300 and a few earlier cycles since 500 B. C.

[*Editor's Note:* Material including Table 2 and Figure 2 has been omitted. The omitted material relates to the 22-year, 80-year and 200-year cycles.]

Table 3. The 200 Year Cycle illustrated by solar and climatic variations.

T   = The period of rise from min. to max. Excess of 5 years in tenths of a year. Based upon Schove 1955 Tables 4,2.

T + U = The probable length of the solar cycle. Excess of 11 years in tenths of a year. Based upon Schove 1955 Tables 4,2. Bracketed figures may be incorrect.

R   = The estimated mean sunspot number. Based on Auroral Numbers in Schove 1961.

C_{14} = Initial radiocarbon after Damon and Long (1963) and Willis et al. (1960).

D   = Droughts in the American Southwest yielding missing rings. Percentage of such droughts of moving three-century totals.

Century	T	T + U	R	C_{14}	D
II B. C.	+ 1	+ 2			
I B. C.	+ 1	(0)			
A. D. I	+ 3	(+ 2)			
	(+ 2)	(+ 1)			
	(+ 2)	(+ 8)			
	− 1	− 6	55		
V	+ 2	+ 7	50		
VI	− 3	− 3	60		
	+ 2	(+ 3)	46	H	
	+ 3	(0)	52		
	− 2	(+ 1)	55		
X	− 6	(− 3)	53		0
XI	(+ 5)	(+ 5)	51	H	70
	− 5	− 5	59	L	20
	+ 1	+ 4	47		60
	+ 1	− 5	51		10
XV	(+ 2)	+ 2	33	H	40
XVI	− 2	0	57	L	30
	+ 4	+ 3	37	H	50
	− 2	− 5	49	L	10
	− 1	+ 8	43		
XX	(− 5)	(+ 1)	(50)		

7. The 27-day Period

The solar rotation period of 27 days is of potential significance in geology now that diurnal varves (Schwarzbach 1963, p. 90) have been recognized. Even to-day terrestrial temperatures show some response (Visser 1959) to this period, and in Palaeozoic times, when the atmosphere was presumably more transparent to ultra-violet light, this cycle, like other solar cycles, was probably recorded significantly in varves. In this connection, the recent claim from sub-varve-counts that there were less than 200 days in the Palaeozoic year is not merely a challenging theory that could be tested elsewhere; it is an opportunity for investigation of the shorter 27 day solar cycle in geological times. In 840, again, two displays seen by the Buddhist monks in China are dated the 22nd day of the fifth moon and the 21st day of the sixth moon, and thus appear to reflect successive rotations of a sunspot or, strictly, an "M-region". The lunar cycle is said to have been detected in geology by KORRINGA 1957, and aurorae, which are naturally visible more readily at New than at Full moon (cf. SCHOVE and HO 1958, p. 303), do respond to this cycle too. However, DIXON (1939) showed that the effective solar rotation period over an auroral cycle was nearly always 27.31 days. In twelfth century Europe and China conspicuous auroral recurrences after 26 days may have been more freqent than after 28 days[1]).

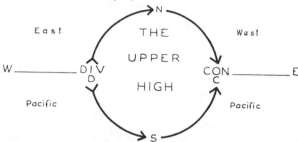

Fig. 3. Inflation of the Upper High by Equatorial easterlies at about 200 mb. The points of upper convergence (C) and upper divergence (D) are in Indonesia and South America and indicate the position of the Eastern and Western Crescents respectively. — Reproduced from SCHOVE 1963 fig. 2, by kind permission of Geofis. Pura e Applicata, Milano.

II. Equatorial Cycles

There is only one regular cycle in meteorology longer than the seasonal year: this is the remarkable alternation of W and E winds in the Equatorial Stratosphere, with a period of 2.2 years. Before we consider this, and before we can discuss solar connections, we must consider the more intermittent cycles of the Equatorial Troposphere.

1. Pressure Persistance

The "principle of equatorial inflation" has been outlined elsewhere (SCHOVE 1963 b) but it is illustrated in fig. 3. The principal Upper High lies in the Tropics over the hemisphere most heated by the sun and it is frequently located near the Indian Ocean. The absolute topography of the

---

[1]) The climatic significance of the lunar cycle, the synodic cycle of 29.53 days from New Moon to New Moon, has been discovered by BRIER (1964, and in progress). In the United States there is a double cycle with the main maximum on the 17th day (just after full moon). The reason for this is because at that day the moon is overhead at 3.0 a. m., so that the lunar tidal forces are strongest at the time normally most favourable for precipitation in that part of the world. In England CRADDOCK (1964) reports systematic relationships in the periods 1891—1903 and 1916—1939, with more rain at New Moon.

319

200 mb level tends to imprint itself on the pressure anomaly pattern at
the surface. When pressure is above normal at both levels the air passing
round the high is re-injected at its eastern extremity by the equatorial
easterlies. The effect of this in the troposphere is to set up a pressure persistence which appears to mask the influence of the 26-month stratospheric cycle. One high pressure year tends to be followed by another, and
a drought year is more likely to be followed by a second drought year
than by a wet year. This equatorial persistence is unique in meteorological
time-series and would seem to provide a simple check by which equatorial
varves could at once be distinguished from sub-tropical varves in early
geological periods.

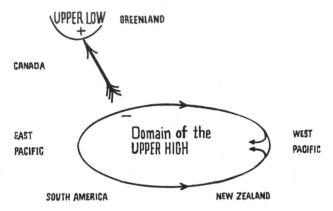

Fig. 4. Pressure changes from one five-year period to the next with a rising
parameter (cf. fig. 5). (Simplified from a detailed unpublished map.)

The persistence, both at the 200 mb level and at the surface, is especially characteristic of the half-year from July to January, and it
would be interesting to re-investigate from this viewpoint the Pleistocene
varves measured in East Africa by Nilsson (In DE GEER 1934) at places
with a double rainy season.

2. The Pressure Parameter

A rough index of equatorial inflation (reversed) in the domain of the
Upper High is provided by WALKER's indices of the Southern Oscillation.
Here we shall use a somewhat similar index, the "first approximation to
the pressure parameter" abbreviated as PP, the reversed pressure in regions near the Indian Ocean (Mean of Cape Town, west India, east India
and south-east Australia). This index was described briefly by SCHOVE
(1961 b, p. 609), and the annual anomalies are now presented in the table
below.

The PP is reflected, after several months delay in tropical temperatures,
and in some tree-rings (cf. DE BOER 1952) and it should correlate with

any equatorial varves, but annual values as such have little significance outside Australia and the Tropics. However, the difference between the mean values of two successive five-year periods is of g l o b a l significance. The parameter change then represents (fig. 4) a transfer of pressure from the domain of the Upper High to the several upper troughs. The see-saw of five-year pressure means between Greenland and the Indian Ocean suggested previously (Schove 1961 b, p. 608) can now be demonstrated by using two long series for western Greenland published by Sestoft in

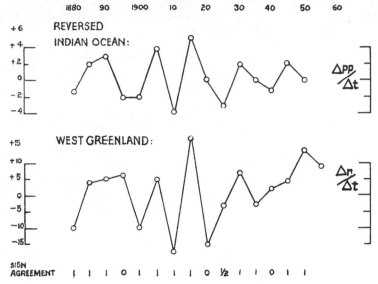

Fig. 5. Opposing barometric trends in the Indian Ocean and the west Greenland areas.

c. 1880 = 1881/5 — 1876/80

U p p e r   C u r v e : Change in the PP, the "First Approximation to the Pressure Parameter". PP = the reversed mean barometric anomaly at Cape Town, west India, east India and south-east Australia.

L o w e r   C u r v e : Barometric means of Upernavik and Jakobshavn (Baur 1962, pp. 772—777).

Sign-agreements are indicated at the base of the figure.

Baur (1962); this is shown in fig. 5. When the trend $\Delta PP/\Delta t$ is positive, the air flows, like the rivers of the geographer, from the high contours to the valleys (troughs), as indicated by the thick arrow. The pressure changes in the troposphere in the two domains are indicated by the + and — signs.

The sign-agreement between Greenland and the PP is sufficiently good to hope that regional changes in five-year means of ice-layers (cf. H. Bader, Köln 1964) will eventually by prepared and will show good sign-agreement with the PP parameter. Indeed, even if equatorial varves fail to cross-date with Scandinavia (as DE GEER had originally hoped), we may yet find in this way a teleconnection between the tropics and the varves of New England and Canada (which TERASME, 1963, has shown contain material for both pollen-analysis and carbon-dating). Approximate dating through palaeomagnetism is possible (cf. GRIFFITHS et al. 1960 and Cox et al. 1963 fig. 1).

The PP-data of Table 4 differ from other meteorological time-series in a second respect: they show a real tendency toward cycles. Although the Upper Trough region also has its characteristic cycle, of 7³/₄ years (cf. SCHOVE 1961 b, p. 608) in the period 1870—1960 (a Northern Oscillation), cycles are the exception in meteorology.

Table 4. Annual anomalies of the PP. Reversed mean pressure anomalies (tenth-mb) in the Indian Ocean countries. S. E. Australian data before 1857 are interpolated.

	Year +									
	0	1	2	3	4	5	6	7	8	9
1840	( )	(− 2)	(+ 2)	(+ 1)	(+ 1)	(− 5)	(− 4)	(+ 3)	(+ 6)	(+ 6)
1850	(0)	(+ 5)	(− 3)	(− 3)	(− 2)	(− 3)	(− 1)	− 2	− 1	− 2
60	+ 1	+ 5	+ 7	+ 5	− 5	− 4	− 4	− 2	− 8	− 3
70	+ 5	+ 2	+ 4	+ 1	0	+ 3	− 2	− 13	+ 1	+ 3
80	− 2	− 5	+ 1	0	− 4	− 7	0	0	− 6	− 1
90	+ 3	− 5	+ 6	+ 5	+ 2	+ 1	− 3	0	+ 5	− 3
1900	0	− 2	− 4	0	− 1	− 4	+ 1	+ 1	− 3	+ 4
10	+ 5	− 2	− 2	− 1	− 10	+ 3	+ 10	+ 8	− 3	− 6
20	+ 3	+ 1	+ 5	+ 8	+ 1	+ 1	0	0	+ 1	+ 2
30	− 6	− 1	0	+ 2	0	+ 1	0	+ 2	+ 4	+ 2
40	− 7	− 6	0	+ 1	− 1	0	+ 3	+ 1	0	0
50	+ 2	(+ 1)	(+ 1)	(− 4)	(− 1)	(+ 4)	(+ 8)	(− 3)	(− 4)	(− 3)
60	(0)									

3. PP Cycles

The PP-data show several non-persistent wave-trains. As DE BOER and BERLAGE (1962) have shown in the case of Djakarta pressure, the wavelength varies with the amplitude of the initial anomaly. Certain wavelengths appear to be preferred.

There are some periods, 1840/75 and 1918/56, when an 11-year wave-train predominates, but in the period 1880—1910 the 3-year cycle was outstanding. A 7-year cycle was important in the period 1908—30 and a 4-year wave-train from 1929 to 1948. The 3-year cycle was apparently generated near Indonesia and the "Eastern Crescent" (cf. SCHOVE 1963, fig. 3) and its essential equatorial aspect is illustrated in fig. 6. Changes outside the tropics are included in the original maps, but, as BERLAGE (1957, p. 35) points out, these are less significant because the standard deviation is so much greater.

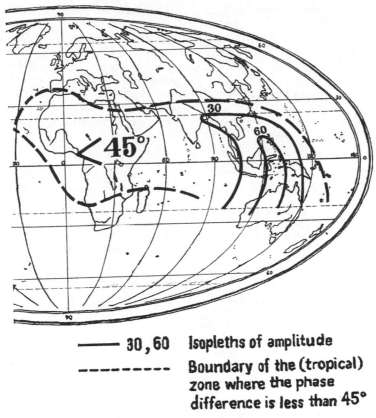

———— 30,60   Isopleths of amplitude

— — — — — —   Boundary of the (tropical)
zone where the phase
difference is less than 45°

Fig. 6. The generation of the 3-year cycle in the Eastern Crescent, 1896—1905.
Extracted from von Schubert's (1924/7) maps in Berlage, 1957, figs. 6 and 29.
(cf. Schove 1963, fig. 3.) Amplitude and phase are indicated for the Tropics only.

The 7-year cycle may be associated with the Western Crecent. Certainly, in South America it can be traced back before 1870 (Berlage 1957, fig. 2 and p. 97), and its exact length, 6.9 years, makes it quite distinct from the 8-year cycle of the Polar Trough. In the Eastern Crescent it tends to break into two cycles of 3.4 years that de Boer (1952, fig. II c) suggests may be the fundamental cycle.

The cycles associated with the domain of the Upper High could be gauged from the spectrum of the data of Table 4. In the meantime, two independent investigations, for the NW and SE extremities of the domain, suggest some cycles that may be expected.

In Australia O'Mahony's (1961) thorough analysis of rainfall cycles in separate months makes it possible to distinguish between significant and insignificant cycles in the range 2 to 7 years. There are clusters of cycles at 2.15, 2.7, 3.0, probably 3.3 and 3.7 and certainly 4.9 and about 7 years. The 3-year cycle is well-marked in Darwin pressure (Walker 1941).

In the American Southwest, tree-ring cycles over the past millenium were critically studied by BRYSON and DUTTON (1961). Cycles of certain lengths — 2.2, 2.73, 3.1, 4.8 (and 5.8) occur in several diagrams (cf. their Table 1, p. 602). An investigation of SCHULMAN's (1956) Rio Grande and Colorado indices might reveal which of these relate to the Upper High.

In the equatorial zone itself, a computer analysis of barometric cycles is being carried out by MITCHELL.

[*Editor's Note:* Material has been omitted at this point.]

Table 5. Real and possible atmospheric cycles.

Years	Character	References
2.2	Equatorial Stratosphere	II 4, 1, 3. III 2. (EBDON)
2.0 to 2.4	Possible varve-cycles	II 3. IV 9, 2, 4, 7, 8.
2.7	Temperate or Tropical (?)	II 3. (BRUNT)
2.6 to 2.8	Possible palaeocycles in varves	IV 2, 9, 7.
3.0/3.1	Upper High (Eastern Crescent)	II 3 (von SCHUBERT)
3.0/3.1	Possible varve cycles	IV 4, 7, 8, 9.
3.4	Upper High (E. Hemisphere)	II 3 (DE BOER)
3.4	Possible Miocene varve-cycle	IV 7
3.7/3.8	Upper High (?)	II 3. III 2. (LOCKYERS)
3.7 to 4.0	Possible varve cycles	IV 2, 4, 9.
4.8/4.9	Upper High (?) and Arctic	II 3 (BROOKS etc.)
4.6/4.9	Possible varve cycles	IV 2, 4.
$5^{1}/_{2}$ to 6.3	Localised: Past and present Probable half-solar period	III 2. II 3. IV 4, 9.
6.8/7.0	Upper High (Western Crescent) Possibly 2×3.4.	II 3 (BERLAGE)
$7^{3}/_{4}$	Upper Trough	II 2 (SCHOVE 1961 b)
7.9 or 8	Palaeovarves	IV 4 and 9.
11.1 (Mean)	Sunspots	III 3
22.2	Double Sunspot period Possible Devonian equatorial	II 4. III 4. IV 4 (ANDERSON)
55/90	"Eighty-year" solar cycle	I 5. III 7. IV 6
200/205	Even century Solar cycle	I 6. III 7. (SCHOVE 1955)
Long	Possible longer cycles	III 9.

This cycle sometimes shows itself in European summers. In the anticyclonic 1530's, the alternation of dry and wet summers is reflected in an alternation of early and late vintages (LADURIE, 1959, 1960) and in narrow and wide tree-rings (HUBER and SIEBENLIST, 1963) at dry sites. It helps to explain (remembering also the normal inverse relations between the west and east sides of an ocean) some mystifying results found by HUSTICH in sub-Arctic tree-rings. In European temperatures since 1761 (LANDSBERG 1962) this cycle, of length 2.17 years, was outstanding.

DE GEER (1940) noted that varve-sequences often showed a series of maxima in odd years followed — often about a decade later — by a series of maxima in even years. He attempted to use "signatures" of this kind — biennial maxima — to teleconnect curves from different parts of the world. BROOKS (1928 b) had correctly concluded that it was a cycle of 2.2 or 2.15 years that gave to curves of varves their characteristic zigzag appearance (cf. however IV 8 below).

The various cycles selected as worthy of mention in this paper are listed in Table 5. Most of the persistent cycles appear to be unimportant away from the Tropics.

[*Editor's Note:* Material on submultiples of the solar cycle has been omitted at this point. Revisions for Table 5 are now possible. In the instrumental period to the 1970s the first five cycles can be confirmed, but there is also a weak cycle of 2.4 affecting the summers of the northern hemisphere. Better values for the next two cycles may be 3.9 and 5.0.]

3. Actual Solar Cycles

An 11-year interval between successive solar maxima is the exception not the rule. Intervals have varied from 8 to 16 years, and in order to investigate solar effects it is necessary to use an objective list of maxima (Waldmeier 1961, p. 18 from 1610; Schove 1955 for pre- 1610).

A thorough investigation of the climatic anomalies at the actual maxima and minima has been carried out for the period 1899—1962 by Willett (1963), who finds many correlations and who notes that the influences are most significant
a) in the late summer season
b) in the Pacific quadrant.

The PP-data of Table 4 suggest that in some solar cycles since 1840 (see fig. 7 below) the pressure is reduced in the domain of the Upper High.

In the semi-arid American Southwest extreme droughts in certain years are so severe that the tree-ring is missing. Absent rings since A. D. 277 are indicated by circles in Schulman's (1956) charts (figs. p. 37, 92/3 and 114. Also p. 110, 118/9, 86 and 82). It was found that they were only half as frequent in the year of sunspot maximum and the subsequent two years as in the following three. Indeed, as would be anticipated from the summer position of the Upper High (cf. also Schove 1961 b, p. 606—607), the contrast was due almost entirely to the indices from the south-east part of the region (The first three references), for which the ratio was 1 : 3. A long-period relationship in the same sense will be mentioned below (III 8 and Table 3).

Five-year moving means of the PP indicate: Maxima (> 0.3 mb) at 1849, 61, 72, 92/4, 1908, 17/8, 22, 37, 48 and 54. Earlier maxima probably occurred c. 1826/9 and c. 1840.

Minima at 1844, 57, 66, 79, 86, 1903, 13, 28/9, 42, 52 and 59. Earlier minima probably occurred at 1823/4 and 1833/36. Comparison with the solar curve helps to suggest the hypotheses that follow.

4. Strong, Medium and Weak Cycles

We may assume:
a) Solar activity causes an eruption of the Upper High so that the peak tends to become a crater.
b) The geographical extent of the eruption varies in such a way that a mean sunspot number between about 85 and 145 results in lower pressure in the domain as a whole.

Some of the evidence is indicated in the figure, although it should be noted that data was available for only two "Strong" cycles. It is clear that

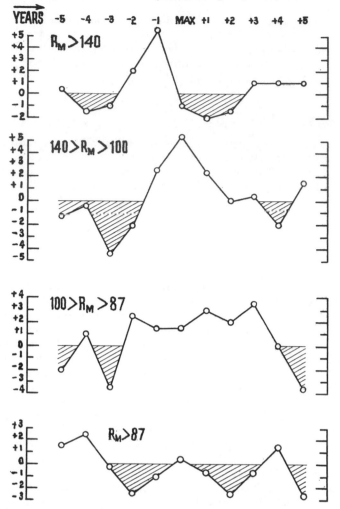

Fig. 7. Mean PP-anomalies (tenth-mb.) in relation to the year of sunspot maximum.
a = Sunspot maxima greater than 141; b = 100 < R_M < 141; c = 87 < R_M < 100; d = Maxima less than 87.

The PP-anomalies used are the reversed mean pressures of Cape Town, west India, east India, Indonesia and south-east Australia (as given in Table 4).

solar-meteorological relationships are not linear, and the failure of weak cycles to register is consistent with the suggestions of BERLAGE (1957) and VISSER (1959). Indeed, the threshold of 85 suggested by VISSER for Djakarta pressure is exactly that required to reconcile the PP and solar curves.

An alternation of cycles greater and less than 85 naturally gives rise to a meteorological cycle of 22 years (cf. WILLETT 1963).

5. Secular Solar Trends and Ten-year means

Curve of overlapping 10 or 11-year sunspot numbers have been discussed previously (SCHOVE 1961 b p. 610 and 1961 a pp. 111—113). In Europe the pressure curve (cf. SCHOVE 1963 b) depends on other factors, but the PP changes are still significant and, in low latitudes, sunspots appear to be more not less important. A preliminary index of the parameter in the Middle Ages (Unpublished) shows a sign agreement both with Nile Flood data and with Auroral Numbers. There are no obvious cycles in the range 22 to 55 years.

[*Editor's Note:* Material has been omitted at this point.]

8. The 200-year cycle

The 200-year drought cycle noted by SCHULMAN (1956) seemed "accidental" (SCHOVE 1961 b, p. 614). However, the "absent ring droughts" referred to above (III 3) suggest that it applies at least to most of the present millenium, and data have been included in Table 3 to illustrate the agreement with the solar data. Thus, there were only four such cases in the eighteenth century compared with 19 in the seventeenth and 12 in the nineteenth, leading to the value $100 \times 4/35$ for the percentage shown. Indeed, it would seem that such droughts are more frequent when sunspots were few, notably in what were previously (SCHOVE 1955, p. 144) termed the "quiescent periods of the odd centuries". Summer temperatures in Finland throughout the same millenium are reflected in tree-ring widths and the seed-bearing years, and there too SIREN (1961) finds the even centuries more favourable than the odd.

[*Editor's Note:* Material including Figures 8 and 9 has been omitted. The omitted material relates to cycles longer than 200 years especially the astronomical cycles of 21,000 (precession), 40,000 (inclination and hence seasonal effects), and 92,000 (eccentricity). For explanation, see the following: Imbrie, V., and K. P. Imbrie, *Ice Ages*, Macmillan, London, 1979, 224p.]

Table 6. Approximate dates of some thermal maxima and minima (cf. fig. 10) in millions of years (KULP 1961 scale).

Warm		Cold	
		c. 560 (?)	Eocambrian (Double minimum)
c. 530 (?)	Mid-Cambrian		
c. 490	Arenigian (SPJELDNAES)	c. 485	Early Llanvirnian (SPJ.)
c. 445	Caradocian/Ashg. border	c. 440	Ashgillian
c. 425	Ord./Silur. border	c. 420	Valentian
		c. 405	Sil./Devonian border
c. 415	Mid. or Late Sil.		
c. 390	Early Devonian		
		c. 300	Namurian or Westph. (Up. Kuttung)
		c. 280	Permo-Carb. (Early Sakmarian)
		c. 245	Late Artinskian
c. 205	Mid-Trias	c. 190	Keuper
c. 180	Rhaetic (Trias-Jura)	c. 175	Mid-Lias (δ)
c. 170	Lias/Dogger border	c. 165	Dogger (δ)-Bathonian border
c. 140	Kimer. or Portland	c. 125	Barremian or Aptian
c. 115	Albian	c. 90	Turonian
c. 80	Senonian	c. 70	Maestrichtian
		c. 60	Palaeocene
c. 55 and 48	Early and Mid. Eocene	c. 25	Oligocene/Miocene border
c. 17	Mid-Miocene		

0.92, .85, .80, .68, .61, .54, .37, .28, .15, .10 (ARRHENIUS scale) .25, .21, .17, .135, .095, .005 (EMILIANI scale)

0.96, .88, .72, .64, .57, .50, .44, .39, .34, .18, .05, .03 (ARRHENIUS scale) .27, .18, .11, .06, .015 (EMILIANI scale)

[*Editor's Note:* Material omitted at this point lists five rules for the determination of palaeoequator positions.]

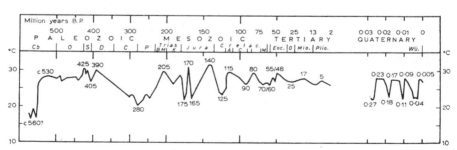

Fig. 10. The temperatures of past equatorial seas. Palaeozoic: Speculative. Cf. MA (1943) and SCHOVE, NAIRN and OPDYKE 1958, p. 220. Mesozoic: Based on BOWEN's 1962/63 data. Post-Mesozoic: Based mainly on EMILIANI (1961). Approximate dates of thermal maxima and minima are indicated in millions of years B. P. (cf also Table 6 and SCHWARZBACH 1963). Note varying scales of time.

[*Editor's Note:* Material including Figures 11 and 12 and the cycles discussed by Anderson (Paper 37) has been omitted.]

10. Conclusions

The table 6 of the various cycles, past and present, mentioned in this paper showed that in general the cycles of the past differed from those of the present. In particular, if the stratospheric cycle (now 2.2) has been a permanent feature of our atmosphere its wave-length must have varied at least from 2 to 2.4. The solar 11-year cycle and its multiples can be demonstrated in the present climate, but its harmonics have so far not been satisfactorarily demonstrated either in the present or the past. Nevertheless, even in the Palaeozoic, cycles of $5^1/2$ and 22 years have been confirmed in particular cases. Equatorial wave-trains to-day of 2.7, 3.0, 3.4, 3.75, 4.85 and 6.9 may be reflections of oscillations of the Upper High. The cycle of $7^3/4$ years is more important near the Upper Trough. Long solar cycles of 22, about 80 and 200 years are sometimes reflected in climatic fluctuations. These various points would be clarified if power spectra could be calculated for varves such as Zechstein 2 or Pleistocene Africa that represent low palaeomagnetic latitudes.

Acknowledgements

Thanks are due, for their reprints and other suggestions, to the authors named in the bibliography. Thanks are also due for helpful suggestions and references to:
Professor C. W. ALLEN, Dr. H. ARNOLD, Prof. H. P. BERLAGE, Dr. R. BOWEN, Miss C. M. BOTLEY, Dr. G. W. BRIER, Dr. J. M. CRADDOCK, Prof. P. E. DAMON, Baroness DE GEER, Prof. C. EMILIANI, Prof. RHODES FAIRBRIDGE, Dr. O. FÜRST, Prof. H. GROSS, Dr. W. B. HARLAND, Dr. M. R. HOUSE, D. J. MURRAY MITCHELL, Dr. A. M. NAIRN, Dr. E. NILSSON, D. J. R. PROBERT-JONES, Dr. H. REMY, Prof. RICHTER-BERNBURG, Prof. W. O. ROBERTS, Prof. M. SCHWARZBACH, Dr. N. SPJELDNAES.

Bibliography

AGER, D. V.: Principles of Paleoecology. — New York and London (Mc Graw Hill) 1963.

ANDERSON, R. Y.: Solar-terrestrial sediments in varved sediments. — Ann. New York Acad. Sci., **95**, 1, 424—439 (1961).

ANDERSON, R. Y. & KOOPMANS, L. H.: Harmonic analysis of Varve Time Series. — J. Geophys. Res., **68**, 877—893 (1963).

ARRHENIUS, G.: Swedish Deep-Sea Exped. 1947—1948, Repts., **5** (3) (1952).

BERLAGE, H. P.: Fluctuations of the general atmosperic circulation of more than one year, their nature and prognostic value. — Kon. Ned. Meteor. Inst., De Bilt, Meded. en Verh. **69** (1957).

Berlage, H. P. & de Boer, H. J.: On the Southern Oscillation ... — Geof. pura e applicata, **46**, 329—351 (1960).
Boer, H. J. de: Kon. Magn. Meteor. Obs., Batavia, Verh. **30** (1947).
—: Tree-ring measurements in Java and the sunspot cycle form A. D. 1514. — Proc. Kon. Ned. Akad., Amsterdam. B **55**, 386 (1952).
Boer, H. J. de & Berlage, H. P.: Geof. Pura e App., **53**, 198—207 (1962).
Borell, R. & Offerberg, J.: Sveriges Geol. Undersökning, Ca **31** (1955).
Bowen, R.: Experientia, Basel. **18**, p. 438 1962.
—: Experientia, **19**, p. 401, 1963.
Bowen, R. & Fontes, J. C.: Experientia, **19**, 268—274 1963.
Bowen, R. & Fritz, P.: Experientia, **19**, p. 461 ff., 1963.
Braitsch, O.: Die Entstehung der Schichtung in rhythmisch geschichteten Evaporiten. — Geol. Rundsch., **52**, 405—417, (1962).
Brier, G. W.: The Lunar Synodical Period and Precipitation in the United States. — J. Atm. Sci., June 1964.
—: Diurnal and Semi-diurnal Atmospheric Tides in relation to Precipitation Variations. — (In progress).
Brooks, C. E. P.: Proc. Roy. Soc., A **105** (1924).
—: Periodicities in the Nile Floods. — Mem. Roy. Met. Soc., II, **12** (1928 a).
—: The problem of the varves. — Quart. J. Roy. Met. Soc. **54**, 64 (1928 b).
Bryson, R. A. & Dutton, J. A.: Ann. New York Acad. Sci., **95**, 1,580—604 (1961).
Charlesworth, J. K.: The Quaternary Era. — 2 vols. London (Arnold) (1957).
Cox, A., Doell, R. R. & Dalrymple, G. B.: Science, **142**, 382—5 (1963) and cf. **143**, 351—2 (1964). (Re-period of reversed magnetisation 1—1.8 mill. B. P.).
Craddock, J. M.: Quart. J. Roy. Met. Soc., **90**, (1964).
Damon, P. E. & Long, A.: Carbon-14, Carbon Dioxide and Climate. — IUGG XIII General Assembly, Berkeley, California (1963).
de Geer, G.: Equatorial palaeolithic varves in East Africa. — Geog. Ann., Stockholm, **11**, 75—96 (1934).
—: Geochronologia Suecia, Principles. — Kungl. Sv. Vet Akad. Handl. **18**, 6. Stockholm (1940).
Dewey, E. R.: The 200-year Cycle in the Length of the Sunspot Cycle. — J. Cycle Research, **9**, No. 2 (1960).
—: Foundation for the study of Cycles. — Research Bulletin (1963) (2) and **1964 (1)**.
Dixon, F. E.: A 27.3 day period in the Aurora Borealis (Scotland, 1858—1938). — Terrestrial Magnetism, **44**, 335—338 (1939).
Eaton, G. P.: J. Geol., **72**, 1—35. (1964).
Ebdon, R. A.: The tropical stratospheric wind fluctuation. — Weather, **18**, 2—7 (1963).
Ellison, M. A.: The Sun and its influence. — London (Routledge and Kegan Paul) (1955).
Emiliani, C.: Trans. New York Acad. Sci., **95**, 1, 521—536. (1961).
Flint, R. F. & Brandtner, F.: Trans. New York Acad. Sci., **95**, 1, 457—461 (1961).
Gleissberg, W.: The Observatory, London. **67**, 123—125 (1945).
—: The eighty-year cycle. — J. Brit. Astr. Assoc. **68**, 148—152 (1958).
—: Zs. Astron., **55**, 153—160 (1962).
Godson, W. L.: Twenty-six-month oscillations in Geophysical Phenomena. — Arctic Meteorology Research Group, McGill University, Montreal. Publication in Meteorology **65** (Jan. 1964), 115—129.

GRIFFITHS, D. H. et al.: The remanent magnetism of some recent varved sediments. — Proc. Roy. Soc., A, **256**, 359—383 (1960).

GRONAU, K. L.: Gesellschaft naturforschender Freunde. (1808.) (Information from Miss C. M. Botley.)

HESS, W.: Himmels- und Naturerscheinungen in Einblattdrucken des XV. bis XVIII. Jahrhunderts. — Leipzig (1911).

HOUSE, M. R.: Bursts in Evolution. — Advancement of Science, London, March 1963.

HUBER, B. & SIEBENLIST, V.: Mitt der Floristisch-soziologischen Arbeitsgemeinschaft, **10**, (1963).

HUSTICH, I.: Correlation of Tree-ring chronologies of Alaska, Labrador and northern Europe. — Acta Geographica, **15**, 3, Helsinki (1956).

KASSNER, C.: Nordlicht in Alt-Berlin. — Met. Zeit **58**, 243.

KORRINGA, P.: Lunar Periodicity. — Mem. Geol. Soc. Amer. **67**, 1, 917—934 (1957).

KULP, J. L.: Science, **133**, 1105—1114 (1961).

LADURIE, E. LE ROY: Histoire et climat. — Annales, Paris. 1959 No. 1, 1—36 (?) (1959).

—: Climat et recoltes. — Annales, Paris. 1960 No. 2—3, 434—446 (1960).

LANDSBERG, H. E.: Biennial Pulses in the atmosphere. — Beitr. Physik Atm., **35**, 184—194 (1962).

LANDSBERG, H. E., MITCHELL, J. M., CRUTCHER, H. L. & QUINLAN, F. T.: Surface signs of the biennial atmospheric pulse. — Monthly Weather Review, **91**, 549—556 (1963).

LINK, F.: Observations et catalogue des aurores boreales apparues en occident de — 626 à 1600. — Inst. Géophys. Acad. Tchécoslov. Sci. **173** (1962).

LONA, F.: Ber. geobot. Inst. ETH, Stiftg. Rübel. Zürich **34**, 64—67 (1963).

O'MAHONY, G.: Time-series analysis of some Australian Rainfall data. — Melbourne. (Bur. Meteor., Met. Study **14**, May 1961.)

MA, T. Y. H.: Past Climate and Continental Drift. — 3 vols. (1943). Proc. Geol. Soc., China, **4**, 91—102 (1961).

MAKSIMOV, I. V.: On the Eighty-Year Cycle of Terrestrial Climatic Fluctuations. — Dokl. Akad. Nauk SSSR **86**, 5, 917—920 (1952). English translation by E. R. HOPE. DRB Canada (1953).

NAIRN, A. E. M. (edit.): Descriptive Paleoclimatology. — New York and London. (Interscience) (1961).

—: Problems in Paleoclimatology. — New York and London. (Wiley) 1964.

POTTER, P. E. & PETTIJOHN, F. J.: Paleocurrents and Basin Analysis. — Berlin (Springer-Verlag) 1963.

PROBERT-JONES, J. R.: An analysis of the fluctuations in the tropical stratospheric wind. — Quart. J. Roy. Met. Soc., **90**, 15—26 (1964).

RETHLY, A. & BERKES, Z.: Nordlichtbeobachtungen in Ungarn, 1523—1960. — Budapest (Hungarian Acad. Sci) (1963).

RIGG, J. B.: Climatic determinism, II. — Weather, **16**, 298—303.

SCHOVE, D. J.: Visions and dated Auroral Displays (A. D. 400—600). — J. Brit. Archaeol. Assoc., (3, **13**, 34—49 (1950).

—: The Sunspot Cycle 649 B. C. to A. D. 2000. — J. Geophys. Res., **60**, 2, 127—146 (1955).

—: Sunspot maxima since 649 B. C. — J. Brit. Astr. Assoc., **66**, 59—61 (1956).

—: Solar cycles and the Spectrum of Time. — Ann. New York Acad. Sci., New York, **95**, 1, 107—123 (1961 a).

—: Ann. New York Acad. Sci., **95**, 1, 605—622 (1961 b).

331

—: The Major Pressure Oscillation, c. 1875—1960. — Geof. Pura e App., Milano, **46**, 233—263 (1961 c).

—: Barometer and Wind . . . 1796—1950. — Nature, **197**, 1101 (1963 a).

—: Models of the S. O. in the 300—100 mb. layer . . .— Geofis. Pura e Appl. **55**, 249—261 (1963 b).

Schove, D. J. & Lowther, A. W. G.: Tree-rings and Medieval Archaeology. — Medieval Archaeology, **1**, 78—95 (1957).

Schove, D. J., Nairn, A. E. M. & Ho, P. Y.: The climatic geography of the Permian. — Geografiska Annaler, **40**, 216—231 (1958).

Schulman, E.: Dendroclimatic changes in semiarid America. — Arizona (1956).

Schwarzacher, W.: J. Geol. **72**, 195—212 (1964).

Schwarzbach, M.: Climates of the Past. — London and New York. (Van Nostrand) (1963).

Seibold, E. & Wiegert, R.: Zs. Geophys. **26**, 87—103 (1960).

Shapiro, R. & Ward, F.: J. Atm. Sci., **19**, 506—508.

Siren, G.: Communicationes Instituti Forestalis Fenniae, **54** (2) (1961).

Spjeldnaes, N.: Ordovician Climatic zones. — Norsk Geol. Tidsskr. **41**, 45—47. Bergen (1961).

Strachow, N. M.: Paläoklimatologische Weltkarten. — Sowjetische Arbeiten z. Paläoklimatogie, **2**, Geol. Inst. Univ. Köln (1963).

Terasme, J.: J. Sedim. Petrol., **33**, 314—319 (1963).

Thorndike, L.: A history of magic and experimental science. IV, New York (1934 f).

Veryard, R. G. & Ebdon, R. A.: Fluctuations in tropical stratospheric winds. — Met. Mag., **90**, 125—143 (1961).

Visser, S. W.: Meded. en Verh. Kon. Nederl. Met. Inst., **75** (1959 a).

—: Geofis. Pura e Appl. **43**, 302—318 (1959 b).

Waldmeier, M.: The Sunspot Activity in the Years 1610—1960. — Zürich (Schulthess) (1961).

Walker, G. T.: Period-hunting in practice. — Quart. J. Roy. Met. Soc., **67**, 15—18.

Weikinn, C.: Quellentexte zur Witterungsgeschichte, 1. Hydrogr., Teil 1. — Berlin (Akademie-Verlag) (1958).

Willett, H. C. & Prohaska, J. T.: Long-term Solar-climatic relationships (Final scientific report. NSF Grant 14077). — Cambridge, Mass., (1963).

Willis, E. H., Tauber & Münnich, K. O.: Variations in the Atmospheric Radiocarbon concentration over the past 1300 years. — Radiocarbon, **2**, 1—4 (1960).

van Woerkom, A. J. J.: In: Climatic Change (ed. H. Shapley) 1953, pp. 147—157, 1953.

Zeuner, F. E.: Dating the Past. 4th ed. (1958).

Zinner, E.: Geschichte und Bibliographie der Astronomischen Literatur in Deutschland zur Zeit der Renaissance. — Leipzig (1941).

39

Reprinted from *Palaeogeography, Palaeoclimatology, Palaeoecology*
25:209–233 (1978)

TREE-RING AND VARVE SCALES COMBINED, c. 13500 B.C. TO A.D. 1977

D. J. SCHOVE

St. David's College, Beckenham BR3 3 BQ (Great Britain)

(Accepted for publication November 2, 1977)

ABSTRACT

Schove, D. J., 1978. Tree-ring and varve scales combined, c. 13500 B.C. to A.D. 1977.
Palaeogeogr., Palaeoclimatol., Palaeoecol., 25: 209—233.

Varve chronologies are of several kinds but glacial varves respond to summer temperature and fluvial varves to summer rainfall.

Tree-ring chronologies near the timberline respond to summer temperature but European oak chronologies respond to summer rainfall.

Certain global features of the weather of the summer season, notably the Biennial Index, are therefore parameters that can be used for cross-dating both varves and tree-rings in different continents.

Characteristic curves for different parameters in the Late Glacial (Zones Ia, Ib and II) and in the past thousand years are presented. A moisture curve is given for the North European Plain (200 B.C.—A.D. 1100). A conversion chart for tree-rings and radiocarbon dates is extended back into the Late Glacial on the tentative assumption that the radiocarbon error was about 950 years at the beginning of the Holocene (III/IV), i.e. that 10300 b.p. should be 11250 B.P. or 9300 B.C.

Supplementary sources of information useful in obtaining approximate dates include: (a) palaeomagnetism; (b) tephrochronology and X-ray analysis; (c) cycle analysis; (d) climatic peculiarities associated with specific radiocarbon centuries; (e) X-ray analysis of specific varves; and (f) new methods of varve-analysis.

The varve-series used for the Late Glacial, if the author's cross-dating between North America and Scandinavia is acceptable, constitute a floating chronology of about 4000 years. Given approximate ^{14}C dates for any long series of varves or tree-rings in one part of the world, it is now possible to obtain cross-dating with any other long series in another part of the world, and it will be easy to replace the tentative '950-year' error by a precise figure determined from a combined varve and tree-ring scale extending back from the present day to (say) the zero of Sauramo's scale for varves in Finland. In the meantime the '950' is mnemonically convenient, as this would make the year on the B.C. scale one thousand less than the year on the b.p. scale.

INTRODUCTION

The Swedish pioneer of geochronology, the Baron De Geer, believed that varve and tree-ring curves could be matched visually across the Atlantic. We now know that visual matching, even within a single locality, is often mis-leading unless confirmed by significance tests at the 99.9% level. Both varves and tree-rings, however, reflect the weather, and meteorological time-series today are not numbers in a random order; certain parameters derived from them now make it feasible to link the Late Glacial varve-chronology with the bristlecone pine tree-ring chronology; the chronology of ice-cores could like-wise be checked against either series. *Glacial varves* like tree-rings from the *upper timber-line*, respond to summer *temperature*. *Fluvial varves*, like many *dry-site tree-rings*, often respond to summer *rainfall*. The layers found in cores from the Greenland ice-caps respond to summer *snowfall*.

The non-random elements in summer weather on both sides of the Atlantic can be separately isolated and often cross-dated. The interrelationships found today provide clues which can be used as far back as the Late Glacial. In the twentieth century for instance:

(a) A *cold* season on one side of the Atlantic (e.g. E Canada) is usually *warm* on the other side (e.g. Scandinavia). This was contrary to the expecta-tion of De Geer.

(b) An abnormal summer on one side is, however, often followed by an abnormal summer of the same kind (e.g. hot) one year later.

(c) A decade of warm summers on one side of the Atlantic nevertheless synchronizes with a decade of warm summers on the other side. Indeed, for any one season, curves of overlapping ten-year means look very similar. Five-year moving means behave in the same manner.

(d) Thirty-year turning points are again similarly dated. Curves of over-lapping hundred year means, if truly representative, would again be parallel.

(e) In this century, prior to the outstanding exception of 1976, summers in odd years were drier than summers in even years (Schove, 1969, p.290). The amplitude of this biennial cycle varies similarly in both E Canada and NW Europe.

(f) There are certain characteristic responses to sunspot cycles of differing amplitude-classes (Schove, 1971c).

The general method used and applied to each parameter in turn involves firstly determining weighted dates of maxima and minima and secondly calculating product moments between different varve-series to find the position of synchronization.

VARVE CHRONOLOGIES 13000—9000 B.C.

These relationships were used in an attempt by the writer to cross-date various varve-series measured by Antevs in North America not only with one another but also with the Swedish and Finnish varve-series of Nilsson, Sauramo and others (Schove, 1971a, b, 1975a, p.10). The following specific equations were proposed:

S Sweden (Nilsson)	— Finland (Sauramo)	= 8226 (Early Zone II)
		8217 (Late Zone II)
N central Sweden	— Finland	= 8130/8155 according to century
N Sweden (Jarnefors)	— Finland	= 1226 (north of Uppsala)
(Döviken scale)		
N Sweden (varve 6923 or Döviken zero)		= 1166± in Canada
Finland	— Canada	= 52—63 according to century
	(Timiskaming)	
Quebec	— Canada	= 83
Espanola (S Canada)	— Canada	= 78, then 59
Burke Falls	— Canada	= 251 (uncertain)
Bracebridge	— Huntsville	= 187/186
N New England	— Bracebridge	= 6939 ± 5
N New England	— S Sweden (Nilsson)	= about 17027, then 16960
N New England	— Finland (varve 1200)	= about 8745

The principal connections claimed are illustrated in Fig.1.

At present no independent confirmation is available, although no counter-claims have been proposed. However, Mörner tells me that Swedish varve-specialists do not accept these teleconnections. Errors were noted in the Swedish countings (Schove, 1971a, p.226 ff) of Nilsson and De Geer, and further errors were suspected, so that the writer used the countings of Sauramo in Finland rather than those of Nilsson in Sweden when preparing Fig.1. When the Finnish curves of Niemalä (1971), Ignatius and Saarnisto (in prep.) have been published on a scale that will enable comparisons with

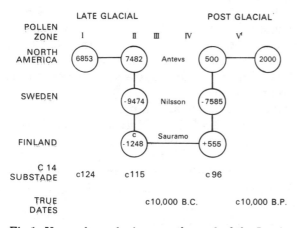

Fig.1. Varve chronologies near the end of the Ice Age (established connections are shown by lines). The Swedish dates 'before Christ' (f.Kr.) are indicated as negative.

Sauramo's countings, some minor errors may be found in the Finnish chrono-logy; Niemalä suggests revised nomenclature of the moraines in S Finland, but I understand from Professor Donner that no counting errors in Sauramo's work have been detected. Ignoring this source of possible error we have in

Fig.1 a count of 3900 varve-years. The radiocarbon ages at the two extremities of this period suggest that about 4200 radiocarbon years are involved (5568 year half-life).

The absolute dates suggested were based on the somewhat arbitrary assumption (Fig.2 and Schove, 1971a, table 5) that the radiocarbon error at the end of the Late Glacial (Zone III/IV interface) was 950 years so that 10300 b.p. is equivalent to 9300 B.C. This assumption was based on Early Holocene counts of non-glacial varves (cf. Stuiver in Olsson, 1970, p. 197 ff), and assuming a constant hard-water ^{14}C error, is near the value of varve countings made at Schleinsee (Geyh et al., 1971), but when the gap between the bristlecone pine and the Canadian varve chronologies has been effectively bridged, a revision will be possible. With this approximation the true year B.C. is a thousand less than the ^{14}C year b.p., so that the 950 is mnemonically convenient. Mörner (in prep.) considers a ^{14}C error of 950 years as impossible and in his paper (1977a, p.137) keeps the Pleistocene/Holocene boundary at 10000 B.P. or 8000 B.C. as was customary in pre-radiocarbon days. The ^{14}C equivalent of the boundary is near 10300 b.p. (8350 b.c.), and to avoid confusion I distinguish the Swedish varve-dates (as in 1971a) using the Swedish abbreviation f.Kr. instead of B.C.

Meanwhile, the Swedish countings in the Early Holocene are being recounted. Dr. Bo Strömberg (written comm., 1977) writes that his preliminary results should mean that the true time-scale 'is some 30 years longer between Stockholm and Uppsala' (roughly the three centuries 8125 f.Kr to 7825 f.Kr). Indeed, the writer found (1971a, fig.8) about 25 more varves in Finland implied by the cross-dating formulae. Strömberg adds that the scale is correct between Uppsala and Gävle, and that his co-worker finds it correct within 20 years between Gävle (say, varve c. 7525) and Bollnäs (say, varves c. 7400).

The use of a densitometer and other methods of microanalysis was suggested (Schove, 1971a, p.231) as a means of testing the various equations, but so far no results have come to the author's notice.

The errors in parts of the Swedish series are small, as in Strömberg's curves (1977, fig.3, p.101) for the range 8690—8213 f.Kr. in Nilsson's chronology, but the Swedish chronology has been assumed to be less reliable than the Canadian or the Finnish counts.

The substades indicated as *c124*, *c115* and *c96* in Fig.1 are radiocarbon centuries dated c. 12450, 11550 and 9650 b.p., these centuries being warm (cf. Schove, 1971a, fig.2) and represented by reliable varve-series. The approximate radiocarbon dates of earlier warm centuries are listed in Schove (1975a, p.433; and cf. Cline and Hays, 1976).

RADIOCARBON ERROR

This error-curve of Fig.2 is highly simplified but it is intended for the correct placement in time of floating tree-ring chronologies of the order of

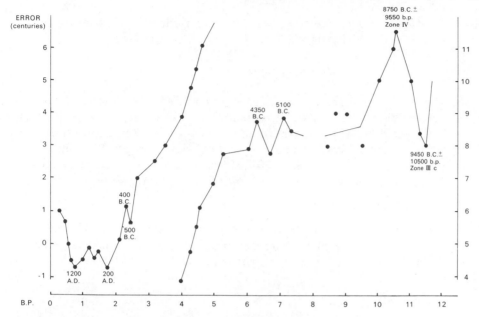

Fig. 2. The ^{14}C 'error' curve: the difference between the conventional radiocarbon date (B.P.) and the true or tree-ring date (B.P.). This is highly simplified; in the unpublished detailed version some 'Suess wiggles' give rise to discontinuities and two or more error values (B.P.—b.p.) can occur for certain tree-ring dates (B.P.). The left-hand part of the diagram is based on the bristlecone pine chronology, the left-hand scale being applicable back to 4500 B.P., and the right-hand scale for the mid-Holocene dates where the 'error' rises to about 800 years. The right-hand part of the curve is unreliable and tentative, as it is based on the assumption that this error is 950 years at the beginning of the Holocene (c. 11,000 B.P.) and on other implications of Schove's (1971a) table 5 where the varve-chronologies are considered as an integrated floating chronology.

several centuries. When there is an irregularity on the ^{14}C curve there should be a discontinuity on our figure, three tree-ring dates sometimes corresponding to one radiocarbon date.

Radiocarbon dates b.p. near 870 (A.D. 1200±), 1620, 2175, 2465, 2625, 2900, 3180, 3370, 3540, 3800, 4465, 4675, 5350, 6030, 6595, 8700, 8900, 10175 and 10990 are suspected of having ambiguous answers in the tree-ring chronology and some of these are being investigated by Suess (1977) and Pilcher et al. (1977). For most centuries the work of Clark (1975) is now accurate but intensive studies of tree-rings within a given century or so reveal the reality of further small Suess-type wiggles. As an example Mook et al. (1972) have demonstrated that stade 47, that is 4750 b.p. or 2800 b.c., is equivalent to 3625 B.C. but stade 49 (4950 b.p. or 3000 b.c.) is 3700 B.C., the errors at the two dates two radiocarbon centuries apart being 825 and 700 years respectively. Floating tree-ring chronologies at other periods (e.g. 3470—3200 B.C. and 1370—810 B.C.) have even been placed in their correct centuries merely because of the characteristic irregularities in the ^{14}C curve (cf. however, Pilcher et al., 1977) during those periods.

PRE-BÖLLING VARVES (European Zone Ia)

Many sets of North American varves were numbered by Antevs. These series can now be shown to overlap one another. However, there is still a gap between the main series for southern and northern New England, series identified by A and B dates respectively. The A dates are pre-Bölling, and, assuming the cold phase (varves 4650/4800) corresponds to the Rosendale readvance in the Hudson Valley about 14800 b.p., the formula:

A + (b.p.) = 19400 ± 100

is a possible improvement on the previous value (in Schove, 1971b, p.932) of 18700 ± 300. Palaeomagnetic measurements recently have not been successful in testing the early claim by Johnson et al. (1948) that declination was as follows:

$$A = 3700 \text{ E (incl. near } 40°)$$

$$= 4200 \text{ W (incl. min. } 45°)$$

$$5100 \text{ E}$$

and 5800 to   6300 W (30°) (incl. measuring to 55°)

The floating New Haven varves were claimed to be strongly westerly about No. 260 and weakly westerly at No. 400 (McNish and Johnson, 1939). If these values prove correct, a check on the formulae will be possible. The palaeomagnetic and climatic fluctuations have been combined for dating purposes for Lake Biwa where a high intensity peak has tentatively been dated 16500 b.p. (Schove, 1975b, p.432).

In order that the varves of Zone Ia can be dated absolutely, it is necessary to find overlaps bridging the glacial readvances near 17200 and 13000 b.p., and tree-rings from trees overwhelmed by glacial advances at these times might provide the missing links.

BÖLLING VARVES (Zone Ib)

The correlation given in Figs.1, 3 and 4 (after Schove, 1971a, and cf. 1977c) between the pollen-zones and the varves is, in my view, consistent with that given by Tauber (Fig.3 in Olsson, 1970, p.187), who regards the varves south of Lake Bolmen as corresponding to the Bölling period. Varves of this period, and even the patterns of recession (Schove, 1971a, Fig.7a), were cross-dated with those in northern New England, U.S.A., and both sets of varves were used to sketch parameters that should prove valid for other Bölling varves known in Austria and elsewhere. Some Swedish writers regard these varves as post-Bölling.

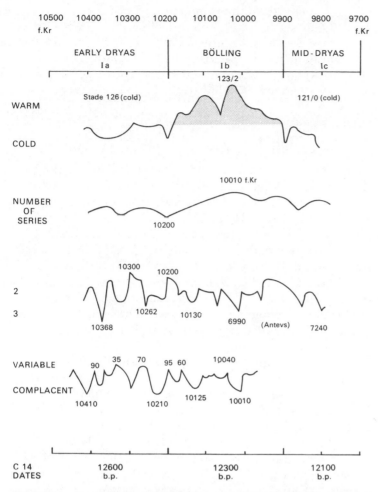

Fig. 3. The chronology of the Bölling, a period which in the strict sense lasted two to four centuries only. From top to bottom: (a) warmth reflected in glacial recession (curves in Schove, 1971a, Fig.7a); (b) warmth reflected in total number of varve-series; (c) biennial and triennial cycles; (d) variable and complacent periods (up to seven years).

In the broad sense the Bölling (Mangerüd et al., 1974, p.119) includes the whole radiocarbon millennium 13000—12000 b.p., but the very hot phase lasted less than two centuries (c. 12300 B.P. or c. 10050 f.Kr). Over a millennium in SE Sweden (cf. now also Rudmark, 1975, pp.75—77) the rate of recession averaged 90 m per year (Ringberg, 1971, p.4). During the hot phase in Vermont (U.S.A.) the rate reached a record value of 335 m per year (Schove, 1971b, after Antevs). The top curve of our Fig.3 illustrates the warmth compared with the cold Mid-Dryas phase c. 12100 b.p. that intervened before the Alleröd (Fig.4).

The frequency of biennial (as opposed to triennial) cycles is likewise indicated, with some dates first in the Swedish and then in the New England varve scales; this curve is influenced by the proportion of proximal varves which are characteristically zigzag and may also reflect varying solar activity (cf. Schove, 1971b) but does not seem to indicate high sunspot activity during the hottest period. Fluctuations in what seems to be the variability of summer climate (in seven-year groups) are sketched in the lowest curve, consistent transatlantic parallels for this parameter being available up to about varve 9980.

A curve of global geomagnetic intensity would be a useful addition to the Bölling curves, as soon as agreement has been reached on the Göteborg magnetic excursion, ascribed by Mörner (1977b) to the period 13750—12350 b.p. and involving a magnetic flip in what I term the mid-Bölling, c. 12375 b.p. near the time of the Fjaras stadial. In SE Sweden varves were investigated by Noel and Tarling (1975), and a series of intensity minima were found at intervals of about 24 years at 10151, 10136, 10113, 10074, 10051 and 10029 f.Kr. (Nilsson scale) with a major maximum at about 9920 f.Kr. (about 12200 b.p.), with further low values following at their date of 9785 f.Kr. They considered that there was a reversal for 30 years from varve 10153 to 10127 f.Kr. Thompson and Berglund (1976) failed to find any Bölling reversal in their Swedish cores, 13000—11000 b.p. and believe that the violent climatic changes of the period produce sediments of very variable mechanical properties, which are often unreliable in reflecting the magnetic elements of the time.

Declination was found by Noel and Tarling to be abnormally westerly during the hot phase (10150—10000 f.Kr.) but east of north again by varve 9820 f.Kr. (Middle Dryas or Zone Ic). Inclination values were high c. 10175, low c. 10100 and high again c. 10030 f.Kr. on the Nilsson scale; they were again low c. 9930 and high c. 9870 f.Kr.

This period, according to the writer's cross-dating, corresponds to the period of the northern New England varves, so that the period of the 'flip' should be near varve 6900 (Schove, 1971a, table 5, p.230), and such varves would repay a fuller investigation than originally carried out in the thirties (Johnson et al., 1948), when only declination (east of North about 10°, decreasing) and inclination (about 55°) were measured (cf. Verosub, in prep.).

Current studies of the palaeomagnetism of varves are being made by Mörner and will be published in the *Journal of Interdisciplinary Cycle Research*, in *Stockholm Contributions to Geology* and in the *Canadian Journal of Earth Science*.

The effects of the Geomagnetic Excursion on ^{14}C dates and on climate are of interest. At the time of the 'flip' [14]C dates would be expected to indicate ages that were much too young on account of the modulation of the cosmic ray flux. The climatic effects may include the abnormal warmth of Bölling summers, which in England were hotter than the summers of today (cf. Coope and Pennington, 1977), although more complex relationships are indicated by Fairbridge (1977). The level of the ice-sheets may have been lowered to the point of no return despite the local readvances of the Middle and later Dryas (Zones Ic and III, respectively).

ALLERÖD VARVES (Zone II)

Varves from the warm Alleröd zone, the radiocarbon millennium 11900—10900 b.p. or the eight centuries 11800—11000 b.p. are dated in the Swedish Nilsson chronology as 9700—8900 f.Kr. Curves of the derived parameters (cf. Fig.4) were compared and cross-dated not only with Finland, but also with the U.S.A. and Canada (Schove, 1971b and cf. 1971a, fig.7b). The teleconnection methods used are sometimes regarded as subjective; nevertheless it may be emphasised that given a correct approximate century on the radiocarbon scale the equations are correct to a significance of over 99%, and similar methods have been successful (Schove, 1976 and 1977a) in fixing the floating chronology of the American Maya in the A.D. time-scale.

One curve in Fig.4 shows the biennial tendency and the curve below indicates whether the hottest summers in Baltic countries were in odd (as at 9530 f.Kr. on the N (Nilsson) scale), or even (as at 9430) years. The latter curve does not incorporate the North American evidence, as the phase pattern shows no consistent agreement across the Atlantic.

The average length of the biennial cycle today is not 2.0 but 2.2 years so that there is often (but not always) a tendency for hot summers in one decade to be in odd years and in the next decade to be in even years. Oddness is measured by using first differences and smoothing the results (Schove, 1974, p.167) and the curves so found have cycles, not necessarily of 20 years. The length of these oddness cycles shows a similarity of pattern on both sides of the Atlantic and is shown in the next curve, whereas the curve above suggests there were also oddness cycles of 30—33 years (e.g. between 9462 and 9428).

A phase (cf. Figs.5 and 6) may persist, and indeed in our own century, until 1975, the odd summers were drier both in England and eastern Canada. Such persistence is maintained despite the average cycle length of 2.2 years, because the cycle is occasionally longer, and in the Alleröd occasional four-year cycles, when the biennial cycle misses a beat, coincide with the periods of phase-persistence.

In our curve (Fig.2) radiocarbon error appears to be at a maximum in the end of Zone IV and at a minimum early in Zone III. Such a result is tentative and depends on the reliability of the varve-counts used; possibly, however, ^{14}C dates are complicated by the effect of the magnetic excursion.

The outstanding palaeomagnetic peculiarity of the Alleröd is the inclination, for this was abnormally high in NW England and SW Sweden (over 70°) but low in the eastern U.S.A.

Varve-series from Zone III need investigation as there is at present an apparent error in the Swedish series, the Finnish series having been used for our Fig.1. Niemalä (1971) suggested revision of the moraines of Zone Ic in Finland as has been noted, but until the unpublished details of the curves

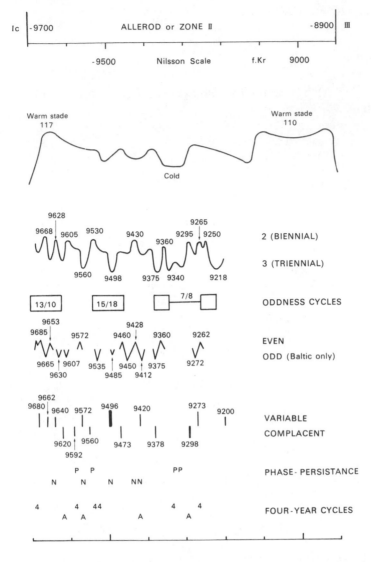

Fig. 4. Chronology of the Alleröd period. Some parameters based on combined North American and Baltic varve-series. From top to bottom: (a) warmth reflected in glacial recession (Schove 1971a, Fig. 7b); (b) biennial and triennial cycles; (c) oddness cycles when alternate anomalies are reversed in sign; (d) the phase of the biennial oscillation maxima (in the Baltic *only*); (e) variable and complacent periods (up to seven years in length); (f) biennial cycles: phase persistence (*P*) or non-persistence (*N*); (g) four-year cycles, present (*4*) or absent (*A*) in transatlantic varve-series.

have been compared with those of Sauramo, the latter's are assumed to be approximately correct.

EARLY HOLOCENE VARVES

My assumption that the end of the Ice Age (10300 b.p.) was dated 11250 B.P. (the Swedish varve date being 10039 before the present or c. 8088 N) was again based on the tentative radiocarbon error of 950 years at that date. Estimates discussed at the INQUA meeting in Birmingham (Aug. 1977) range from 0 to 1000 years. We badly need measured varves (as distinct from the counts made at the Lake of the Clouds, etc.) that will bring the varve chronology back to the point at which it can be teleconnected to the tree-ring chronologies.

Varves being investigated in northern Sweden appeared to fit those in Canada, but the significance level of the author's formula, in which the Döviken zero corresponds to the Canadian varve c. 1166, would be low if that zero is incorrectly linked to Zone IV varve-date near 6923 f.Kr. (cf. Lundqvist, in prep.). Nevertheless, the several Canadian series (Schove, 1975, p.10 links the Quebec and Espanola series with the Timiskaming) carry us through the Preboreal to the Cockburn cold stade, about 8250 b.p. roughly 7250± B.C., almost reaching the bristlecone pine period (cf. Fig.2).

Varve counts have been made in Germany through the period 9200 to 5700 b.p., and the use of X-rays and carbonate analyses promise exactitude soon (cf. Geyh et al., 1971). Connections with the north Swedish varves, which already go back to c. 8800 b.p. and the bristlecone pine chronology, which now begins well before 7500 b.p. on the radiocarbon scale, should not then prove difficult. Glacial varves, if good series can be found, would be more easily dated to the year, and more readily teleconnected with the upper timberline trees of the bristlecone pine, both being sensitive to a short summer season.

Field work involving microanalysis of varves is now necessary. X-ray densitometers need to be used to analyse the pattern of variation within specified varves, and meteorological sequences in each of the three summer months can then be used, both to clinch the formulae suggested and to eliminate errors due to double or missing varves. Whereas the author's methods need at least three centuries of good varves and knowledge of the approximate millennium on, say, the radiocarbon time-scale, X-ray plots could be used to date series which have so far not been teleconnected at all.

The palaeomagnetic studies cited suggest that, at least for eastern U.S.A. and NW European sites, declination changes are in unison in this period, with westerly maxima near stades of Zone III, say, 110—105, easterly near 100—95, west near 88—83 and east near 79—75 (i.e. roughly near 7750 b.p.). More precision is expected soon, as further well-dated cores are analysed.

In Switzerland, the well-known varves of Faulensee may be genuinely annual for the radiocarbon range 7660—5700 b.p., and palaeomagnetic analysis might usefully extend the bristlecone time-scale. Dates were originally given for these varves by Welten (cf. Vogel in Olsson, 1970, Fig.1, p.317) as 4250—2300 v.Kr; if these varves are consistently 2300 years older than Welten thought this would mean that the radiocarbon error remains about 850 years for several centuries further back than the c. 5500 B.C. beginning of the bristlecone (middle) curve of our Fig.2, the tree-ring ages of these varves becoming 6560—4600 B.C., thus the Swiss varves overlap the zone where floating tree-ring chronologies are being developed in Ireland by Hillam and Baillie and in Germany by Becker. Such an assumption, *if correct*, would justify connecting up the bristlecone and the Canadian right-hand varve curves c. 8000 B.C. in our Fig.2. The German varve-counts lead to a similar conclusion.

A long series of measurements from Dalmatia, received from Professors Beug and Seibold (cf. 1958), is more promising; measurements were made separately for the dark winter layer and the light summer layer. Over two centuries of these measurements were indeed published under dates 1667—1945; a ^{14}C date subsequently proved that they belonged to a period near 5300—5000 b.p., but the chart (Seibold, 1958, Fig.2) will still prove useful for cross-dating both varves and tree-rings.

The fluvial varves counted by Liden but never published in Sweden are regarded with justifiable suspicion, but for more continental climates there are less grounds for uncertainty.

TEPHROCHRONOLOGY

In Scandinavia and the Faroes, debris from Icelandic volcanoes has been found (Persson, 1971) in peat deposits, and despite the prevailing west winds, similar deposits may perhaps be found in the Greenland ice-cores.

Persson's important findings suggest that volcanic ash layers might be dated in varves as follows:

24th Century B.C.	(1880 b.c.)	3730 b.p.
11th Century B.C.	(870 b.c.)	2820 b.p.
A.D. 400—440	(a.d. 390)	1560 b.p.
A.D. 1362	(a.d. 875 etc.)	1075 b.p.
c. 1375—1620		
1875		

The Aegean eruption of Thera (Santorin) about 1500± B.C. should be identifiable in tree-rings of the Bronze Age being investigated in Asia Minor (Kuniholm, 1978). The effects of the eruptions of A.D. 1783 (but not that of 43 B.C.) have been traced by peculiar colours of the tree-rings (Schove, 1954).

A useful list of volcanic eruptions giving dust-veils has been published by Lamb (1970, 1977) for the A.D. period and some effects will be found in both varves and tree-rings.

THE GAP: 9000—8253 B.P.

The gap between the long-floating varve chronologies of the Early Holocene and the shorter but anchored varve chronologies of recent centuries can be bridged as soon as we can agree on rough dates for some intervening varve chronologies. Once an approximate dating is obtained for a good varve-series, cross-dating with tree-ring chronologies since 7104 B.P. (and shortly c. 8253 Ferguson, pers. comm.) will become practicable.

An approximate date may be obtained by: (a) palaeomagnetic studies of small groups of consecutive varves; (b) tephrochronological evidence of volcanic ash in specific varves; (c) spectral analyses of cycles in varve thickness and derived parameters; (d) climatic peculiarities associated with specific centuries.

These topics will be considered in turn before we discuss the chronologies of the Later Holocene.

PALAEOMAGNETISM

The three palaeomagnetic parameters — intensity, declination and inclination — have now been investigated throughout the Holocene for certain regions (Creer et al., 1976) but few applications to varves have so far been published. A pioneer study by Griffiths (1955) of fluvial varves was dated according to Liden's 'Swedish Chronology', but it can now be corrected provisionally by comparing the declination curve found with the trends as now known. We can ignore any 'Westward Drift' in our first approximation, as on both sides of the Atlantic declination was easterly near 3600 b.p. (c. 2000 B.C.), westerly near stades 29/26 (1200—950 B.C.) but easterly again at 1600 b.p. (A.D. 400). Moreover, in southern Finland a short westerly 'excursion' is provisionally dated c. 300 B.C. by Tolonen and his colleagues (1975). The best match necessitates correcting the varve-dates of Liden over his span from 1000 f.Kr.—A.D. '700' to about 1400 B.C.—A.D. 300. However, varves have already been counted through this period in Finland and a definitive solution will be available as soon as they have been studied palaeomagnetically. Such palaeomagnetic studies could be supplemented with X-ray analyses of any unusual varves likely to prove significant horizons chronologically.

CYCLES

The biennial oscillation

The biennial cycle of 2.2 years has already proved useful in teleconnecting transatlantic varves (Schove, 1971a, b). A biennial index is readily constructed (Schove, 1974). Typical curves of its amplitude are shown in Fig.5 for regions responding to temperate winter rainfall, monsoon rainfall and temperate summer rainfall respectively. The phase of the rings in the New and Old World is shown in Fig.6. Since these figures were drawn a pilot analysis has been made of summer temperature indices based on timberline trees in the U.S.A. (using especially the work of LaMarche and Stockton, 1974), Norway (using especially the spruce data in Schove, 1957) and Siberia (using the work of Lovelius for the River Tana region). In these three regions the amplitude fluctuations are linked, but the phase in the mountains of the American southwest tends to be opposite to that in Siberia and (often) Norway. In the last two regions *even* summers are hotter e.g. c. 1502, 1522; *odd* summers are hotter c. 1530/1, c. 1562, c. 1580, c. 1640, c. 1680, c. 1730s, 1770s, c. 1860 (but changing to *even* c. 1869 except in Norway). The American trees behave in the reverse sense so that the bristlecone pine chronology may thereby eventually be correlated with Old World tree-ring chronologies to the exact year. This method thus made it possible to cross-date the floating tree-ring chronology in Asia Minor with the bristlecone series (see below).

In the three curves of our Fig.5 there is little apparent similarity, and,

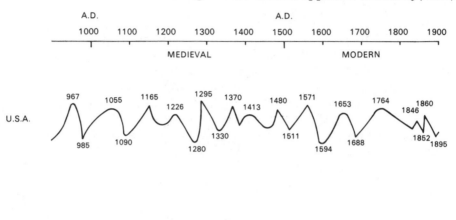
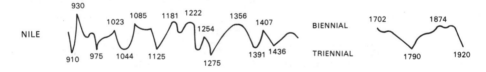
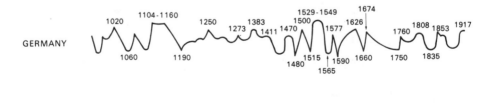

Fig.5. Biennial and triennial cycles, A.D. 910—1929. From top to bottom: (a) the American southwest: tree-rings reflecting a winter rainfall pattern; (b) the Nile: river levels reflecting summer Ethiopian rainfall; (c) Germany: tree-rings reflecting especially a summer rainfall pattern. (The peaks correspond to biennial phases.)

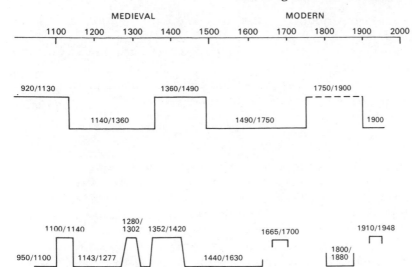

Fig.6. Biennial oscillation: the phase in north temperate latitudes, A.D. 920—1948. Above: winter pattern in American Southwest trees. Below: summer pattern in German trees. (The peaks correspond to even maxima.)

except for the German curve, no obvious similarities with solar fluctuations. Comparisons of the original data, nevertheless, suggested that the amplitude peaks in the summer Nile data preceded those in the German summer tree-rings (by about three years) and in the American winter-sensitive tree-rings (by about 11 years). The phase of the biennial cycle was found to have significant persistence for a half-century, but the apparent cycle of 80—110 years in the American amplitude curve does not correspond to any regular change of phase. The generation of biennial cycles in strong solar cycles has been demonstrated previously (cf. Schove, 1971c), but the length between successive maxima may be another important factor in their development.

Triennial cycles

Three-year cycles are not shown separately in our figures but these cycles, characteristic of sunspot minima and weak sunspot maxima, warrant separate study as they too have provided useful clues to teleconnection. Summer-sensitive tree-rings in Nevada through a period of nearly two millennia, as LaMarche and Stockton (1974) have demonstrated, have significant cycles at 2.2, 3.3, 7 and 27 years but the amplitude and phase-patterns of cycles of either 3 or 3.3 years are not yet understood.

Other short cycles

Many of these were included in a previous list (Schove, 1964), but not all oscillations important in the instrumental period of the past 100 years can be expected to be significant in earlier centuries.

The autocorrelation periodogram and autocovariance functions have been applied (by Dr. J. Pochobradsky) to the writer's date-lists of outstanding varves in the Late Glacial, and a 6-year cycle appeared that is not in existence in summer weather today. 'Pilot doses' of other parameters are now available for similar collaboration, and it is hoped that a program has been devised which will 'unscramble' the solar cycle from the results.

The solar cycle

The 11-year solar cycle is conspicuously absent in most meteorological time-series, and other parameters now need to be studied. The solar cycle (Schove, 1967) of 11.1 years has nevertheless persisted in auroral activity in approximately the same phase for 11½ centuries; over that *whole* period biennial cycles in German tree-rings have shown a similar phase. This method of detecting the solar cycle phase has not yet been applied to earlier millennia and might not be successful: it assumes that the cycle-length is one-ninth of a century.

A strong sunspot maximum tends to give rise to cycles in tree-rings, especi-ally a biennial cycle about 10 or 11 years afterwards generally, that is, at the subsequent sunspot maximum. The peak amplitude of triennial cycles is centred two years before the strong maximum, normally just after the preced-ing sunspot minimum.

Irregular solar oscillations. Moving 30-year means of auroral and sunspot numbers enable us to determine what we will call A- (or S-) max and A-min, conspicuous at the central dates given in Table I.

The 80-year solar cycle. The 80-year Gleissberg cycle is well-established, although its weakness, and the variability of its length, make it difficult to trace; since A.D. 200 the maxima have been near years of form:

$$78\, q + 48$$

where q is an integer. Meteorological or tree-ring effects through this whole period have not yet been investigated, but the cycle exists in the ice-core layers at Dye 3 in southern Greenland (cf. Hibler, in prep.) since c. A.D. 600.

175—180 years. A cycle of this length is found in the Greenland ice-cap since A.D. 1200 (Dansgaard et al., 1971, fig.11, p.47) but it may not occur in earlier centuries. The planetary cycle of this period is not reflected in any corresponding solar cycle.

The 200-year solar cycle. A cycle of about 200—205 years in the sunspot cycle since A.D. 300 (Schove, 1955; Zhukov and Muzalevskii, 1969) is consistent with a similar cycle through the same period in the radiocarbon error curve. The intervals between tree-ring dates of wiggles on this curve

TABLE I

Maxima (A-max) and minima (A-min) of auroral activity or sunspot numbers (S-max, S-min) as indicated by the central dates of 30-year moving means

A-max	A-min	A-max A.D.	A-min A.D.
c. 205 B.C.	?	1090—1126	1154
c. 170 B.C.	c. 150 B.C.	1200	1237
c. 100 B.C.	c. 75 B.C.	1263	1325—35
		1371	1418
c. A.D. 310	A.D. 335	1450	1480—85
365	415	1541—58	1600
440	470		
510	540	1675	1695
575	600	1725—30	1755—60
760	810	1775	1810 S-min
835—75	915	1849	1900 (c. 1888—'89, '89—'90, 1910)
930	950		
975	1050	1955 (?)	
		(cf. also S-max at c. 1859, 1882—1883)	

suggest that the cycle existed also before A.D. 300 and that its average length is 203 years.

Sometimes the 200-year cycle is suppressed, possibly when it coincides with a minimum of the 80-year cycle, so that then we have the 400-year cycle noted by Link (1968).

Tree-ring 'evidence', which I produced (1964) in support of meteorological effects of the 200-year cycle, does not reach a level of significance (as Professor Fritts first pointed out to me) but the sunspot-rich even centuries in the A.D. period have usually been warmer than the odd centuries.

260-year moisture cycle. A cycle in the texture of Danish raised bogs has been found by Aaby (1976), who notes increases of wetness at dates which, converted to tree-ring chronology, are: c. 2660 B.C., c. 2040 B.C., 1500 B.C., 600 B.C., A.D. 250, 500, 1025 and 1525 (with possible increases at 2370 B.C., 1050 B.C., 325 B.C., 50 B.C., and A.D. 1275).

Some support for this cycle is found in Fig.7, drawn before I heard of Aaby's work, and in lists, for the North European Plain as a whole, of centuries classified as follows (Schove, 1966/67, now converted to absolute dates):

Dry: 4760 B.C., 3730 B.C., 2670 B.C., 1710 B.C., 400 B.C.
    A.D. 185, 470 and 990
Wet: 1975± B.C., 880 B.C., 205 B.C.
    A.D. 60, 535, 815 and 1170

There is no associated temperature fluctuation.

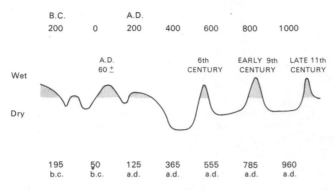

Fig. 7. Wetness of the North European Plain, 200 B.C.—A.D. 1100 reflected in low humification of peat. True dates (B.C./A.D.) above correspond to radiocarbon equivalents (b.c./a.d.) given below the curve.

A 350—400-year cycle. A persistent Holocene cycle of 350 years is noted in the Greenland ice-core (Dansgaard et al., 1971, Fig.5, p.47), but this period is in effect based on the radiocarbon time-scale and the additional 950 years tentatively added to the Holocene in our Fig.2 would increase the length of the cycle to 380 years. It seems probable that this is essentially the same as Link's 400-year cycle.

A 1500-year cycle. In an early version of the present Appendix (cf. Schove, 1967) I concluded that the only possibility of a significant cycle in the range 450—2000 years in the Holocene was one of c. 1265 radiocarbon years. This is perhaps equivalent to a cycle noted by Bray (1972) of 1325 radiocarbon years, but, if not accidental, either cycle must be nearer 1500 years on the absolute time-scale.

The 3100-year magnetic cycle. In the same article, however, Bray pointed out that a double cycle of 2700 radiocarbon years is more important, and cycles of 2775—2800 ^{14}C years have since (Creer et al., 1976) been found in palaeomagnetism, in declination in NW England and in inclination in the Lebanon and Black Sea. The true length of such a cycle would be near 3100 years.

The 2400-year cycle in [18]O used in one calibration of the Camp Century time-scale (Dansgaard et al., 1973, Fig.10) may prove to be this same cycle, as the conventional radiocarbon chronology for the Würm could be 30% too short (cf. Shackleton et al., 1977).

6000—7000 years. Dansgaard's cycles of 2000 and 3000 years (Dansgaard et al., 1971, Figs.4—8) can likewise be adjusted to different time-scales. In my own conversion of Lake Biwa chronology to radiocarbon equivalents (Schove, 1975b), the magnetic intensity cycles averaging about 5000 radiocarbon years would likewise correspond to a period of 6000—7000 years if a true cycle is involved. The Holocene peak of magnetic intensity is thought to be near the thermal maximum. This is probably accidental, but the errors in the radiocarbon age would be affected by any major cycles in terrestrial magnetism.

Firmly established evidence for the reality and precise period of any one of these longer cycles would be very helpful in the correct placement in time of both ice-cores and floating varve-chronologies.

CLIMATIC SUBSTADES

In the belief that the temperature changes from century to century were similar in sign in different parts of the globe, an attempt was made — as part of the Spectrum of Time Project — to collect and classify radiocarbon dates of palaeoclimatic significance. The results were weighted, partly as a result of discussions with the field workers providing the samples dated in *Radiocarbon*. The classifications included indications of (usually summer) warmth, cold, warming, cooling. Histograms were formed with the *radiocarbon century* as the time unit, and the radiocarbon centuries were abbreviated by deleting the last two figures of the mean radiocarbon date. The 'First Little Ice Age' after the thermal maximum, dated c. 4575 b.p., thus became substade 45. Comparison with pollen diagrams was often helpful when the radiocarbon date seemed off-centre. A selection of these substades is given in the Appendix.

The first long list was published in Schove (1966/67, pp.52—53 and errata p.234), but the evidence was not published at the time for lack of space. Much better evidence now confirms the ^{14}C chronology for certain regions (cf. Denton and Karlen, 1973), notably northern Sweden (cf. Karlen, 1976). Good evidence is now becoming available also for true centuries, notably of summer temperature in the American southwest (LaMarche and Stockton, 1974). Some quotations from the 1966/67 list are now qualified or supplemented from the sources cited. A detailed Late Glacial list (Schove, 1971a, table 5) and lists of summer subpluvials in the African Sahel (cf. Schove, 1975c) have since been published. 'Winter' subpluvials affecting December— April rainfall from SW Ecuador to the snow-sensitive glaciers of the Colorado Front Range are also being studied, and they seem to indicate lower pressure in the E and NE Pacific and in Canada in the periods c. 8400—7500 and c. 5000—3500 b.p.

The substades selected for inclusion here are those likely to help in the placement of floating varve and tree-ring chronologies, especially in the Middle Holocene. The tree-ring series are expected to indicate the duration of abnormal phases: some substades lasted much longer than a 'radiocarbon century': others probably lasted about fifty years only.

LATE HOLOCENE VARVES

There is no varve-series comparable or even cross-dated with the bristlecone pine chronology, and the still unpublished Swedish series is thought to contain both missing and double varves. There is, however, the pioneer Soviet study at Lake Saki or Sakskij on the western shores of the Crimea, a record of measurements ostensibly going back to 2394 B.C. Agreement with rainfall at nearby Sevastopol seemed significant back to 1842. The mineral portion of the silt layer depends essentially on precipitation but each year the action of micro-organisms affects the silt when precipitated and results in varves that are clear and readily counted. These claims, and the data published by Schostakowitsch (1936), urgently need verification by cross-dating with a new core, but so far even Soviet scientists have not taken the work seriously. Curves were given also for Lake Pert, on a plain in south Karelia, 50 km northwest of Petrozarodsk, and the varves are being studied by S. I. Kostin. President Dr. Dzotsenidze of Tiflis tells me that new varve-series are being made in the semi-arid regions of Soviet southwest Asia.

In the U.S.A. new methods are already proving successful on modern varves. The work of Soutar and Kipp (in WMO, 1975) proves that Californian submarine varves can be correlated with both rainfall and temperature records and Miss Kipp tells me that measurements are being made back through the historical period.

In the continental climate of the Interior the varves (as at the Lake of the Clouds) are often too thin for measurement or consistent counting, but the current work of Swain (1974) is of special interest, as correlations with pollens are being made for each decade; Swain is at present working on vegetation and climatic reconstructions through the past 2000 years. Moreover, Bryson tells me that cores from small American lakes are successfully preserved by being frozen (cf. Saarnisto et al., 1977) before measuring. Glacial varves, being found today, in many mountain areas of continental interiors would certainly repay study. Mayr has reported (pers. comm.) an Alpine site in Europe with varves back to A.D. 1300.

Varve-series usually need to be cross-dated locally so that regional standardized series can be developed before they can be convincingly cross-dated with tree-rings, and, to facilitate this, early exchanges of print-outs and publication of data would be helpful.

LATE HOLOCENE TREE-RINGS

Tree-ring series for several millennia are now becoming available not only for North America but also for several parts of Europe. In England the oak-chronology series goes back only to A.D. 491 (Schove, 1974, 1978) but a floating chronology for drought-sensitive oak from Roman Britain has now been fixed by comparison with German series (Schove, 1977b). The phase of the biennial cycles may prove useful in checking that a single ring is not omitted. However, the phase, as distinct from the amplitude, of the biennial cycle is no help in transatlantic cross-dating of Late Glacial varves (cf. Fig.6).

The biennial cycle in the summer-sensitive bristlecone pines from the upper timberline of North America has been discussed above and alpine studies (cf. Fritts, 1977), e.g. of the stone pine, promise to provide a summer temperature index back to A.D. 500.

Winter-sensitive series for the Mediterranean are not yet available in standardized form, but they too (Waisel, etc.) could soon be coordinated into the general scheme. A series for Turkey going back to before 1500 B.C. (Kuniholm, 1977) could no doubt be teleconnected to the bristlecone pine chronology and to the long series that are being developed in Ireland (Ballie, Hillam and others), Germany (Hollstein, Eckstein, Becker, etc.) and the Danube Lands. In these three regions tree-ring series are expected shortly to become continuous over the last 6000 years, and possibly to 8000 years in the Danube area.

Since the above was written the writer has successfully teleconnected the Saki varve-curves from the north side of the Black Sea with the tree-ring curves from the south side. The varves are usually too old by 1 or 1½ centuries, the error in the period 1600—1050 B.C. being 146 years, as the tree-ring curves have been teleconnected with both the Irish oak and the bristlecone series for this period (1000 on Kuniholm's Asia Minor scale = 6470 or 1531 B.C. on the American scale and 6445 ± 20 on the Irish scale of Pilcher et al., 1977).

CONCLUSIONS

Floating chronologies of both varves and tree-rings in the Holocene are needed to complete the gaps between Late Glacial varves and mid-Holocene tree-rings. Approximate dates can be obtained through palaeomagnetism, tephrochronology, climate substages and oscillations of certain other derived parameters which can then be used for cross-dating series in different continents. An absolute chronology of the Late Glacial and Holocene can thereby be obtained, and the controversial estimate of 950 years radiocarbon error at the boundary can be replaced by a precise value. The solution would be a valuable contribution to the Spectrum of Time and the writer would be pleased to help in the cross-dating of long series of replicated varves or tree-rings.

ACKNOWLEDGEMENTS

Thanks are due to the authors cited for the reprints and other information and to many others who have sent material for the so-called Spectrum of Time project. The figures were kindly drawn by the Cartography section of the Soil Survey of England and Wales.

REFERENCES

Aaby, B., 1976. Cyclic palaeoclimatic variations and the future. Newsl. Stratigr., 5(1): 66—69.

Bray, J. R., 1972. Cyclic temperature oscillation from 0—20,300 yrs b.p. Nature, 237: 277—279.

Clark, R. M., 1975. A calibration curve for radiocarbon dates. Antiquity, 49: 251—266.

Cline, R. M. and Hays, J. D. (Editors), 1976. Investigation of Late Quaternary palaeo-oceanography and palaeoclimatology. Geol. Soc. Am. Mem., 145.

Coope, G. R. and Pennington, W., 1977. The Windermere Interstadial of the Late Devensian. Philos. Trans. R. Soc., Lond., B 280: 208.

Creer, K. M., Gross, D. and Lineback, J., 1976. Origin of regional geomagnetic variations recorded by Wisconsian and Holocene sediments from Lake Michigan, U.S.A. and Lake Windermere, England. Geol. Soc. Am. Bull., 87: 531—540 (cf. also Quaternary Res., 7(1977): 411).

Dansgaard, W., Johnson, S. J., Clausen, H. and Langway, C., 1971. Climatic record revealed by the Camp 'Century' ice core. In: K. T. Turekian (Editor), The Late Cenozoic Glacial Ages. Yale University Press, New Haven, Conn., pp. 37—56.

Denton, G. H. and Karlen, W., 1973. Holocene climatic variations. J. Quaternary Res., 3(2): 155—205.

Fairbridge, R. W., 1977. Global climatic change during the 13,500 yr b.p. Gothenburg geomagnetic excursion. Nature, 265(5593): 430—431.

Fritts, H. C., 1977. Tree-rings and Climate. Academic Press, London, 567 pp.

Geyh, M. A., Merkt, J. and Muller, H., 1971. Sedimentological pollen-analytical and isotopic studies of annually laminated sediments in the central part of the Schleinsee, Germany. Arch. Hydrobiol., 69(3): 366—399.

Griffiths, D. H., 1955. The remanent magnetism of varved clays from Sweden. Mon. Notes R. Astron. Soc. Geophys. Suppl., 7: 103—114.

Hillaire-Marcel, C. and Occhietti, S., 1977. Fréquence des datations au ^{14}C de faunes marines post-glaciaires de l'Est du Canada et variations climatiques. Palaeogeogr., Palaeoclimatol., Palaeoecol., 21(1): 17—54.

Johnson, E. A., Murphy, T. and Torreson, O. W., 1948. Pre-history of the earth's magnetic field. Terr. Magn., 53: 349—372.

Karlen, W., 1976. Lacustrine sediments and tree-limit variations as indicators of Holocene climatic fluctuations in Lappland: N Sweden. Geogr. Ann., 58A: 1—34.

Kuniholm, P.I., 1977. Dendrochronology at Gordion and on the Anatolian Plateau. Thesis, University of Pennsylvania, Philadelphia, Pa.

LaMarche, V. C., Jr., 1974. Palaeoclimatic inferences from long term tree-ring records. Science, 183 (March): figs. 4—6.

LaMarche, V. C. and Stockton, C. U., 1974. Chronologies from temperature sensitive bristlecone pines. Tree Ring Bull., 34: 21—45.

Lamb, H. H., 1970. Volcanic dust in the atmosphere. Philos. Trans. R. Soc., 266B: 425—539.

Lamb, H. H., 1977. Climate: Past, Present and Future, Vol. II. Methuen, London, 835 pp.

Link, F., 1968. The 400-year cycle. J. Br. Astron. Assoc., 7B: 195—205.

Mangerüd, J. et al., 1974. Quaternary stratigraphy of Norden. Boreas, 3: 109—126.

McNish, A. G. and Johnson, E. A., 1939. Secular variations in declination in New England. Glacial varves. Int. Assoc. Terr. Magn. Elect., Washington Assembly, pp. 1—7.

Mook, W. G., Munaut, A. V. and Waterbolk, H. T., 1972. Determination of age and duration of stratified prehistoric bog settlements. Proc. 8th Int. Conf. on Radiocarbon Dating, Wellington, p.1.

Mörner, N.-A., 1977a. A 10,000-year temperature record from Gotland, Sweden. Palaeogeogr., Palaeoclimatol., Palaeoecol., 21: 113—138 (see p.133).

Mörner, N.-A., 1977b. The Gothenburg Magnetic Excursion. Quaternary Res., 7: 413—427.

Mörner, N.-A., 1977c. Varve chronology on Södertörn. Geol. För. Stockholm Förh., 99: 423—425.

Niemalä, J., 1971. Quaternary Stratigraphy of the clay-layers between Helsinki and Hameenlinna in south Finland. Geol. Surv. Finland., Bull., 259 (in German).

Noel, M. and Tarling, D. H., 1975. The Laschamp geomagnetic event. Nature, 253: 705—707.

Olsson, I. (Editor), 1970. Radiocarbon Variations and Absolute Chronology. Proc. Nobel Symp., 12., Uppsala. Wiley, New York, N.Y., 652 pp.

Persson, C., 1971. Tephrochronology. Sver. Geol. Undersök., C, 65(2): 3—34.

Pilcher, J. R., Hillam, J., Baillie, M. G. L. and Pearson, C. W., 1977. A long sub-fossil oak tree-ring chronology from the North of Ireland. New Phytol., 79: 713—729.

Pisias, N. C., Dauphin, J. P. and Sancetta, C., 1973. Spectral analysis of Late Pleistocene—Holocene sediments. Quaternary Res., 3(1): 3—9.

Ringberg, B., 1971. Glacial geology and the deglaciation of E. Blekinge, S.E. Sweden. Sver. Geol. Undersök., 65(7).

Rudmark, L., 1975. The deglaciation at Kalmarsund, southeastern Sweden. Sver. Geol. Undersök., C, 69(5).

Saarnisto, M., Hottunen, P. and Tolonen, K., 1977. Annual lamination of sediments in Lake Lovojarvi, southern Finland, during the past 600 years. Ann. Bot. Fennici, 14: 35—45.

Schostakowitsch, W. C., 1936. Geschichtete Bodenablagerungen der Seen als Klima-Annalen (Rainfall of Europe since 2880 B.C.). Meteorol. Zeit., 53: 176—182.

Schove, D. J., 1954. Summer temperature and tree-rings in N. Scandinavia, A.D. 1461—1950. Geogr. Ann., 36: 49—80.

Schove, D. J., 1955. The sunspot cycle, 649 B.C. to A.D. 2000. J. Geophys. Res., 60: 127—146.

Schove, D. J., 1964. Solar cycles and equatorial climates. Geol. Rundschau, 54(1): 448—477.

Schove, D. J., 1966/7. World climate 8000—0. B.C. R. Meteorol. Soc. Proc. Int. Symp. World Climate, pp. 52—53 and errata, p. 234.

Schove, D. J., 1967. Sunspot cycles. In: R. W. Fairbridge (Editor), Encyclopaedia of Atmospheric Sciences and Astrogeology. Reinhold, New York, N.Y., pp. 963—968.

Schove, D. J., 1969. The biennial oscillation, tree-rings and sunspots. Weather, 24: 390—397.

Schove, D. J., 1971a. Varve teleconnection across the Baltic. Geogr. Ann., 53A: 214—234.

Schove, D. J., 1971b. A varve teleconnection project. In: M. Ters (Editor), Etudes sur le Quaternaire dans le Monde. Paris, pp. 927—935.

Schove, D. J., 1971c. Biennial oscillations and solar cycles. Weather, 26: 201—209.

Schove, D. J., 1974. Dendrochronological dating of oak, A.D. 650—906. Med. Arch., 18: 165—172. (The Old Windsor dates should be reduced by 160 years.)

Schove, D. J., 1975a. Tree-ring teleconnection in Europe and transatlantic varve teleconnections. Masca Newsl., Philadelphia, Pa., 11: 9—10.

Schove, D. J., 1975b. World climatic chronology and Lake Biwa. In: S. Horie (Editor), Palaeolimnology of Lake Biwa and the Japanese Pleistocene. 3, pp. 429—437.

Schove, D. J., 1976. Maya chronology and the Spectrum of Time. Nature, 261: 471—473.

Schove, D. J., 1975. Note. Geogr. J., 141(2): 202.

Schove, D. J., 1977a. Maya dates A.D. 352—1296. Nature, 286: 670.

Schove, D. J., 1977b. Summer weather in S. E. England 54 B.C. Weather, 32(8): 34.

Schove, D. J., 1977c. Note. Philos. Trans. R. Soc., B280: 120—181.

Schove, D. J., 1978. Anglo-Saxon tree-rings. Tree-Ring Newsl., July.

Seibold, E., 1958. Jahreslagen in Sedimenten der mittleren Adria. Geol. Rundschau, 47(1): 110—117 (see Fig. 2).

Shackleton, N. J., 1977. The Oxygen Isotope record of the Late Pleistocene. Philos. Trans. R. Soc., 180—181.

Soutar, A. and Crill, P. A., 1977. Sedimentation and climatic patterns in the Santa Barbara Basin during the 19th and 20th centuries. Geol. Soc. Am. Bull., 98: 1161—1172.

Strömberg, B., 1977. Rückzug des Inlandeises am Billingen. Z. Geomorphol., N.F., Suppl., 27: 89—111.

Swain, A. M., 1974. A history of fire and vegetation in N. E. Minnesota as recorded in Lake Sadimento. Quaternary Res., 3(3): 383—396.

Thompson, R. and Berglund, B., 1976. Late Weichselian geomagnetic 'reversal' as a possible example of the reinforcement syndrome. Nature, 263 (No. 5577): 490—491.

Tolonen, K., Siiriäinen, A. and Thompson, R., 1975. Prehistoric field erosion sediment in Lake Lojarri, S. Finland and its palaeomagnetic dating. Ann. Bot. Fennici, 12: 161—164 (see Fig.3).

WMO, 1975. Proceedings of the WMO/IMAP Symposium on Long-Term Climatic Fluctuations. Norwich, August, 1978. Geneva, No. 421.

Zhukov, L. V. and Muzalevskii, Y. S., 1969. A correlation spectral analysis of the periodicities in solar activity. Sov. Astron., 13(3): 473—479.

APPENDIX

Radiocarbon centuries or substades (The probable error is usually about fifty years; Quotations refer to the 1966/7 list)

Sub-stade	^{14}C date b.p.	True date B.C.	
87	8750	8th millennium B.C.	temperature above modern normals 'and continued above that level until c. 4750 b.p.'; in N Sweden, however, stade 73 may have had cooler summers
c. 78±	7850±	8th/7th	winter subpluvial: stades 84/70 in NW of S America
73	7350±	7th	cold; very dry summers in both mid-U.S.A. and NW Europe
68/64	6850—6450	6th, up to 5350 B.C.	hot summers: relatively moist in U.S. Plains; dry in African Sahel
63/62	6315±	5215±	relatively cooler; dry again in mid-U.S.A.
60	6050±	4975±	hot summers
59/58	5900	4750	very dry in Europe; now moist in Sahel as if high pressure belts further north again
55	5550	4450	hot summers again; Neolithic subpluvial in Sahara
53	5325	4225	'thermal maximum (last and greatest of several)'; in N Sweden but not in Canada the tree-line had been higher in the earlier hot centuries long dry phase sets in NW Europe soon afterwards
49	4925	3700	dry century in several parts of the world
46	4625	3450	dry to moist (NW Europe)
c. 46/45 and c. 42	4600—4250	3430—2970	winter subpluvials (W Americas) with two maxima at stades indicated
45	4575	3375	'First Little Ice Age; short but severe'
41/40	4100	2750	December—March rains often high in subtropics, etc.

APPENDIX (*cont.*)

Sub-stade	¹⁴C date b.p.	True date B.C.	
40	4050	2675	dry again in NW Europe (mid-sub-boreal) and many other regions
39	3930	2500	cold summers (White Mountains, SW of U.S.A.)
36	3650	2030	last very warm centuries in most parts of the globe
35	3575	c. 1975	very moist in North European Plain
35/34	3500	c. 1900	rapid fall of summer temperature; mild winters in NW Europe
33	3375	1750—1700	dry North European Plain; mild winters continue
32	3225	1600—1550	'Second Little Ice Age: longer but less severe'
			still dry in NW Europe
c. 30	c. 3100—2900	1500—1200	warmer interval
28/27	2800	c. 1030	'Third Little Ice Age: severe'; the 11th century B.C. was indeed very cold; in N Europe it was also dry
26	2650	c. 880	moist in North European Plain
24/22	2300±	6th and especially 5th century	'Fourth Little Ice Age: less severe than 11th century B.C. but more prolonged'
22	2250	400 B.C.	drier North European Plain
21	2150	c. 205± B.C.	moist North European Plain
21/16	2150—1650	200 B.C.—A.D. 400	warmer (Roman period)
19	1960	A.D. 50±	moist North European Plain (but short dry spell A.D. 60—90)
18	1845	A.D. 190±	dry North European Plain
18/17	1780	200	change to moist in Denmark
15/14	1550	c. 430±	dry again but in Denmark change to wet by A.D. 500
14	1425	565	moist in NW European Plain
13	1350	7th century	cold century? (rise of Islam)
12	1160	c. 805±	moist NW European Plain
10	1050	10th century	dry NW European Plain; cold then warm
10/9	1000	990±	dry NW European Plain
9	950	1030	moist NW European Plain
8	850	1170±	warm moist NW European Plain

ERRATUM

Page 224, line 12, ". . . German tree-rings . . ." should be ". . . American tree-rings. . . ." Page 228, the English tree-ring dates before A.D. 800 are incorrect. I relied too much on skeleton plots and archaeological and radiocarbon evidence, and rejected the obvious computer result that the "Old Windsor" series begins in A.D. 419.

40

Varve-chronologies and their teleconnections, 14000-750 B.C.

D. J. SCHOVE
St David's College, Beckenham, Kent, England

1 PREPARATION OF DATA

Varve-chronologies can be accurately dated and it will be possible to bridge the gap between Holocene varves and tree-rings c 6000 B.C. and glacial varves c 7500 B.C. The recommended procedure forms the topic of this paper.

Charts of varves or tree-rings cannot be matched by skeleton plots or visual comparison except in particular localities, although this method is useful for glacial varves in the west-east direction parallel to the ice-front. For long-distance comparisons we must search for and isolate those non-random elements found in weather to-day. Cross-correlation with other varves or tree-rings can then be applied.

Late Glacial varves in North America and Europe through a 3900-year period have thus (Schove 1969,1971,1978) been fitted into a coherent transatlantic chronology, but this is still a floating one fixed only in Radiocarbon or b.p. time (c 12450/8250 b.p.).

Holocene varves can be counted back from the present-day, but the series available at present are not accurately counted or dated; the Lake Saki varves (USSR) have been redated here and placed in the absolute B.C. time-scale provided by the bristlecone pines (USA).

Collaboration is needed so that long series can be constructed, replicated and cross-dated with tree-rings especially in the period of the missing millenium, 6300-8250 tree-ring years B.C.

1.1 Primary Series

Time series of measurements should, wherever possible, be made of the thickness not only of each complete varve but also of each part of the couplet (e.g. winter clay and summer silt) separately.

Peculiar colours or structural features of unusual varves should be noted. X-ray densitometer plots may prove essential for correct numbering.

Replicated series should then be cross-dated against one another by using skeleton plots or the standard tree-ring programs as developed in Belfast and Hamburg. The measurements can be normalized so that the standard deviation is $\pm$ 1 and the (running) mean is zero. However, proximal (summer) and distal (mainly winter) varves are best made into separate series, and for some parameters (e.g. turning points of thirty-year running means) the original data may be preferable.

The standardized series are amalgamated into regional means and plotted on the usual scales of 2 years per cm and 10 years per inch. The author would be grateful for any charts of long series in these forms.

1.2 Secondary Series

From a primary series a, b, c ...... secondary series can be obtained to highlight non-random elements. An 'Oddness Index' useful for regional teleconnections is derived from first differences of ring widths with alternate signs reversed (e.g. b − a, b − c, d − c, d − e) and then using running means in groups of two (Schove 1974: 167). This series is A, B, C ...... where

$$B = \frac{2c - b - d}{2}$$

A Biennial Index for inter-continental teleconnections is obtained by using three-year moving means of the squares of this index:

$$\beta = \frac{A^2 + B^2 + C^2}{3}$$

This Biennial Series β, γ, σ reflects the amplitude (linked with sunspot activity) but not the phase of the Biennial Oscillation.

Alternatively, it may suffice to note the central dates of biennial cycles which are characteristic of specific decades (e.g. the 1530s A.D.) or centuries. Triennial cycles should also be noted, as these are often simultaneous in different continents, as in the period A.D. 1880/1910 when sunspot cycles were weak.

1.3 Other Non-random features

Other useful parameters (Shove, 1971a) include the turning-point years of ten- and five-year means. The moving five-year means reflect what I term the pressure parameter effect (Shove 1976); this is the tendency for air to flow from regions where (in the season in question) the atmosphere bulges (e.g. the Upper High over the Indian Ocean) to regions of the Upper Troughs (e.g. the Davis Strait). A clue to the phase of the eleven-year sunspot cycles may sometimes be derived indirectly in this way, but statisticians have shown that running means give rise to spurious cycles and five-year running means yield pseudo-cycles of 10-15 years. Years of very wide and very narrow rings are noted in the 'Decile' test. Years of sudden changes up or down with phases of abnormal variability or complacancy are useful especially in lower latitudes.

The central years of the abnormal features should be noted and each abnormality weighted on a scale − 4 to + 4, ± 1 being regarded as loosely corresponding to the standard deviation. They are conveniently plotted on scales of 10 years and of 100 years to 1 cm as skeleton plots. Cross-correlation with all possible lags usually yields the correct position of any extensive overlap between two series.

Curves can be drawn freehand, and in particular curves for the biennial versus triennial tendencies may make it possible to obtain approximate synchronization visually without further arithmetic. The Swedish and North American glacial varves were cross-dated first in this way. Some abnormalities are nevertheless time-transgressive: two-year cycles thus occur (in

the period A.D. 800 − 1978) about 11 years earlier in German summer-sensitive rings and in the Nile Flood levels they occur several years earlier still.

2 LATE GLACIAL

Code-letters used for the algebra of well-known glacial varves are as follows:
D   N. Sweden (Döviken zero)
V   Central Sweden (Swedish dates f.Kr)
N   South Sweden (Nilsson's f.Kr. dates)
F   Finland (Relative to Sauramo's zero)
C   Canada (Timiskaming)
Q   Quebec
Br Bracebridge-Huntsville
A   S. New England
B   N. New England

2.1 Trans-Baltic Solutions

Swedish varves have usually been assigned a year 'before Christ' or 'f.Kr.' on a Swedish time-scale (N or V). Finnish dates (F) differ from such negative years by a number of years which varies from 8226 in the Late Glacial (F − N) to 8140 in the Early Holocene (F − V). The errors appear to lie in the Swedish rather than the Finnish counts, and no corrections to Sauramo's numeration have yet been reported by Niemala.

2.2 Trans-Atlantic Solutions

Wide and narrow proximal varves often correspond to warm and cold summers, but warm seasons in eastern Canada more often correspond to cold seasons in north-west Europe so that the year-to-year tendency (especially in distal varves where the winter portion is more important) may be in opposition on the two sides of the Atlantic.

The Finnish and Canadian series in the Early Holocene (Pollen-zones IV and V) were thus found to be related by the formula (varying according to century):
F − C = 52 to 63
The writer's main Canadian series is built up especially from the Timiskaming curves of Antevs (1925). In relation to the zero of the Timiskaming series, Antevs' Quebec series (from Maniwaki, north of Ottawa 47°N 76°W) was found to be 83 years earlier and the Espanola series (from north-east of Lake Huron 47°N 82°W) was found to be first 78 (for Espanola 3/4) and later (for Espanola 5, separated by an unconformity) 59 years earlier.

The main Canadian series thus extends from − 83 to about + 2000 and was shown to correspond with the period c 8175 to c 6063 'B.C.' (f.Kr.) in the N. Swedish varve-chronology (V). This is known to correspond to radiocarbon dates c 10300/8250 b.p. or c 8350/6300 b.c. It is expected to correspond with tree-ring dates $9350/7250 \pm 200$ B.C. on the extrapolated bristlecone scale.

Current dates in N. Sweden are linked to the Döviken zero (6923 f.Kr. on the V-scale or about 9050 b.p. on the C14 scale) and this year seems to correspond to year 1166 ± 1 in the Canadian series of Antevs (1925) and Hughes (1965).

No satisfactory series of varves for the cold period of European pollen-zone III could be identified for north America, but the zero of Antevs' 'Burke Falls' series ($45\frac{1}{2}$ N $79\frac{2}{3}$ W) is about 251 years older than the first set of the Espanola series ($46°$N $82°$W) which together constitute a chronology that is not yet linked with Europe. An approximate C14 date is needed.

The Huntsville and Bracebridge series from east of Lake Huron ($45°$N $79°$W) were on the other hand teleconnected first with one another (the Huntsville zero is 187/186 years later than that of Bracebridge), and the combined series was cross-dated with the U.S.A. (B) series for northern New England, the postulated zero of this is c 6939 ± 5 years earlier than that of Bracebridge. Cross-dating with the Baltic was made for each of these series, which were shown to commence in the Boelling period and to continue in the Alleroed, thus corresponding to the Two Creeks period in the wide sense. These years were connected with the conventional Swedish varve-years (N in pollen-zones Ib and II) by the formula:

$$Br - N = 10,030 \pm 4$$

The United States varves were discussed separately (Schove 1969) but no link with Europe has yet been found for the A-series. New series from Poland, The USSR and Esthonia are needed for this.

3. HOLOCENE

Glacial varve series are now being counted backwards from the present day, notably at Wisconsin and in Finland, using new techniques of obtaining cores from modern lakes which freeze every winter. Varve-structure is preserved by refreezing after coring.

Other types of varves are being studied and are found even at the sea bottom off the California coast and in the Red Sea. The ice-layers found in Greenland and Antartica have annual layers which are being dated by tephrochronology in Denmark back at least to A.D. 1783.

The well-known varves of Faulensee, Switzerland and of Yugoslavia need to be re-investigated as the dates originally assigned to them were incorrect.

Some varves have been counted but not yet measured. The Lake of the Clouds varves are thus too thin for measurements and in any case are not weather-sensitive. German varves studied at Schleinsee, S.W. Germany by Geyh and his collaborators (1971) will be of special interest as they cover the period 3250/ 7200 B.C. and almost span the missing millenium. Measurements would be invaluable.

The chronology of Holocene varves has nowhere been firmly established. This applies to the best-known series from Lake Saki in the Crimea, USSR; the methods adopted for glacial teleconnections were used to find the correction necessary at c 1500/1000 B.C.

3.1 Lake Saki varves since 2000 B.C. (d)

The silt deposits of the western shore of Lake Saki, Crimea, USSR were measured and counted by Perfilev in 1929. The mineral portion depended especially on precipitation and the micro-organisms in the hot season affected the silt deposited so that one varve could be distinguished from its neighbours. Annual measurements were published by Schostakovitch (1934).

One possible dating check is the sawtooth or zigzag signature found in the 1530s in tree-rings in Germany and the USSR as far east as Siberia (and also in vintages in France). Certainly it occurs in the Saki varves at the right decade and Dr S. Eddy tells me that the counting may have been correct back to Roman times. In our algebra this series is termed d.

3.2 Tree-ring series c 2000 − 750 B.C.

Four tree-ring series are available for comparison through the first third of the Saki period:
a. USA: Bristlecone Pine 5142 B.C.+..
b. Ireland: Oak c 4000/1000 B.C.
c. Asia Minor: Juniper c 1600/750 B.C.
e. Central Europe: Oak c 2000/1000 B.C.
The first of these series was dated in absolute chronology (Ferguson 1969) and the last series (Becker) is not yet published.

The Irish and Asia Minor series have mean-while been cross-dated with the Bristlecone series using the author's methods described above.

3.3 Asia Minor Tree-Rings 1562 - 767 B.C.(c)

Archaeological timber from the Hittite and Archaic period at Gordion, Turkey, has been used (Kuniholm 1977) for a coherent float-ing chronology by Kuniholm. These ring-widths reflect especially winter and spring rainfall, and were therefore expected to cross-date the Lake Saki series from the opposite side of the Black Sea. The cor-rect dates were found by tests which in-cluded direct comparison with the Bris-tlecone (α) Series, as the curve of bien-nial and triennial cycles showed good agreement.

In the Decile Test a lag of 5470 gave 51 points with the same signs in extreme years; a lag of 5469 also gave a non-random result with 19 similar and 51 oppo-site signs. This is because the preceding year's weather also affects the Asia Minor trees. The year numbered as 1000 (estima-ted archaeologically by Kuniholm as c 1540 B.C.) was therefore dated (Schove in progress) 1531 B.C. and the series was dated 1562 - 767 B.C.

3.4 Irish Tree-Rings c 3980 - 1070 B.C.(b)

This important series has been published (Pilcher et al. 1978) as a floating chro-nology estimated as c 3998 - 1088 $\pm$ 10 B.C. (dated by comparison with Bristlecone C14 dates). This series had a higher serial correlation than the bristlecone and did not readily yield a more precise direct cross-dating. Cross-dating was neverthe-less eventually possible with the Saki varves (d) using e.g. dates of maxima and minima of five-year means.

3.5 Central Europe

Professor Becker's tree-ring series are not yet published but are mentioned be-cause of their potential value for dating European varves at Faulensee, Lake Saki and elsewhere.

3.6 Interrelationships

Lake Saki curves (d) were at first assumed to be within a century of the correct dates but this failed to lead to any cross-dating. Wide-aperture tests (Biennial, Triennial etc.) with the other series on the assumption that the error was less than 700 years led to the conclusion that the error lay between 100 and 200 years and narrow-aperture tests eventually led to solutions 145/153 years in the period c 1600/1000 B.C. Through most of this pe-riod the assumption of a 145-year error made sense of visual matching with the 30-year and 5-year running means of Asia Minor curves. If the Saki 'B.C.' date is regarded as 'minus d' the following for-mulae applied:

$$b - d = 2676/2684$$
$$a - d = 8149 \pm 5$$
$$c - d = 8130 \pm 5$$

The tree-ring series were related as follows:

$$a - b = 5470$$
$$c - b = c5448 \pm 2$$
$$a - c = \quad 22 \pm 2 \text{ (Partly inferred)}$$

3.7 Tephrochronological prospects

The Lake Saki varves would repay further investigation in this period, and detection of the year with Santorini dust would yield the date of the Thera eruption in the Aegean sea. Babylonian observations of Venus were missing in a year regarded as 1635 B.C. and the hypothesis of Weir that there was an eruption in that year would be tested.

3.8 Climatic implications c 1500/1000 B.C.

The several series examined above suggest:

1. There was a warm moist period c 1550-1300 B.C. but occasionally prolonged droughts as in 1429/24, 1416/11 and c 1352 B.C. affected both sides of the Black Sea.

2. Dryness prevailed in the Black Sea area as the global temperature fell in the 13th century B.C., and severe droughts occasionally affected both the Saki varves and the Asia Minor tree-rings (e.g. c 1295, c 1265).

3. Moist (but globally colder) conditions returned in the period c 1220/1140 B.C., notably in the period c 1200/1190 conven-tionally associated with the end of the Bronze Age and the Trojan Wars.

4. Further periods of drought in N.W. Asia Minor occurred c 1150, c 1130, 960s, 924/1, 888, 845/839 B.C. The climate was globally cold to c 950 B.C. and then be-came warmer, and the last of these droughts synchronized with the period of Ahab in Palestine.

3.9 Future work

Collaboration in the collection, replication and dating of long series of varves and tree-rings, especially through the missing millenium c 8250/9200 B.P. or c 6300/7250 B.C. is requested. The approximate substade (Schove 1978a) or radiocarbon century and palaeomagnetism of any floating chronology provides the first clues to the dating and Oddness and Biennial indices will normally lead to an approximate dating. For precise dates it is helpful to know the meteorological factors and seasons involved and if modern varves are available for comparison 'Response Functions' as used in tree-ring analysis should be calculated.

REFERENCES

Antevs, E. 1922, The recession of the Last Ice Sheet in New England, Amer. Geogr. Research Series No. 11 (and other papers).

Ferguson, C.W. 1969, A 7104-year Annual Tree-Ring Chronology, Tree-ring Bulletin 29 (3-4): 3 - 29.

Geyh, M.A., J. Merkt & H. Muller 1971, Sedimentological pollen-analytical and isotopic studies of annually laminated sediments in the central part of the Schleinsee, Germany, Arch. Hydrobiol. 69: 355-399.

Hughes, O.L. 1965, Surficial geology of part of the Cochrane District, Ontario, Canadian Geol. Soc. of Amer., Special Paper 84: 535-565.

Kuniholm, P.I. 1977, Dendrochronology at Gordion on the Anatolian Plateau, Pennsylvania Univ. Doctoral Dissertation.

Pilcher, J.R., J. Hillam, M.G.L. Baillie & G.W. Pearson 1977, A long sub-fossil oak tree-ring chronology from the north of Ireland, New Phytologist, 79: 713-729.

Schostakowitsch, W.B., Bodenablagerungen der Seen und periodische Schwankungen der Naturerscheinungen, Memoires de l'Institut Hydrologique, Leningrad, 13: 95-140.

Schove, D.J. 1969, A varve teleconnection project. In M. Ters (ed.), Etudes sur le Quaternaire dans le Monde, pp. 927-935, Paris.

Schove, D.J. 1971, Varve teleconnection across the Baltic, Geogr. Annaler, 53A: 214-234.

Schove, D.J. 1974, Dendrochronological dating of oak ... A.D. 650-906 (recte A.D. 490-746), Medieval Archaeol. 18: 165-172.

Schove, D.J. 1976, In S. Bezzaz, D. Dalby (eds.) Drought in Africa, 2. London, International African Institute.

Schove, D.J. 1978a, Tree-ring and Varve Scales combined, Palaeogeography, Pal. Pal., 25 (1978): 209 - 233.

Schove, D.J. 1978b, Tree-ring and Varve Teleconnections, c 4000 - 750 B.C. (Submitted to Nature, London).

APPENDIX

The figures below have been kindly drawn for me by H. Bolliger (EAWAG-ETH, Switzerland) and help to illustrate this paper and that which has now appeared in Pal. Pal. Pal., 25 (1978): 209-233. Progress in measuring and dating (back to the Middle Ages) ice-varves in Greenland is meanwhile reflected in papers by C.U. Hammer (cf. J. Glaciol., 20, 1978, 3-26) and varves back to 8500 B.C. have been studied by Degens and Kempe (cf. this volume).

The period for which varve or tree-ring measurements are most needed is illustrated in Figure 2. The way in which 2- and 3-year cycles can be used for first approximate synchronization is illustrated in Figure 3. Decile Tests use weighted outstanding maxima and minima from each series (cf. Schove, 1971, p. 224 and 231-2) and help in exact synchronization. A sample of very wide tree-rings in Turkey, when 5570 years were added to the Asia Minor Scale (as in Table 1), was found to synchronize with 15 wide rings (over 125) but only 5 narrow (less than 75) in the U.S.A.

Thirty year tests show sometimes good agreement as shown in Figure 5 Annual curves occasionally do, as in Figure 6.

In general, no single test has high significance but once many independent tests lead to the same correlation the results can be trusted. The correlations shown in Figure 4 are approximations covering a wider time-range than those of Table 1.

[*Editor's Note:* The Decile Test results were more complex than stated in section 3.3 on page 322, the scores being as follows for 5470: Wide Gordion = 29 Wide USA but only 5 Narrow; Narrow Gordion = 22 Wide USA but only 4 Narrow. Preceding lag 5469 showed an opposite behavior.

Dr. Pilcher reports that the absolute dates of the Irish tree-rings are known back to 5000 B.C. and confirms the dating in this paper except for the obvious mistake of 3998/3980 B.C. on our p. 361 (personal communication of April 1983).]

Table 1. The four scales c 1500/1000 B.C. with equivalent dates on the Bristlecone scale + 65007000.

- 1500	1400	1300	1100	- 1000	
1501 B.C.				1001 B.C.	
+6500	U.S.A. Scale			+7000	Bristlecone
c6479	Irish Scale			c6979	add 21
+1030	Asia Minor Scale			+1530	add 5570
−1645	U.S.S.R. Scale			−1145	add 8145

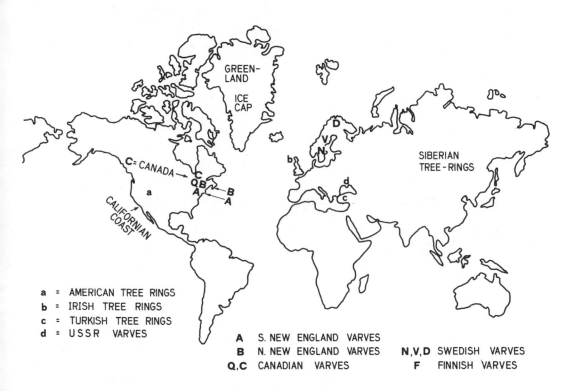

a = AMERICAN TREE RINGS
b = IRISH TREE RINGS
c = TURKISH TREE RINGS
d = USSR VARVES

A   S. NEW ENGLAND VARVES
B   N. NEW ENGLAND VARVES   N,V,D SWEDISH VARVES
Q,C CANADIAN VARVES   F   FINNISH VARVES

Figure 1. Map showing sites of the four scales (a to d) and of the principal Late Glacial varve series.

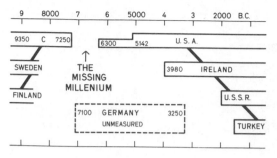

Figure 2. The gap c 7250/6300 B.C. and the
S.W. German varves that may provide an
overlap.

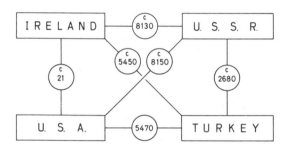

Figure 4. The approximate interrelations
of the four chronologies with the differences in years between each pair.

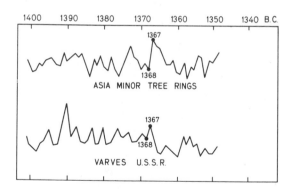

Figure 6. Tree-rings in Asia Minor and
Varves in the USSR, 1400/1340 b.C. It is
assumed that the error in Lake Saki varves
is 145 years (Note the thirty-year wet
period 1391/62 B.C. in the tree-rings, cf.
Fig. 4).

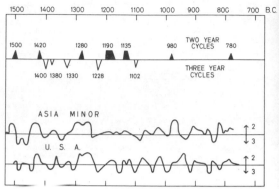

Figure 3. Two and three-year cycles common
to most tree-ring and varve series, with
curves for Asia Minor and the U.S.A.

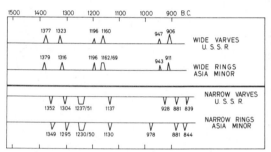

Figure 5. Central dates of thirty-year
WIDE (wet) and NARROW (dry) periods common
to Lake Saki varves and Asia Minor Tree-
Rings.

ERRATUM

Page 323, column 1, line 1 from the bottom, ". . .
A.D. 490–746 . ." should read ". . . A.D.
419–675. . . ." Page 323, column 2, line 14 from
the bottom, 5570 should be 5470. Table 1,
column 4, 5570 should be 5470. Page 325, Fig.
5, in "Narrow Varves USSR," 881 should be
865. The Saki error was then about 135. Fig. 6,
the lower curve is plotted one year off, but
the dates are correct.

CONCLUSION

In this book we have discussed sunspot cycles—short (11-year), medium (80-year), and long (200-year)—and we have demonstrated the irregularities of their periods and the unreliability of their effects. With an awareness of the complexities of these cycles we now need to plan a program that will explain the complexities by using the evidence from tree-rings and varves.

Solar excursions can be dated through most of the Holocene. In A.D. times the sunspot and auroral observations tell the same story as the radiocarbon changes. In B.C. times we have just the radiocarbon evidence by itself, but, following especially Eddy and Stuiver, we can derive the true dates of the strong and weak periods (see Table I in Eddy, 1977; Fig. 1 in Paper 36).

Table 1. Prolonged Solar Maxima and Minima

Solar Maxima		Solar Minima	
Modern Maximum	A.D. 1780	Maunder Minimum	A.D. 1646–1699
Late Medieval Maximum	A.D. 1360–1380	Spoerer Minimum	A.D. 1400–1510
Medieval Maximum	A.D. 1120–1210	Wolf Minimum	A.D. 1280–1350
Late Roman/Byzantine Maximum	A.D. 4th and 6th centuries	Norman Minimum	A.D. 1010–1090
		Dark Age Minimum	A.D. 660–740
Alexandria Maximum	350–200 B.C.	Greek Minimum	c. 400 B.C.
	c. 1800 B.C.		c. 800 B.C.
	23rd century B.C.		c. 1400 B.C.
	28th century B.C.		c. 2300 B.C.
	33rd century B.C.		c. 2900 B.C.
	36th century B.C.		3350–3250 B.C.
	3670–3650 B.C.		3500–3450 B.C.
	4250–3950 B.C.		3620–3600 B.C.
	49th and 47th centuries B.C.		45th–44th centuries B.C.
	54th century B.C.		53rd century B.C.

Conclusion

Sunspot cycles averaging 11.1 years have been dated through the last 2000 years and we have learned something of the way in which they and the several longer cycles affect our climate. We have found that within certain limits we can predict both sunspot activity and its climatological consequences.

The papers selected explain the how but not the why of sunspot cycles—their cause remains an enigma. Despite the hope of Wolf in 1852, no close parallels have yet been found between the stars and sunspot cycles (see Bray and Loughhead, 1964, p. 282ff; Godoli, 1976; Wilson et al., 1981). Certainly Babcock's emphasis on the significance of differential rotation (see White, 1977, p. 449) is generally accepted. Beyond that, however, we cannot at present proceed.

History, it has been said, is too important to be left to the historians. Solar physics, on the other hand, is too important to be left to the scientists. To predict sunspot cycles scientifically we must study them in *geology* and in *history,* for their past is the key to their future.

REFERENCES

Bray, R. J., and R. E. Loughhead, 1964, *Sunspots,* Chapman & Hall, London, 304p. (Textbook of Sunspot Physics.)

Eddy, J. A., 1977, Climate and the Changing Sun, *Climatic Change* **1:**173–190.

Godoli, G., 1976, Stellar Activity of the Solar Type, in *Basic Mechanism of Solar Activity,* V. Bumba and J. Kleczek, eds., Reidel, Dordrecht, Holland, pp. 421–446.

White, O. R., ed., 1977, *The Solar Output and Its Variation,* Colorado Associated University Press, Boulder, 586p.

Wilson, O. C., A. H. Vaughan, and D. Mihalas, 1981, The Activity Cycles of Stars, *Sci. American* **244:**82–91.

APPENDIXES

Appendix A. Sunspot minima 653 B.C. to A.D. 1501[a]
(Dates in parentheses are uncertain.)

Period	0	1	2	3	4	5	6	7	8	9	Mean Remainder for the Century
−700+					(48) 653 B.C.	(59) 642 B.C.	(70) 631 B.C.				4
−600+	(3)[b]										3
−500+	(3) 498	16 485	27 474	38 463	(49) 452	c				102	4
−400+	2 399	(13) 388	25 376	(36) 365	47 354	57 344 B.C.					3
−300+					59 242	59 242	71 230	82 219	90 211	101 200 B.C.	4
−200+	1 200	13 188	23 178	34 167	46 155	59 142	71 130	81 120	92 109 B.C.	103 98 B.C.	3
−100+	3 98	13 88	22 79	32 69	43 58	55 46	68 33	80 21	92 9 B.C.	104 4 A.D.	2
0+	A.D. 4	17	28	38	49	60	( )	( )	( )	( )	c. 5?
100+	( )	( )	22	(34)	46	57	68	(80)	(92)	(105?)	c. 1?

(continued)

Appendix A. (continued)

Period	0	1	2	3	4	5	6	7	8	9	Mean Remainder for the Century
200+	(5)	(15)	(26)	(38)	(51)	(63)	(74)	(87)	98	107	c. 6?
300+	7	16	26	37	48	57	67	79	91	103	c. 4
400+	3	15	26	37	46	57	70	82	95	107	c. 3
500+	11	21	32	45	56	65	78	(85)	(98)	110	10.6
600+	10	18	(32)	(44)	(54)	(64)	73	84	(97)	(107)	(9.0)
700+	(7)	(19)	(32)	42	(54)	(63)	77	87	96	106	(8.3)
800+	6	19	27	39	51	61	(72)	(86)	(96)	(104)	(6.8)
900+	(4)	(16)	25	37	(53)	(62)	(74)	(86)	(96)	103	(6.3)
1000+	3	15	(28)	41	(54)	67	77	87	97	111	8.1
1100+	11	19	28	37	49	61	72	84	93	102	6.4
1200+	2	(15)	(27)	39	(51)	61	75	86	(97)	108	6.1
1300+	8	(18)	(27)	(40)	52	61	71	82	(91)	(104)	(5.7)
1400+	(4)	(16)	(28)	(39)	(50)	(60)	(74)	(85)	(93)	106	(5.9)

[a]This table is revised slightly from Table 3 in Paper 20. For a continuation of this table see Table 1 in our Introduction.
[b]for example, −597 or 598 B.C.
[c]Livy's dates corrected by 3 years

Appendix B. Sunspot maxima c. 649 B.C. to A.D. 1506[a]
(After A.D. 300, parentheses indicate probable error greater than 1.)

Period	0	1	2	3	4	5	6	7	8	9	Mean Remainder for the Century
−700+					(52) 649	(63) 638	(74) 627			(107) 594	(7)
−600+	(7) 594									(107) 494	(7)
−500+	(7) 494	20 481	31 470	42 459	(53) 448	b					(8)
−400+	6 395	(17) 384	29 372	(40) 361	51 350	61 340					(7)
−300+						64 237	77 224	86 215	95 206	108 193	9
−200+	8 193	17 184	29 172	37 164	(51) 150	65 136	75 126	87 114	96 105	106 95	7.4
−100+	6 95	16 85	27 74	37 64	46 55	59 42	71 30	84 17	96 5 B.C.	109 A.D.9	5.0
0+	9	21	33	42	53	64	( )	( )	( )	103	10?
100+	3	15	27	(38)	(48)	60	(72)	86	96	( )	5?

(continued)

Appendix B. (continued)

Period	0	1	2	3	4	5	6	7	8	9	Mean Remainder for the Century
200+	( )	22	( )	(42)	55?	67	78	91	101	111	12?
300+	11	20	32	43	52	61	70	86	96	(109)	8.3
400+	(9)	(21)	(29)	(41)	51	(63)	(76)	(89)	99	111	(9.2)
500+	11	21	32	45	56	65	78	(85)	(98)	110	10.6
600+	10	18	(32)	(44)	(54)	(64)	73	84	(97)	(107)	(9.0)
700+	(7)	(19)	(32)	42	(54)	(63)	77	87	96	106	(8.3)
800+	6	19	27	39	51	61	(72)	(86)	(96)	(104)	6.8
	(4)	(16)	25	37	(53)	(62)	(74)	(86)	(96)	103	(6.3)
1000+	3	15	(28)	41	(54)	67	77	87	97	111	8.1
1100+	11	19	28	37	49	61	72	84	93	102	6.4
1200+	2	(15)	(27)	39	51	61	75	86	(97)	108	6.1
1300+	8	(18)	(27)	(40)	52	61	71	82	(91)	(104)	(5.7)
1400+	(4)	(16)	(28)	(39)	(50)	(60)	(74)	(85)	(93)	106	(5.9)

[a]This table is revised slightly from Table 3 in Paper 20. For a continuation of the table see Table 2 in our Introduction.
[b]Livy's dates corrected as in Appendix A.

APPENDIX C: SUNSPOTS AND RADIOCARBON SINCE A.D. 5

Solar and auroral cycles of medium length correspond to radiocarbon changes, and selected strong and weak phases and their effects are indicated by boldface and italics respectively in the table in this appendix. The preliminary results of Stuiver and Quay (1980, 1981) have been read from their charts for columns e and f, the production rate they plotted at the end of each decade being tabulated here for the mid-decade to give the best fit with auroral numbers; the reduction factor used to correct for geomagnetism is based on the dashed line in their (1981) chart. The crude conversion formula then leads to a calculated auroral index (column g) based on radiocarbon evidence alone. The Stuiver and Quay data were based partly on European oak; Stuiver's American results have now been used for columns c and d, and columns e, f, and g will need revision up to A.D. 1170 to fit. When interlaboratory agreement is reached on the ^{14}C levels to be used for column c the dependent variables (c, d, f, and g) will need a further slight adjustment.

The auroral evidence by itself is indicated in column h. This differs from Paper 19 in that newly discovered evidence has made slight revisions possible. The effective lag between my auroral numbers (N) and the radiocarbon level is twelve years to fifteen years. An empirical model suggests adding the level for the decade in question (R_0) to weighted levels for subsequent decades:

$$\frac{R_0 + 3R + 2R_2 + R_3}{7} = f(N).$$

This model ignores the effects of radiocarbon more than thirty-five years old. Some discrepancies between calculated and observed auroral indices are presumably explained, as in the sixteenth century, by north-south movements of the auroral girdle in Eurasia.

An independent index, presumably linked to the frequency of coronal holes, is provided by the time of arrival at the solar surface of the effective sunspot minimum. The *remainder*, as defined in Paper 21, is thus regarded in column h as a continuous function (11.1, not 11, being used as the divisor). There are some periods (in parentheses) when alternative interpretations are still possible but the agreement of this index with radiocarbon production (column f) suggests that it usually reflects the weakness of the solar wind and the strength of the cosmic rays impinging on our upper atmosphere.

REFERENCES

Stuiver, M., and P. D. Quay, 1980, Changes in Atmospheric Carbon-14 Attributed to a Variable Sun, *Science* **207**:11–19.

Stuiver, M., and P. D. Quay, 1981, Atmospheric ^{14}C Changes Resulting from Fossil Fuel CO_2 Release and Cosmic Ray Flux Variability, *Earth and Planetary Sci. Letters* **53**:349–362.

A.D.[a]	B.P.[b]	b.p.[c]	B.P.-b.p.[d]	C14[e]	Q_M[f]	AURORAE CAI[g]	OAI[h]	PHASE OF MINIMA Spots[i]	Phase[j]
5	1945	1970	−25						4
15	1935	**2000**	**−65**						5
35	1915	*1920*	−5						6
45	1905	**1960**	**−55**						5
85	1865	*1870*	−5						5
125	1825	**1890**	**−65**						(6)
135	1815	*1820*	−5						(1)
165	1785	**1840**	**−55**						(2)
185	1765	*1800*	−35						(2)
205	1745	**1840**	**−95**						(3)
245	1705	1770	−65	−3					(5)
255	1695	1760	−65	−1					(5)
265	1685	*1695*	−10	+2	+6	44			(6)
275	1675	1705	−30	−2	−11	61			(7)
285	1665	**1750**	**−85**	−8	−1	51			(7)
295	1695	1720	−65	−12	−15	65	50		(9)
305	1645	1760	−115	−12	**−17**	**67**	**75**		8
315	1635	**1765**	**−130**	−8	−9	59	65		7
325	1625	1725	−100	−7	−13	63	50		5
335	1615	1725	−110	**−13**	−2	52	30		5
345	1605	1690	−85	−12	−4	54	40		4
355	1595	1685	−90	−11	0	50	60		3
365	1585	*1665*	*−80*	−10	−2	52	55		2
375	1575	**1700**	**−125**	−12	−6	56	65		**1**
385	1565	1675	−110	−14	−4	54	50		**1**
395	1555	1655	−100	**−16**	−7	57	60		2
405	1545	1655	−110	−8	0	50	(45)		3
415	1535	1625	−90	−3	−2	52	*(35)*		4
425	1525	1620	−95	*−1*	*+14*	*36*	*(35)*		*(4)*
435	1515	1555	*−40*	−5	+10	40	(50)		*(4)*
445	1500	*1550*	−45	−9	+1	49	(55)		4
455	1495	**1570**	−75	−10	0	50	(60)		3
465	1485	1565	−80	−8	+1	49	(40)		2
475	1475	1555	−80	−9	0	50	(55)		(3)
485	1465	*1550*	−85	−11	−1	51	(50)		(4)
495	1455	1555	−100	−14	**−15**	**65**	(55)		(4)
505	1445	**1570**	−125	−16	−14	64	**70**		(6)
515	1435	*1565*	−130	**−16**	−12	62	60		7
525	1425	**1595**	**−170**	−15	−11	61	55		**6**
535	1415	1545	−130	**−16**	−2	52	50		**6**
545	1400	1505	−100	−14	−5	55	50		7
555	1395	1505	−110	−13	−1	51	50		7
565	1385	1490	−105	−12	−2	52	70		7
575	1375	1485	−110	−10	0	50	60		7
585	1365	*1470*	*−105*	−12	−6	56	70		6
595	1355	1475	−120	−11	−1	51	60		(5)
605	1345	1445	−100	−8	0	50	35		(4)
615	1335	*1435*	−100	−8	−5	55	**65**		(4)
									(3?)

(continued)

A.D.[a]	B.P.[b]	b.p.[c]	B.P.-b.p.[d]	C14[e]	Q_M[f]	AURORAE CAI[g]	OAI[h]	PHASE OF MINIMA Spots[i]	Phase[j]
625	1325	1440	−115	−11	**−10**	**60**	(60)		**(3)**
635	1315	**1460**	**−145**	**−15**	+7	43	(40)		(4)
645	1305	1385	−80	**−15**	+8	42	(55)		(5)
655	1295	1360	−65	−14	+9	41	(45)		(5)
665	1285	1340	−55	−13	+13	37	(35)		4
675	1275	1310	−35	−11	+15	35	55		2
685	1265	1295	−30	−8	+21	29	45		2
695	1255	*1260*	−5	−4	+6	44	(25)		2
705	1245	**1290**	**−45**		9	41	(50)		(2)
715	1235	1270	−35		13	37	(50)		(2)
725	1225	*1240*	*−15*		+6	44	(40)		(3)
735	1215	1245	−30		−1	51	(35)		(5)
745	1205	1260	−55		**−15**	**65**	**(60)**		5
755	1195	**1290**	**−95**		−11	61	50		4
765	1185	1280	**−95**		+4	46	**70**		5
775	1175	1245	−70	−12	+18	32	60		5
785	1165	*1165*	0	−12	−1	51	50		(4)
795	1155	**1205**	**−50**	−10	+2	48	*50*		3
805	1145	1185	*−40*	−10	*+3*	47	55		2
815	1135	*1180*	−45	−10	0	50	40		2
825	1125	1195	−70	**−18**	−4	54	55		2
835	1115	1180	−65	−15	0	50	**70**		2
845	1105	**1205**	**−100**	−12	**−16**	**66**	60		2
855	1095	1195	−100	**−18**	−12	62	65		1
865	1085	1180	−95	**−18**	−8	58	50		2
875	1075	1165	−90	−12	−9	59	75		1
885	1065	1160	−95	−12	−1	51	50		(2)
895	1055	1130	−75	−11	*+5*	*45*	*35*		(1)
905	1045	1100	*−55*	−5	+4	46	55		0
915	1035	*1095*	−60		−13	63	40		0
925	1025	1140	−115		**−14**	**64**	**65**		0
935	1115	**1145**	**−130**		−6	56	55		0
945	1005	1105	−100		−10	60	45		(2)
955	995	1110	−115		−11	61	35		3
965	985	1105	−120		−16	65	70		(3)
975	975	**1110**	**−135**		−0	50	65		3
985	965	1055	−90		+3	47	55		2
995	955	1040	−85		*+13*	*38*	*35*		1
1005	945	990	*−45*	( )	−1	51	70		0
1015	935	**1015**	**−80**	−9	+13	38	55		0
1025	925	965	−40	−11	+12	39	35		2
1035	915	950	−35	−10	*+21*	*30*	45		2
1045	905	*905*	0	−9	+8	42	40		3
1055	895	925	−30	−4	+9	41	*20*		4
1065	885	*05*	−20	−6	+9	41	45		5
1075	875	895	−20	−8	0	50	55		5

(continued)

A.D.[a]	B.P.[b]	b.p.[c]	B.P.-b.p.[d]	C14[e]	Q_M[f]	AURORAE CAI[g]	OAI[h]	PHASE OF MINIMA Spots[i]	Phase[j]
1085	865	905	−40	−7	−10	59	50		5
1095	855	930	−75	−6	**−16**	**66**	**75**		5
1105	845	**945**	**−100**	−11	−15	64	50		(5)
1115	835	935	−100	−16	−10	59	75		4
1125	825	910	−*85*	−16	−11	60	75		2
1135	815	*905*	−90	−14	**−32**	**80**	**80**		1
1145	805	**965**	**−160**	−15	−10	59	*45*		**0**
1155	795	905	−110	**−21**	+2	*48*	50		0
1165	785	*860*	−*75*	−16	−6	56	45		1
1175	775	870	−95	−15	**−11**	**60**	65		2
1185	765	**875**	**−110**	−14	−4	54	60		1
1195	755	845	−90	−16	−7	57	50		0
1205	745	860	**−115**	(−16)	+4	46	**70**		**(−2)**
1215	735	830	−95	(−16)	−1	51	50		(−1)
1225	725	805	−*80*	−14	−2	52	**55**		(0)
1235	715	*795*	−*80*	−12	−5	55	*30*		1
1245	705	800	−95	−14	**−8**	**57**	45		1
1255	695	**805**	**−110**	**−16**	+7	44	50		1
1265	685	*745*	−60	−10	−2	52	45		(2)
1275	675	755	−80	−12	+9	42	50		(3)
1285	665	705	−40	−7	+21	31	35		3
1295	655	655	0	−1	+24	*28*	*25*		3
1305	645	620	+25	0	+19	33	50		(2)
1315	635	635	0	0	+18	34	*35*		(1)
1325	625	605	+20	0	+28	*25*	40		(1)
1335	615	*555*	+*60*	4	+13	39	30		(2)
1345	605	575	+30	1	−3	53	40		2
1355	595	605	−10	−4	−3	53	50		3
1365	585	595	−10	−4	**−23**	**70**	65		2
1375	575	**650**	**−75**	**−11**	−15	63	**85**		1
1385	565	630	−65	−9	+1	49	65		0
1395	555	560	−5	−4	−3	53	*25*		**(−2)**
1405	545	555	−10	−4	−8	57	50		(−1)
1415	535	520	+15	0	+13	39	*20*		(0)
1425	525	490	+35	+4	+20	33	30		0
1435	515	455	+60	+5	+*29*	*25*	35		1
1445	505	420	+85	+8	+*31*	*23*	40		1
1455	495	390	+*105*	+12	+23	30	40		0
1465	485	385	+100	+10	+20	33	30		(+1)
1475	475	375	+100	+10	+16	36	*15*		(+2)
1485	465	370	+95	+9	+19	33	25		(+1)
1495	455	345	+*110*	+*12*	+16	36	30		+1
1505	445	350	+95	+12	+8	43	35	38	2
1515	435	345	+90	+10	+12	40	40	30	2
1525	425	320	+105	+12	+15	*37*	60	45	2
1535	415	295	+*120*	+*14*	+9	42	65	58	1

(continued)

A.D.[a]	B.P.[b]	b.p.[c]	B.P.-b.p.[d]	C14[e]	Q_M[f]	AURORAE CAI[g]	OAI[h]	PHASE OF MINIMA Spots[i]	Phase[j]
1545	405	300	+105	+12	+4	46	60	58	−1
1555	395	305	+90	+10	−5	54	65	59	−1
1565	385	320	+65	+6	−6	55	70	40	0
1575	375	320	+55	+4	−4	54	75	60	0
1585	365	330	+35	+5	−20	67	60	45	−1
1595	355	330	+25	0	−14	62	45	31	−1
1605	345	335	+10	+1	−20	67	50	39	−2
1615	335	350	−15	−3	−13	61	50	48	−2
1625	325	330	−5	−1	+2	48	60	51	−1
1635	315	280	+35	+4	+5	46	50	29	0
1645	305	255	+50	+6	+9	42	25	29	1
1655		230	+65	+7	+13	39	30	15	0
1665	285	205	+80	9	+25	28	25	12	0
1675	275	155	+120	15	+19	33	30	24	1
1685	265	160	+105	13	+24	29	25	18	0
1695	255	105	+150	17	+22	31	15	7	0
1705	245	105	+140	+17	+19	44	35	35	0
1715	235	95	+140	17	+1	49	50	34	1
1725	225	130	+85	12	0	50	65	61	1
1735	215	150	+65	8	−17	65	65	61	0
1745	205	175	+30	2	−25	71	40	42	0
1755	195	180	+15	0	−19	66		37	0
1765	185	170	+15	1	−25	71		54	−1
1775	175	185	−10	−2	−29	75		71	−3
1785	165	200	−35	−5	−33	78		71	−5
1795	155	200	−45	−8	−16	64		36	−4
1805	145	155	−10	+2	+1	49		28	−1
1815	135	105	−30	3	−3	53		21	0
1825	125	105	+20	2	−9	58		27	1
1835	115	110	+5	0	−10	59		67	0
1845	105	110	−5	−2	−11	60		57	−1
1855	95	120	−25	−3	−16	64		43	0
1865	85	125	−40	−6	−18	66		49	1
1875	75	110	−35	−6	−17	65		51	1
1885	65	100	−35	−6	−13	61		38	1
1895	55	80	−25	−4	−14			45	1
1905	45	75	−30					36	1
1915	35	110	−75					39	2
1925	25	135	−110					42	1
1935	15	150	−135					51	0
1945	5	165	−160					72	0
1955	−5							92	−1
1965	−15							61	−2
1975	−25							62	−1

[a]Date of mid-decade of form 240–249 (except in col. f)
[b]Tree-ring date Before the Present
[c]Conventional radiocarbon date before the present NBS oxalic acid standard (Stuiver, M., 1982, Radiocarbon **24**:1–26; cf. Suess, H. E., 1980, Radiocarbon **22**:779)

(continued)

[d]Radiocarbon deviation, corresponding to C14 content per mil for the decade in question

[e]C14 content per mil (after Stuiver and Quay, 1980, *Science* **207**:11-19)

[f]C14 Production rate during decade, calculated using Reservoir Model. (After Stuiver and Quay, 1981, *Earth and Planetary Sci. Letters* **53**:349-362)

[g]Calculated Auroral Index based on $50 - a \Delta Q_M$ where a is reduction factor to correct for magnetic field intensity: a is 0 near A.D. 700, 0.95 near A.D. 200, and 1000 and 0.8 near A.D. 2000

[h]Observed Auroral Index based on Paper 19 (decades of form 01-10) with slight revision

[i]Mean sunspot numbers (decades from 1500-1509) based only on aurorae to A.D. 1609

[j]Phase of sunspot minima defined by remainders (when last two digits of years are divided by 11.1) and treated here as a continuous function

APPENDIX D: CYCLE INDEX

Weather and sunspot oscillations are usually transient and subject to phase changes, and some of those mentioned may be statistical accidents. The radiocarbon (presumably solar) cycles in the 100-year to 2000-year range have been analyzed through 6000 years in the following: Neftel, A., H. Oeschger, and H. E. Suess, 1981, *Earth and Planetary Sci. Letters* **56:**127–147.

Years	Oscillation	Benchmark Papers
2	biennial oscillation in weather (shorter when linked to strong solar activity)	33, 34, 39, 40
2.17	quasibiennial oscillation (QBO) in weather	33, 34, 39, 40
2.7	weather (eigenvector) cycle of 32 months = $\frac{1}{4} \times 10.8$ years	38
3	triennial weather cycle, A.D. 1880–1910	38
3.3	(sunspots weak)	38
3 to 10	other alleged cycles in weather	38
5.0	half sunspot cycle (20th century)	38
6.0	half sunspot cycle (A.D. 1870–1900)	38
10	strong sunspot cycles	20
11.1	mean sunspot cycle	20, 30, 31, 32, 37, 38
12	weak sunspot cycles	20
20 to 23	Hale's double sunspot cycle (heliomagnetic)	24, 37, Figs. 2 and 13–16 in our Intro.
42 to 46	quadruple sunspot cycles (occasional)	Fig. 9 in our Intro.
55 to 62	medium sunspot, auroral, and radiocarbon cycles	Fig. 9 in our Intro.
78	Gleissberg cycle in sunspot phase	25, 26
91 to 95	medium solar cycles A.D. 500–1600; phase change at A.D. 930	26, Fig. 9 in our Intro.
133	radiocarbon cycle (A.D. 1–1000), aurorae, sunspots	28
150	radiocarbon (2000–1 B.C.) and, with phase change at 2000 B.C., especially strong 4000–2000 B.C.	28
180	sunspot phase and radiocarbon cycle	27

(continued)

Years	Oscillation	Benchmark Papers
200	the long sunspot cycle (A.D. 200+); radiocarbon cycle is strong before 2000 B.C. and from 500 B.C. with phase lag of 70 years c. 1800–500 B.C.	20, 27, 40, and Fig. 9 in our Intro.
260	possible rainfall cycle	40
286	radiocarbon cycle (4000–1 B.C.)	40
500, 1000, and 2300	possible radiocarbon cycles	40
1500	alleged climatic cycles	40

AUTHOR CITATION INDEX

Aaby, B., 353
Abbot, C. G., 314
Abetti, G., 26, 188
Adamnan, 121
Ager, D. V., 329
Agterberg, F. P., 300
al-Bitrūjī, 47
Allen, J. H., 27, 238
Altheim, F., 104
Ancient Sunspot Records Research
    Group, 69
Anderson, A. O., 116
Anderson, C. N., 196, 235
Anderson, R. Y., 301, 314, 329
Angell, J. K., 272
Anner, G. E., 215
Antevs, E., 362
Appian, 133
Arakawa, H., 266
Archibald, E. D., 271
Aristotle, 130, 132
Arrhenius, G., 329
Ashbrook, J., 210
Aurelius, V., 133
Avicenna, 49

Baillie, M. G. L., 355, 362
Bain, W. C., 21, 26
Bandeen, W. R., 235, 295
Banerjee, I., 300
Barrell, J., 314
Barrett, A. A., 99
Bartlett, M. S., 210
Baur, F., 78
Baxter, M. S., 26
Becker, B., 26, 196
Bell, B., 15, 291
Belmont, A. D., 272, 282
Berglund, B., 356
Berkes, Z., 331
Berlage, H. P., 273, 282, 329, 330
Berson, A., 271
Berson, F. A., 235, 273
Bicknell, P. J., 34, 38
Bielenstein, H., 69
Bigg, E. K., 215
Bigourdan, M. G., 187
Biraben, J-N., 235
Blackman, R. B., 215, 223, 314
Bohn, G., 187

Bonov, A. D., 15
Borchert, J. R., 295
Borell, R., 330
Bossolasco, M., 271
Botley, C. M., 99, 104, 140, 147
Bowen, R., 330
Boyle, R., 170
Bracewell, R. N., 215, 235
Bradley, W. H., 314
Braitsch, O., 330
Brandtner, F., 330
Bray, J. R., 26, 34, 69, 187, 188,
    295, 354, 366
Brecher, K., 127
Bridge, D., 35
Brier, G. W., 330
Britton, C. E., 108, 115, 116
Brooks, C. E. P., 235, 282, 314,
    330
Brown, G. M., 78
Bruzek, S., 99
Bryson, R. A., 330
Bucha, V., 188
Buli, U., 38
Bumba, V., 78
Burritt, E. H., 187

Callendar, G. S., 249, 266
Callisthenes, 132
Carrington, C., 78
Cary, M., 104
Casiri, 47
Cassini, G., 187
Cassini, J., 187, 188
Chambers, G. F., 78, 104
Chapman, S., 129, 188
Charlesworth, J. K., 330
Charmander, 132
Chavannes, E., 104
Chen, M., 26, 99, 129, 141
Chernosky, E. J., 78
Chizevsky, A., 104
Cicero, 133
Cimino, M., 210
Cissoko, S-M., 291
Clark, D. H., 69
Clark, R. M., 354
Clarke, A. M., 173, 174, 187
Clarke, H., 235
Clausen, H. B., 34, 354

Clayton, H. H., 235
Cline, R. M., 354
Cole, T. W., 26
Collingwood, R. G., 147
Cook, A. F., 168
Coope, G. R., 354
Copernicus, 49
Cousins, F. W., 26, 78
Cox, A., 330
Creer, K. M., 141, 354
Crill, P. A., 355
Crooker, N. U., 196
Cross, C. I., 266
Crutcher, H. I., 314, 331

Dai, N., 26, 99, 129, 141
Daimachus, 132
Dalby, D., 291
Dall'Olmo, U., 26, 99
Dalrymple, G. B., 330
Damboldt, T., 78
Damon, P. E., 188, 197, 223, 224,
    330
Dansgaard, W., 34, 354
Dartt, D. G., 272, 282
Dauphin, J. P., 237, 355
Davis, J. M., 188
Davis, N. E., 273
Davison, C., 168, 235
de Boer, H. J., 168, 330
Deehr, C. S., 78, 99
Deevey, E. S., 314
de Geer, E. H., 115
De Geer, G., 314, 330
Degens, E. T., 301
De Jong, A. F. M., 26, 27, 196
Delambre, J. B. J., 47
de Mairan, J. J. d'O., 78, 99
Denton, G. H., 295, 354
DeVries, H., 188, 196, 223
Dewey, E. R., 140, 235, 266, 330
Dicke, R. H., 78, 196, 235, 236
Diels, 104
Dio, 133
Diodorus Siculus, 132
Dionysius of Helicarnassus, 132
Dio-Xiphilinus, 133
Dixon, F. E., 330
Dobson, M., 238
Doell, R. R., 330

Donnelly, R. F., 147
Douglass, A. E., 188, 267, 314
Drake, S., 47
Dubief, J., 291
Dubs, H. H., 38, 42, 69, 104
Duncombe, R. L., 210
Durrant, C. J., 99
Dutton, J. A., 330

Easton, C., 210
Eather, R. H., 99, 147
Eaton, G. P., 330
Ebdon, R. A., 330, 332
Eberhard, W., 104
Eddy, J. A., 34, 69, 147, 187, 196,
    223, 295, 366
Einhard, 34
Ekdahl, C. A., 223
Ellison, M. A., 330
Ellyett, C. D., 238
Emiliani, C., 330
Ephorus, 132
Ettmueller, M. E., 147
Everitt, W. L., 215
Ezekiel, 103

Fairbridge, R. W., 27, 34, 122,
    141, 236, 238, 301, 304, 354
Fan, C. Y., 188
Farrington, B., 104
Feller, W., 210
Ferguson, C. W., 223, 362
Fergusson, G. J., 224
Feynman, J., 141, 196
Flint, R. F., 330
Fontes, J. C., 330
Freeman, K., 104
Fritts, H. C., 236, 354
Fritz, H., 41, 78, 99, 104, 114,
    168, 188, 196
Fritz, P., 330
Frobesius, J. N., 99
Frolow, V., 314

Galen, 50
Galle, J. G., 104
Gartlein, C. W., 104, 115
Gates, W. L., 188, 295
Geyh, M. A., 354, 362
Giaconni, R., 188
Gilbert, G. K., 314
Gildersleeves, P. B., 237
Gilman, P. A., 34
Ginzel, F. K., 104
Glasspoole, J., 282
Gleissberg, W., 2, 15, 26, 78, 94,
    112, 140, 168, 188, 197, 202,
    205, 210, 215, 330
Gnevyshev, M. N., 78, 197, 236
Godoli, G., 366
Godson, W. L., 330
Goldberg, R. A., 236

Goldstein, B., 47, 69
Goudie, A. S., 291
Granar, L., 314
Granger, C. W. J., 236
Grant, R., 187, 188
Gregory of Tours, 118, 119
Grey, D. C., 188, 197, 223
Griffiths, D. H., 331, 354
Groissmayr, F. B., 210
Gross, D., 354
Grotrian, W., 210
Grove, A. T., 292
Groveman, B. S., 236
Gundel, W., 131
Gunderstrup, N., 34
Guo, C.-S., 15

Hagan, M. P., 78
Hakluyt, 39
Hale, G. E., 187, 197, 198
Halley, E., 173, 188
Hammer, C. U., 34
Hancock, D. J., 236
Hanslik, S., 236
Harang, L., 168
Hari, P., 301
Harris, P. R., 301
Harrison Church, R. J., 291
Hays, J. D., 354
Heastie, H., 282
Heath, T. L., 104
Hellman, C. D., 104
Henkel, R., 15, 147
Herman, J. R., 236
Hermelink, H., 49
Herodian, 133
Herr, R. B., 34
Herring, R. S., 291
Herschel, R. A., 187
Hess, W., 115, 331
Hevelius, J., 187, 188
Hibler, D. J., III, 237
Hibler, W. D., 236
Hillaire-Marcel, C., 354
Hillam, J., 355, 362
Ho, P.-Y., 34, 109, 110, 111, 129,
    141, 188, 332
Hoang, P., 104
Holtet, J. A., 78, 99
Hottunen, P., 355
House, M. R., 331
Houtermans, J., 17, 26, 197,
    224
Hsü, J. C. H., 26
Huber, B., 331
Hughes, A. D., 236
Hughes, O. L., 362
Hundhausen, A., 188
Hustich, I., 331
Huygens, C., 147
Hydatius, 116

Iba, Y., 120
Ibn, Guma, 39
Imbrie, J., 236
Imbrie, K. P., 236, 327
Imbrie, V., 327

Jenkinson, A. F., 291
Jeremiah, 103
Jevons, W. S., 236
Jiang, Y., 69, 148
Johnsen, S. J., 34, 236
Johnson, A. I., 291
Johnson, E. A., 354
Johnson, F. S., 295
Johnson, M. J., 78
Johnson, S. J., 41
Jose, P. D., 15, 215

Kageorge, M., 315
Kanda, S., 69, 112, 113, 131, 147,
    168, 188, 197
Kandaurova, K. A., 15
Kane, R. P., 27, 78, 147
Karlen, W., 295, 354
Kassner, C., 331
Keeling, C. D., 223
Keimatsu, M., 27, 35, 99, 122,
    295
Kempe, S., 301
Kendall, M. G., 210
Kennedy, E. S., 48, 50
Kepler, 47
Kiepenheuer, K. O., 27, 79
Kimura, H., 210
Kindelberger, C. P., 236
King, H. C., 187
King, J. W., 24, 236
King-Hele, D. G., 15, 147
Kirkland, D. W., 314
Kleczek, J., 78
Koehl, H. W., 27
Kopecký, M., 79
Köppen, V., 314
Korn, H., 314
Korringa, P., 331
Korshover, J., 272
Kotrc, P., 79
Kratoschwill, F., 210
Kraus, E. B., 267
Krauss, F. B., 104, 130, 134
Krieger, A. S., 188
Kroehl, H. W., 238
Kuklin, G. V., 79
Kulkarni, R. N., 235, 273
Kulp, J. L., 331
Kuniholm, P. I., 354, 362

Ladurie, E., 282, 295
LaLande, J., 79, 187, 236
LaMarche, V. C., Jr., 236, 354
Lamb, H. H., 27, 236, 291, 295,
    354

Lampridius, 133
Landsberg, H. E., 147, 169, 267, 282, 291, 314, 331
Langway, C. C., Jr., 236, 354
Latter, J. H., 35
Leighton, R. B., 187
Lerman, J. C., 188, 224
Libby, W. F., 224
Lieber, A. D., 127
Lin, Y. C., 188
Lincoln, J. V., 168
Lineback, J., 354
Lingenfelter, R. E., 188, 197, 224
Link, F., 27, 35, 99, 108, 134, 205, 331, 354
Lippert, L., 47
Livy, 132
Lona, F., 331
Long, A., 188, 223, 224, 330
Loughhead, R. E., 26, 34, 187, 366
Lowe, E. J., 173
Lowther, A. W. G., 129, 267, 332
Lucan, 133
Lycosthenes, C., 104
Lydus, 132
Lynch, H. B., 267

Ma, T. Y. H., 331
Maas, C., 188
McClelland, L., 35
McCormac, B. M., 236
Macdonald, N. J., 295
McGuire, R. H., Jr., 314
McInnes, B., 140
Mackin, J. H., 314
McNish, A. G., 168, 354
Maksimov, I. V., 331
Maley, J., 292
Mandeville, J. E., 292
Mangeröd, J., 354
Manilius, 133
Manley, G., 188, 249, 282
Maran, S. P., 235, 295
Marcellinus Comes, 116
Marcia, E., 301
Margoliouth, H. M., 147
Markson, R., 236
Marshall, J. R., 295
Martin, H., 314
Mathews, W., 108
Matsushita, S., 188
Maunder, E. W., 69, 79, 147, 169, 187
Meadows, A. J., 188
Meeus, J., 47, 79
Meko, D. M., 27, 295
Merkt, J., 354, 362
Messerli, P., 236
Meyermann, B., 210
Michael, H. N., 224

Michelson, A. A., 210
Mihalas, D., 366
Miles, M. K., 236, 237
Milsom, A. S., 168
Mintz, Y., 188, 295
Mitchell, J. M., Jr., 22, 27, 237, 295, 314, 331
Mock, S. J., 237
Moffitt, B. J., 282
Monin, A. S., 301
Mook, W. G., 26, 27, 196, 224, 354
Mörner, N-A., 79, 301, 354
Muller, H., 354, 362
Munaut, A. V., 354
Munk, W., 26, 197
Münnich, K. O., 332
Murphy, T., 354
Murray, R., 282
Muzalevskii, Yu. S., 16, 28, 215, 356
Myrach, O. v., 380

Nairn, A. E. M., 301, 331, 332
Nakayama, S., 188
Nallino, 49
Namias, J., 292
Needham, J., 35, 69, 111, 129
Neftel, A., 197, 378
Ness, N. F., 295
Newell, R. A., 292
Newhall, C., 35
Newton, H. W., 27, 79, 140, 168
Newton, R. R., 27, 187, 188
Nicholson, S. B., 315
Nicolini, T., 134, 147, 168
Nicolson, S. E., 292
Niemalä, J., 354
Nihusius, B., 104
Noel, M., 354
Nook, W. G., 188
North, J., 35
Norton, A. P., 79
Nupen, W., 315

Obsequens, J., 104, 132, 133
Occhietti, S., 354
Oeschger, H., 197, 378
Offerberg, J., 330
Ogilvie, R. M., 148
Ol', A. I., 236, 237, 301
Olausson, E., 198, 238
Olson, R. H., 295
Olsson, I. U., 188, 224, 355
Olympiodorus, 132
O'Mahoney, G., 331
Opdyke, N. D., 301
Öpik, E., 188
Orosius, 132
Ovid, 133

Pankhurst, R., 292
Parker, B. N., 18, 19, 27, 79, 237
Parker, E. N., 188
Parker, G., 237
Paulus Diaconus, 132
Pearson, C. W., 355
Pennington, W., 354
Pepin, E. O., 197
Perfilief, B. W., 315
Persson, C., 355
Pettijohn, F. J., 331
Pfister, C., 236
Philoponus, 132
Pico, 49
Pilcher, J. R., 355, 362
Pilgram, A., 99, 197
Pingré, A. H., 187
Pisias, N. G., 237, 355
Pittock, A. B., 27, 237, 295
Pliny, 130, 132, 133, 134
Plummer, H. C., 104
Plutarch, 38, 132, 133, 188
Potter, P. E., 331
Probert-Jones, J. R., 331
Procter, R. A., 187
Prohaska, J. T., 332

Quay, P. D., 24, 27, 372, 377
Quinlan, F. T., 331

Rafter, T. A., 224
Ralph, E. K., 224
Ramaty, R., 188, 197, 224
Ranyard, A. C., 188
Reeh, W., 34
Rees, M. J., 187
Renan, E., 50
Rethly, A., 331
Riccioli, G. P., 237
Richter-Bernburg, G., 301
Rigg, J. B., 331
Ringberg, B., 355
Riveroll, D. D., 315
Rizzo, P. V., 148
Roberts, W. O., 295
Robertson, K. A., 140
Rodewald, M., 267
Rognon, P., 292
Roelof, E. C., 188
Rosen, E., 47
Rostow, W. W., 237
Rubashev, B. M., 79, 148, 197, 202, 237
Rudmark, L., 355
Ruf, K., 79

Saarnisto, M., 355
Sabra, A. I., 49
Sancetta, C., 237, 355
Sarton, G., 35, 40, 47
Sayili, A., 48, 50
Schaefer, E. H., 111

Scheiner, C., 35, 187
Scherhag, R., 282
Scherrer, P. H., 295
Schneider, S. H., 188
Schoening, G., 99, 129
Scholz, B. W., 35
Schove, D. J., 12, 13, 15, 23, 27,
    34, 35, 38, 40, 69, 79, 89, 94,
    99, 100, 104, 108, 112, 114,
    115, 122, 129, 131, 134, 140,
    141, 148, 168, 187, 188, 197,
    205, 206, 210, 215, 224, 237,
    238, 249, 250, 267, 273, 282,
    292, 295, 301, 304, 331, 332,
    355, 362
Schroeder, W., 79, 100, 148
Schulman, E., 249, 267, 273, 315,
    332
Schwabe, H., 86, 187
Schwarzacher, W., 332
Schwarzbach, M., 332
Seibold, E., 315, 332, 355
Seliga, T. A., 236
Seneca, 130
Shackleton, N. J., 355
Shapiro, R., 295, 332
Shapley, A. H., 27, 238
Shaw, P. K., 292
Shea, W. R., 35
Short, T., 115
Schostakowitsch, W. B., 315,
    355, 362
Siebenlist, V., 331
Siebert, L., 35
Siiriäinen, A., 356
Sikes, S. K., 292
Silius Italicus, 132
Silk, J. K., 188
Silverman, S., 141
Simkin, T., 35
Simpson, J. A., 188
Siren, G., 301, 332
Siscoe, G. L., 8, 9, 27, 100, 197
Sisenna, 133
Sleptsov-Shevlevich, B. A., 238
Smith, L. M., 237
Smythe, C. M., 187
Sokolov, S. S., 249
Sorokin, P. A., 282
Soutar, A., 355
Sparkes, J. R., 238
Spjeldnaes, N., 332
Spoerer, F. W. G., 69, 79, 148,
    169, 187
Steinhauser, F., 249
Steinschneider, M., 47, 49, 50
Stephenson, P. M., 282
Stetson, H. T., 238, 315
Stewart, B., 198
Stiftar, V. Th., 104
Stockton, C. W., 27, 236, 237, 295
Stormer, C., 188

Stothers, R., 100, 134, 135, 148
Stowe, 173
Strachow, N. M., 332
Street, F. A., 292
Strömberg, B., 355
Stuiver, M., 24, 27, 136, 141, 188,
    198, 224, 372, 376, 377
Suess, H. E., 26, 188, 197, 198,
    224, 376, 378
Sugawara, K., 315
Suter, H., 47, 48, 50
Svalgaard, L., 295
Svenonius, B., 198, 238
Swain, A. M., 355

Tabony, R. C., 238
Taffara, L., 210
Takahashi, K., 238
Talbot, C. H., 247
Tannehill, I. R., 267
Tans, P. P., 27
Tarling, D. H., 354
Tauber, H., 332
Taulis, E., 282
Tchijevsky, A. L., 238
Terasme, J., 332
Tertullian, 133
Thompson, L. M., 295
Thompson, R., 356
Thorndike, L., 332
Timothy, A. F., 188
Titow, J., 267
Titus, 133
Tolonen, K., 355, 356
Torreson, O. W., 354
Toussoun, Prince Omar, 292
Trotter, D. E., 34
Tuckerman, B., 47, 49
Tukey, J. W., 215, 223, 314
Turner, H. H., 198, 210

Udden, J. A., 315

Vaiana, G. S., 188
Vail, O. E., 315
Van Benmelen, W., 271
Van Woerkom, A. J. J., 210, 332
Vasilev, O. B., 15
Vaughan, A. H., 366
Vercelli, F., 38
Vergil, 133
Veryard, R. G., 332
Vestine, E. H., 188
Visser, S. W., 141, 267, 332
Vitinsky, Y. I., 35, 79, 148, 198
Vogel, J. C., 188, 224
von Guericke, O., 148
von Humboldt, A., 35, 79
von Oppolzer, T. R., 188
Vulis, I. L., 301
Vuorela, A., 271
Vyssotsky, A. N., 38

Wagner, A. J., 28, 238, 301
Wagner, T., 267
Walawender, A., 108
Waldmeier, M., 3, 4, 28, 79, 148,
    187, 215, 224, 332
Walker, G. T., 332
Wallick, E. I., 188, 223
Walton, A., 26
Wang, J. H. C., 69
Wang, J. R., 188
Ward, F. W., Jr., 315, 332
Watanabe, N., 141
Waterbolk, H. T., 354
Webster, J. B., 292
Weikinn, C., 108, 332
Weiss, J. E., 148
Weiss, N. O., 148
Welton, M., 315
White, O. R., 366
Whitelock, D., 100
Wiegert, R., 315, 332
Wilcox, J. M., 295
Willett, H. C., 20, 28, 238, 315,
    332
Williams, G. E., 28
Williams, J., 104
Willis, E. H., 136, 141, 332
Wilson, A., 187
Wilson, O. C., 366
Wing, V., 188
Winstanley, D., 292
Wittmann, A., 28, 79, 148, 198
Wolbach, J. G., 15
Wolf, R., 28, 35, 79, 80, 148, 168,
    187, 188, 198, 210
Wolfer, A., 80, 198
Wood, C. A., 238
Wood, K. D., 215
Woodbridge, D., 295
World Meteorological
    Organization, 356
Wright, P. B., 238, 273, 292

Xanthakis, J., 15
Xu, Z., 69, 148

Yamamoto, T., 267
Yarger, D. N., 236
Yoshino, M. M., 238
Young, C. A., 187
Yunnan Observatory, 35, 100,
    148, 198

Zacharias of Mitylene, 117
Zacharius, 103
Zeuner, F. E., 38, 332
Zhukov, L. V., 16, 28, 215, 356
Zinner, E., 49, 332
Zombeck, M., 188
Zonaras, 132, 133
Zumbühl, H. T., 236

SUBJECT INDEX

aa-index (geomagnetic), 76
Academies
    Paris, 88, 172–173
    U.S., 199
Africa. See also Nile
    aurorae, 39
    climate, 231–232, 253, 283–292, 284 (figs.), 291
        (fig.)
    varves, 320, 329
Alcuin, 96, 115
Alleroed (zone II), 341–342 (fig.). See also Late
        Glacial
Alps, 230, 231, 272. See also Switzerland
A-max, 348–349 (table). See also Aurorae
Americans, 192
Amplitude. See Magnitude, of cycles (R_M)
Andes. See Ecuador
Angels (aurorae and comets), 120, 247
Annals, Irish, 115, 126
Anomalies, meteorological, 230, 242–249, 250
Arctic, 94. See also Greenland
Armenia, 126. See also Lake Van (Armenia)
Asia, south-east, 32, 231–232
Asia Minor. See Turkey
Astronomer Royal, 176
Astronomische Mitt., 4, 75
Asymmetry, 3, 7, 12 (fig.), 92, 93 (fig.), 202n
Atlantic, 252, 334, 341, 359
Augsburg, 33
Aurorae
    Africa, 39, 97, 144, 218
    A-max (30 years), 349 (table)
    ancient, 130–135, 144
    calculated, 372, 373 (col. g)
    catalogs, xii, 13, 78, 97, 130–133
    Chinese, 69, 109–112, 125 passim, 129n, 144
    colors, 37, 114–115, 125
    cycles, 13 (fig.), 97
        weak, 279 (fig.)
    first dated (649 B.C.), 144
    frequency, 8–9 (figs.)
    international. See Aurorae, tropicalis
    medieval, 113–129, 317
    numbers
        calculations, 372
        observations, 373 (col. h)
    observed, 8–9 (figs.), 15 (fig.)
    tropicalis, 39, 99, 102, 116
Auroral armies, 108, 113, 317
Auroral battles, 113, 317

Auroral birds, 111, 126
Auroral blood, 40n, 96, 111, 115, 118, 130n
Auroral chasm, 130
Auroral clouds, 116, 124
Auroral fire, 102, 126, 128
Auroral frequency, 8–9, 202–205
Auroral hell, 122
Auroral milk-rain, 130n, 135
Auroral minima, 13 (fig.)
Auroral oval, 141, 372
Australasia, 19 (fig.), 229–231
Australia, 18 (fig.), 25–26, 189, 231, 253, 260, 286,
        323 (fig.), 326 (fig.)
Autocorrelation, 97, 207, 209 (fig.), 212 (fig.), 348

Baghdad, 31, 44
Baltic, 108, 342 (fig.), 359
Bamboo books, 103, 144
Bankruptcies, 241
Beam, aurorae, 131
Before Present, bp and BP, 337 (fig.), 351, 356, 373
Benchmark, 26, 32
Bern, Switzerland, 75, 175
Bias, 139, 159, 194, 305 (fig.)
Biennial cycles, 231, 268–282, 270 (fig.), 300, 339
        (fig.), 347 (fig.), 359, 364 (figs.). See also
        Quasibiennial oscillations (QBO)
Black Sea, 361
Blood. See Auroral blood; Sunspots, blood
Blood rain, 130n
Bölling (zone Ib) 339 (fig.). See also Late Glacial
Bogs, peat, 349, 350 (fig.)
Boulder, Colorado, 78, 145
Bracknell, England, Meteorological Office, xii,
        231, 273, 289
Brazil, northeastern, 18, 229, 298
Bristlecone, 195, 300, 363 (map), 364 (figs.)
Bristlecone year (B.Y.), 300
British Astronomical Association (BAA), 108, 172
British Isles, 20, 115, 128, 192
Bronze Age, 23, 222–223 (figs.), 284 (fig.), 344,
        361
Brussels, 78
Buddhist, 97, 319
Business cycles, 227, 241
Butterfly diagram, 1, 77, 91 (fig.)

Calibration, 182, 372
Camera, pinhole, xi, 71

Canada, 17–19 (figs.), 232, 256, 292, 334–335
    (table), 343. *See also* North America
Cape Town, 253, 326 (fig.)
Carbon dioxide, 181, 216–219, 372
Catalogs
    aurorae, xii, 13, 78, 97, 130–133
    sunspots, 32, 34, 52, 53–61, 97, 99 (refs.)
Cenozoic, 305 (fig.), 328 (fig.)
Centuries, 22, 349, 356–357 (table)
    even and odd, 140, 163–167
Ceremonies, 97
Chin (A.D. 265–420), 32, 62, 64 (fig.)
China, 31–37, 53–69, 108
    climate, 112n
    11th century, 125 passim
    4th century, 109–112
    locusts, 235
    17th century, 69
    7th century B.C., 144
Chinese emperor, 65, 116
Chromosphere, 177
Chronicles, 107
    Anglo-Saxon, 96, 126, 128
    medieval, 88, 98, 116
Chronology, 50, 114. *See also* Catalogs; Eclipses
    ancient, 38, 103, 111, 134n
    climatic, 250–267, 356–357
    errors of, 98, 122, 260
    Greek, 38, 131
    Maya, 50, 233
    Roman, 103, 131, 144, 153
    varve, 335, 359
Clairvoyance, Celtic, 122
Classes, amplitude, 12 (fig.), 13 (fig.), 231
Climate
    equatorial, 232, 316–332
    1500/800 B.C., 345, 361 (figs.), 364 (fig.)
    4th century A.D., 112
    geological, 93–94, 316–332 (fig.)
    Holocene, 234, 283, 343–345 (figs.)
    indoor, 22
    Late Glacial, 333–357
"Clock," in sun, 25, 167–168, 179
Clouds, 83, 123
    auroral, 116, 124
    in sun, 178
Collaboration, west-east, 98
Saint Columba, 98, 116, 120
Comets, 30, 31, 38, 41, 43, 98, 103, 105 (fig.), 128, 131, 234
Conjunctions, planetary, 64 (fig.), 233
Copenhagen, 174
Corona
    auroral, 119, 317
    solar, 174, 177, 183
Correlation, 358
Crises, 227, 237 (ref.), 241, 284, 285
Curves, sunspot, 3–7
Cycles, and wave trains, 378
    2.0-year. *See* Biennial cycles
    2.2-year. *See* Biennial cycles; Quasibiennial
      oscillations (QBO)

2.4-year (summer), 325n
2.7-year, 299, 324, 325n
3-year, 230, 279–282, 290, 325n, 339, 347, 364,
    399. *See also* Triennial cycles (3-year)
3.3-year, 299, 325n
4-year, 231, 299, 322, 325n, 342 (fig.)
5-year, 232, 299
7-year, or 7½-year, 233, 253, 299, 323
8-year, 26, 308 (fig.)
10-year, 82, 86, 163, 194, 227, 241, 268, 299
11-year, 2–9 (figs.), 12–13 (figs.), 74–93, 149–168
    shape, 2–7 (figs.), 12 (fig.), 92 (fig.)
    3000–1000 B.C., 25
12-year, 163, 211, 215, 268
13-year, 26, 308 (fig.)
18-year, 283
22 year, 20-22 (figs.), 92–93, 192–201, 229,
    232–233, 292, 294, 295, 306, 308 (fig.)
46/72-years, 16 (fig.), 192–195
80-year, 11, 16 (fig.), 22 (fig.), 93 (fig.), 94, 140,
    161–164, 193–195, 200 (fig.), 283
72/95-years, 16 (fig.), 94, 164, 193, 204 (table),
    207 (table), 215 (fig.), 299, 310 (fig.), 348
133/135-years, 97, 194, 378
143/155-years, 162 (table), 191, 196, 378
200-year, 16 (fig.), 164, 193–195, 214, 222,
    259–260, 318 (table), 327
240/286-years, 195, 221 (table), 349 (list)
300/1250-years, 196. *See* Megacycles
irregular, 1, 221 (table), 234, 311, 322
long, 94, 97, 311 (fig.), 379
lunar, 128, 319n
medium, 221, 234, 378
Palaeozoic, 25, 311 (fig.), 328 (fig.)
Precambrian, 25
radiocarbon, 378
solar rotation, 114, 128, 319
trade, 226, 241
weather, 299, 323–324, 349, 378

Dante, 122
Data
    daily (1828+), 76, 84–85
    monthly (1749+), 76
    proxy, 230, 364 (figs.)
Davis Strait, 231–232, 253, 259 (table), 288. *See*
    also Upper trough
Dawn, northern, 97
Days, in year, 318
Decades
    auroral, 138 (table), 242, 247, 372–375
    radiocarbon, 25, 372 (table)
    weather, 230, 243–244 (figs.), 255, 264 (table),
    289, 334
Declination, magnetic. *See* Magnetic declination
de Geer, 322, 333
Dendrochronology, 16, 17, 362, 364n
Dendroclimatology, 230. *See also* Tree-rings
Denmark, 179
    Horrebow, 74
Densitometer, 336, 343
Descartes, 97, 101

Deviation, standard, 98
Devonian period, 298, 306–307, 328 (fig.)
De Vries fluctuation, 181, 195, 216
Diaries, 107–108
Dicuil, Irish scholar, 42
Direction, of aurora, 125, 247, 317
Disaster, Skylab, 11, 183, 189
Discoveries
  auroral cycles, 1760s, 74, 97
  auroral frequency, 1733, 101
  fluctuations (LaLande), 1771, 74, 79
  fossil fuel (Suess effect), 1955, 195, 216
  Gleissberg cycle, 1952, 194, 202
  Gnevyshev gap, 1967, 77
  Hale cycle, 1913, 192, 200
  law of latitudes (Spoerer), 1889, 77, 200
  length of cycle (Wolf), 1852, 88, 147
  magnetic cycle (Sabine, etc.), c. 1850, 76, 88
  medium cycles (Pilgram), 1788, 97, 193, 206
  medium cycles (Wolf), 1862, 97, 193, 206, 207
  radiocarbon wiggles, 1958, 216
  smoothed numbers (Wolf), 1879, 4 (fig.), 76
  solar corona, 1715, 184
  solar rotation, 1612, 70
  solar wind (Descartes), c. 1630s, 97
  sunspot cycle (Schwabe), 1843, 75, 82, 86
  sunspot numbers (Wolf), c. 1850, 75, 86
  sunspots (naked eye), B.C., 37n, 38, 42, 53
  sunspots (telescope), 1610/1613, 1, 33, 69n, 70, 188
  200-year cycle, 1955, 17, 163, 194, 196
  wiggles as solar effect, 1961, 195, 218
  zone leap (Carrington), 1858, 76, 91
Double maximum, 13, 146, 165, 194
Douglass, astronomer, 11
Dragons, 37, 96
Drawings, 33–34, 179 (fig.), 187n, 317
Droughts, 257
  African, 283–292
  ancient, 300
  Australian, 292
  biblical, 361
  Chinese, 110, 112
  cycles, 22 (fig.), 259, 318 (table)
  prediction, 230–231, 292
  tropical, 227, 260, 283–292
  U.S., 230–231, 294, 318
  USSR, 36
Dryas (zone I), 338, 339 (fig.). *See also* Late
    Glacial
Duration, of sunspots, 52, 70, 90–91
Dust, 36, 51
Dust-bowl, 265, 294

Earthquakes, 88, 107, 152, 234
Easter Island, 229. *See also* Southern oscillation
    (S.O.)
Eastern Crescent, 232, 286, 322, 323 (map)
Eclipses, 39, 41, 45, 131, 174, 183
Ecuador, 94, 271, 275, 290
Eigenvectors (principal components), tropical
    pressure, 290. *See also* Pressure parameter

Einhard (biographer), 31, 41
Elatina formation, 25
Encyclopedias, oriental, 32, 53, 116
England. *See also* British Isles; Greenwich
    Observatory
  aurorae, 96, 173 passim
  climate, 1620+, 243–244 (figs.)
  diaries, 107
  economy, 227, 241
  summers 1205+, 272, 278–279 (figs.)
  sunspots, 33, 227
  temperature cycles, 21
  tree-rings, 364n
Eocambrian, 328 (fig.)
Epidemics, 103, 105 (fig.), 122, 127, 234
Epochs. *See* Turning points, of sunspots, 1500+
Equation, personal, 75. *See also* Formulae
Equator, 102, 275, 283, 319–320 (figs.)
Equatorial inflation, 271, 275 (fig.), 286, 319 (fig.).
    See also Upper high
Errors, 146, 228, 249, 267, 282, 298, 305 (fig.),
    325n, 362n, 364
  aurorae, 103n, 105, 113n, 127, 129, 131, 145
  auroral numbers, 138–139 (table)
  chronicles, 98, 122, 260
  cycle-lengths, 145
  cycles, 152, 298, 305 (fig.)
  of 11 years, 149n, 163
  hard-water, 336
  probable, 138, 159, 228, 305
  sunspot cycles, 122, 149n, 159–163
  sunspots, 39n, 117, 122, 145, 149n, 151, 176
  transits (really sunspots), 43–47
  tree-rings, 38, 267, 364n
  varves, 336
Eruption hypothesis, 146
  of air, 325
Ethiopia, 231, 276, 285, 291 (fig.), 292. *See also*
    Nile
Europe, 243–247 (table), 257, 281, 350 (fig.), 356
  central, 13 (fig.), 19 (fig.), 27 (ref.), 88, 230
  northern, 13 (fig.), 18 (fig.), 20 (fig.), 228
  pressure, 18–20, 23, 230, 245 (table)
  visions, 113–122, 123–129
  weather, 229–230, 287
  wind, 243–244 (figs.), 287

Faculae, 83, 177
Famines, 96, 107, 228, 257, 280–281, 285, 292
  prediction of, 257. *See also* Droughts
Faroes, 344
Faulensee, 343, 360
Finland, 18 (fig.), 94, 250, 313, 327, 335 (table and
    fig.), 345, 352
Fire, auroral, 102, 126, 128
Fires, 36, 126
f.Kr. (Swedish), 336, 359
Flares, 179
Floods. *See* Nile; Rain
Flux, cosmic ray, 24, 372n
Flying saucers, 114, 128. *See also* Aurorae
Forecasting. *See* Predictions

Formulae
  asymmetry, 7, 12, 92, 110, 162
  cycle, 80-year, 348
  floating numbers, 230
  intervals, 7, 12 (fig.), 92, 160–164
  odd cycles, 193
  radiocarbon, 218, 372
  spot numbers, 75, 239
  teleconnection, 335, 361
  varves, 335 (tables)
Fossil fuel (Suess effect), 372
Fourier. *See* Spectral analysis
France, 118, 128, 317. *See also* Paris
Frequency, auroral, 8–9, 202–205
Frosts, 31, 41

Gaps, in sunspots, 2 (fig.), 66, 77
  in aurorae, 13 (fig.), 134, 138, 165-166. *See also*
    Gnevyshev gap (lull in sunspots at
    maximum); Maunder minimum (17th
    century)
Geography, 98, 325. *See also* Maps
Geology, 299, 328 (fig.). *See also* Varves
Geomagnetism. *See* Magnetism
Germany, 31, 74, 76, 77, 171, 192, 317, 326, 343,
  346, 347 (fig.), 353
Glaciers, fluctuations, 230, 299, 338. *See also* Late
    Glacial
Gleissberg cycle. *See* Cycles, and wave trains,
  80-year
Gnevyshev gap (lull in sunspots at maximum), 2
  (fig.), 146
Gordion (Gordium), 362n, 364 (fig.). *See also*
  Turkey
Gradient, pressure, 243–244 (figs.), 264 (table).
  See also Wind
Greece, 103, 132, 344
Greenland, 20 (fig.), 94, 179, 253, 321, 348
  temperature, 24 (fig.)
Greenwich Observatory, 75, 77
Groups, sunspots, 75, 81–86, 239

Hail, 152, 229
Hale cycle, 1, 199–201
  minima, 233. *See also* Cycles, and wave trains,
    22-year
Han (206 B.C.–A.D. 220), 30, 37, 42, 65, 102
Harmonic dial, 231
Haze, 36, 42, 51
Hell, auroral, 122
Hemisphere, northern solar, 176
Historians, 102, 106, 115, 144
Holocene, 23, 223 (fig.), 234, 283, 294 (fig.), 349
  (list), 350–351 (table)
  varves, 299 passim
Horizons, marker, 344
Horrebow (Danish scientist), 74
Hydrogen, 233

Ice, in North Atlantic, 254
Ice Ages, Little, 25, 185, 216, 220, 259, 349
Ice-cores, 24 (fig.), 25, 34 (ref.), 233, 321

Iceland, 18 (fig.), 179, 254, 259 (table)
Inclination, 340. *See also* Magnetic meridian
Index, oddness, 232, 341
India, 289, 326 (fig.)
Indian Ocean, 18 (fig.), 23 (figs.), 93, 230–231, 234,
  254 (table), 255, 270, 279, 283 passim, 326
  (fig.). *See also* Upper high
Indian-type, 251, 286 passim
Indisposition, of sun, 240
Indonesia (Java), 270 (fig.), 323 (fig.), 326 (fig.)
Institute of Historical Research, London, 108, 122
Intensity. *See* Magnitude
International Quaternary Association (INQUA),
  25
Iona, 120
Iran, 48, 228
Ireland, 120, 126, 311, 317, 360, 362
  annals, 115, 126
  tree-rings, 360, 364n
Isopleth, of zero, 230
Italy, 32, 33, 41, 134 passim, 226, 256

Japan, 56, 78, 256
Jupiter, 178, 233
Jurassic varves, 306 (fig.), 311 (fig.), 328 (fig.)

Korea, 57–66, 117, 122n

Lag, 113, 230–231, 348
  aurora, 139
  biennial cycles, 230
  corpuscles after radiation, 113
  17-year radiocarbon (30 years), 141, 196
  13-year radiocarbon (decades), 372
Lake Biwa, 350
Lake Chad, 284 (fig.)
Lake Pert (Finland), 313, 352
Lake Saki (USSR), 303, 352, 360, 363 (map), 364
  (figs.)
Lake Van (Armenia), 299, 301 (ref.)
Late Glacial, 283, 299, 305, 336–342, 351, 355
  (refs.)
Latitude
  low, 22, 227, 319
  solar, 33, 76, 90, 169, 171
Law
  of latitudes, 1, 77, 91 (fig.)
  of polarity, 1, 199–201
  of zones, 1, 76, 91 (fig.)
Length, of cycles, 1, 12 (fig.), 25
  mean, 25, 75, 147 (table), 152, 163
Letters, 33, 70–71, 183
Level, of radiocarbon, 318, 373 (col. e)
Lights, northern, 7. *See also* Aurorae
Livy, 103n, 132–134, 144, 369n
Locusts, 107, 112, 235
Loess, 42
Louis XIV (1645–1715), 177
Lull. *See* Gaps, in sunspots
Lunar month, 128, 319

Magnetic declination, 88, 340, 343
Magnetic excursions, 182, 340, 345
Magnetic meridian, 123, 247
    11th-century China, 123, 125, 139, 343
    11th-century England, 139, 141n, 247
    9th-century, 317
Magnetic moment (A.D. 100), 182
Magnetic reversal
    earth. *See* Magnetic excursions
    sun, 1, 199
Magnetism
    of earth, 24 (fig.), 76, 88, 350
    late glacial, 338–340
    of sun, 93, 192, 200 (fig.), 233
Magnitude, of cycles (R_M), 1, 14 (table), 15
        (table), 93 (fig.), 159 passim
    ratio, 7,12 (fig.)
    sum, 7, 12, (fig.)
Manuscripts, xii, 47–50, 107, 232, 290
Maps, 102, 228, 229 (figs.), 363 (fig.)
Maunder minimum (17th century), 7, 12, 69, 97,
        101n, 145, 169–188, 221 (table), 227–229
    earlier analogs, 8 (fig.), 9 (fig.), 23, 32, 134, 138,
        149, 166, 331, 365 (table)
Maxima, 153 passim, 166
    auroral, 8–9, 13 (fig.), 103n, 122, 124, 278 (fig.),
        281
    biennial, 324, 342–347 (figs.), 364 (fig.)
    double, 4–6, 19
    Holocene, 364 (fig.), 365 (table)
    minor, 6, 21
    numbered 4–7, 14 (table), 208 (table)
Maximum
    Medieval, 8 (fig.), 186
    secondary (auroral), 13 (fig.), 124
Maya, 50, 233, 341
Mediterranean, 109, 117, 228, 232, 344
Megacycles, 26, 234, 298, 316, 327, 379
Mercury, 31, 32, 43–50, 101
"Merry dancers" (auroral), 174
MESA. *See* Spectral analysis
Mesozoic, 94, 305 (fig.), 328 (fig.)
Meteorite, 38, 69
Meteorology. *See* Bracknell, England,
        Meteorological Office; Cycles, and wave
        trains
Meteors (shooting stars), 98, 117n, 124, 144, 234
Mexico, 19 (fig.), 262, 270 (fig.)
Microvarves, 26, 343
Middle Ages
    aurorae, 2 (fig.), 9 (fig.)
    sunspots, 32, 63 (fig.)
Migrations, steppe-sown, 112, 283
Milk-rain, auroral, 130n, 135
Ming (A.D. 1368–1644), 33, 67(fig.), 108
Minima
    auroral, 13 (fig.)
    Hale, 193, 233
    Medieval, 9 (fig.), 68
    prolonged, 8 (fig.), 68, 117n, 169–188, 365
        (table)
    solar, 12 (fig.), 14 (table), 159 passim, 240, 368

    (table). *See also* Maunder minimum (17th
        century); Spoerer minimum (15th century)
Missing millennium, 344, 356, 364 (fig.)
Models
    cycle, 3 (fig.), 146, 175
    equatorial, 275 (fig.), 319 (fig.), 320 (fig.)
    pluvials, 283
    pressure parameter, 288 (fig.), 321
    QBO, 271
    reservoir, 195, 219
    southern oscillation, 286, 319 (fig.), 320 (fig.)
Modulation, 24, 215, 340
Moon, red, 42
Muhlberg, battle of, 33

Nature, 25, 78
Neolithic, 256, 283, 284 (fig.), 344
Netherlands, 13, 33
New Zealand, 18 (fig.), 19 (fig.), 23 (fig.), 31, 229,
        288 (fig.). *See also* Australasia
Nile, 22, 94, 127, 231–232, 246 (fig.), 284–285, 291
        (fig.)
    cycles, 232, 283
Nomads, African, 283, 286
Nordlicht, 97, 107. *See also* Aurorae
North America, 192, 255, 342 (fig.)
    Alaska, 251, 288, 289
    climate, 244, 250–267
    Northward March of Civilization, 22
Norway, 174, 232, 346. *See also* Scandinavia
Number, of solar cycle, 14 (table), 208 (table)
Numbers, 17 (fig.)
    of aurorae, 8–9 (figs.)
    auroral, 15 (fig.), 83–97, 150
    international (R), 1, 11, 78, 83–87
    pre-1100 B.C., 26
    smoothed, 4, 15 (fig.), 17 (fig.), 76 (fig.)
    sunspot, 10 (table), 17 (fig.), 75, 375 (col. *i*)

Observations
    ancient, 30, 36–38, 41, 53, 63–64 (figs.), 115,
        130–134
    medieval, 8–9 (figs.), 41, 64–68 (figs.), 115–121
    solar
        Chinese, 38, 51–69, 109–112
        daily, 81, 84–85
        1826/1850, 74, 81–87
        medieval, 40–69, 63 (fig.), 64 (fig.)
        c. 1610, 32, 33, 61, 69n, 70–71
Observatory, 108
    Greenwich, 75, 77, 176
    Tokyo, 180
    Yunnan, 32, 52, 61
    Zürich (Swiss), 2–7, 175
Oddness. *See also* Biennial cycles
    cycles, 232, 272, 342
    index, 232, 358–359
    ratio, 269, 277
    since 1205, 272
    summers, 268, 273, 287, 334
Oligocene, 305 (fig.)
Openings (Humboldt), 1, 130

Oscillations, 253. *See also* Cycles, and wave trains; Quasibiennial Oscillations (QBO); Superoscillation, tropical pressure
Oval, auroral, 141, 372
Ozone, 271, 273 (ref.)

Pacific, 94, 229, 351
   ocean, 23 (figs.), 234, 252, 270
   pressure, 23 (fig.), 258 (list), 288 (map). *See also* Superoscillation, tropical pressure
   type, 251, 255, 286
Palaeomagnetism, 123, 125, 139, 141n, 247, 317, 343, 345
Paleozoic, 318, 328 (fig.)
Parameter. *See* Pressure, barometric
Paris, 88, 118, 170, 177
Particles, 1, 113, 216–223
Patterns, 20, 23, 165, 229, 302–305
Peking (Beijung), 108, 125
Periodicities. *See* Cycles, and wave trains
Permian, 298, 305 (fig.)
Phase, 137 (fig.), 163–164 (table), 213 (fig.), 372 (table). *See also* Remainder (crude phase index: 11-year cycle).
Physics, solar, 11, 26–28 (refs.), 223
Pillar, aurora, 130
Planets, 74, 177, 233
   positions, 47n, 177, 233
   records, 64 (fig.)
Pleistocene, 299, 305 (fig.), 320, 328 (fig.). *See also* Late Glacial
Pluvials, and Subpluvials, 283–284 (types), 351, 356–357 (list)
P-max. *See* Pressure, barometric
Poland, 108, 179
Polarity, of spots, 93, 192, 200 (fig.)
Pollen, 112, 352
Portents, 31, 65, 103, 104 (refs.), 106
   and politics, 65, 110
Power spectra. *See* Spectral analysis
Precambrian, 25, 305 (fig.), 328 (fig.)
Predictions, xi, 2–12, 26, 79 (refs.), 146, 231
   crisis, 241
   satellites, 11, 189, 231
   weather, 15 (table), 21, 26, 146, 231, 257, 292
Pregnancy, 116
Prehistory, 23, 361, 365, 378–382
Pressure, barometric, 17, 146, 230, 232, 243–247 (table), 254 (table), 258 (table), 287
   global maps, 18–20 (figs.), 23 (figs.), 288 (figs.), 320 (fig.)
   gradient, 242, 252
   Greenland, 20 (fig.), 23 (figs.), 253, 321
   persistence, 320
   p-max (10 years), 243n, 244n, 256, 262, 262 passim, 267n, 289
   P-max (30 years), 23 (fig.), 242, 250, 258 (list), 267 passim, 289
Pressure parameter, 20, 139, 232, 288 (map), 320 (fig.), 321 (fig.), 322 (table), 326 (table)
Prices, 17, 227, 240
Proxy-data, 229, 313. *See also* Ice-cores; Tree-rings; Varves; Vintages (proxy data for summer temperature)

QBO. *See* Quasibiennial oscillations (QBO)
Quantification, 75, 83–87, 136–141, 149–168, 294, 359, 373 (table)
Quasibiennial oscillations (QBO), 271 (fig.), 273n
   and sunspots, 231, 271 (fig.)
Quaternary, xii, 299, 328 (fig.)

Radiocarbon, 17 (fig.), 68 (fig.), 135n, 181, 217 (fig.), 373 (table)
   ambiguous dates, 337 (list)
   centuries, stades, 284 (fig.), 337, 343, 356 (list)
   deviation, 135n, 152, 161 (fig.), 167 (table), 233, 294 (fig.), 337 (fig.), 343, 353, 372
   Late Glacial, 213 (fig.), 337 (fig.), 341, 353
   production of, 24 (fig.), 194, 365 (table), 372n, 373 (col. f)
Rain
   blood, 40n, 130n
   milk, 130n
Raininess, 112n, 234, 243n, 244n, 254, 261 (fig.), 262–266, 284 (fig.), 349–350 (fig.), 364 (fig.)
Ratio, of magnitude, 7, 12 (fig.)
Relative anomalies, floating, 230, 245 (table), 249
Relative numbers, 187. *See also* Sunspots, numbers
Religion. *See* Sources; Visions
Remainder (crude phase index: 11-year cycle), 124, 137 (fig.), 152, 161 (fig.), 163 (table), 167 (tables), 233, 300, 368 (table), 370 (table), 372
Research, new, 25, 50, 77–78, 98, 145, 283, 292, 295
Revolutions, 226, 228, 281
Rivers, 284, 303
R-max (rainfall), 242. *See also* Thirty-year changes
Rome, 41, 103, 134 passim, 285
Rosa Ursina. *See* Sunspots, drawings
Rotation, solar, 31, 70–71, 77, 92, 114, 170, 179 (fig.), 293, 319, 366
Royal Astronomical Society (R. A. S.), xii, 75, 76
Royal Society, 170, 176
Russia, 36, 114, 287. *See also* Lake Saki

Sahel, 286–287, 351. *See also* Africa; Nile
Saki. *See* Lake Saki; USSR, Saki, Crimea
Salinity, 298
Satellite, *See* Skylab, U.S. satellite
Saturn, 178
Scandinavia, 13, 37, 179, 232, 250, 287, 288 (fig.), 334
Scotland, 108, 174, 250
Sector boundaries, 293. *See also* Solar wind
Sediments, 302–315. *See also* Varves
Shang dynasty, 26
Shape
   of auroral cycle, 13 (fig.)
   of sunspot cycle, 2 (fig.), 3 (fig.), 92 (fig.)
Siberia, 288 (fig.), 346
Sign, in sun, 30, 37, 39
Signature, 256, 278
Sign-leap, polarity, 192, 340
Skylab, U.S. satellite, 11, 183, 189
Smoothing, 4 (fig.), 76 (fig.)
   secular, 15 (fig.), 203, 207
S. O. *See* Southern oscillation (S.O.)

Solar air. *See* Wind, solar
Solar corona. *See* Corona, solar
Solar dynamo, 175
Solar excursions, 234, 365
Solar halo, 66, 110–111
Solar latitude, 33, 76, 90, 169, 171
Solar minima, 12 (fig.), 14 (table), 159 passim, 240, 368 (table). *See also* Maunder minimum (17th century); Spoerer minimum (15th century)
Solar observations. *See* Observations, solar
Solar physics, 11, 26–28 (refs.), 223
Solar rotation, 31, 70–71, 77, 92, 114, 170, 179 (fig.), 293, 319, 366
Solar wind, 97, 101, 186, 233
Sources
   ancient, 37, 53, 105 (fig.), 132–133, 285
   Buddhist, 97, 319
   Chinese, 37, 53, 65, 108, 129n
   chronicle, 107–121, 260, 285, 317
   Islamic 43–44, 285 (list)
   Korean, 53, 120, 122n
   oracle bones, 26
   religious, 97, 317
South America, 18 (fig.), 19 (fig.), 232, 254–255, 279
Southern oscillation (S.O.), 231, 251, 279 (fig.), 286 (list), 292, 320
   reversal, 270. *See also* Equatorial inflation
South Sea Bubble, 241
Spectral analysis, 16 (fig.), 21 (fig.), 193, 207, 209 (fig.), 212 (fig.), 221–222. *See also* Cycles, and wave trains
Spectrum of Time, xii, 7, 40, 98, 105 (fig.), 106 (fig.), 108, 176, 188
Spoerer minimum (15th century), 8–9 (figs.), 176, 221 (table), 317
Spoerer's Law, 77
Stades. *See* Radiocarbon, centuries, stades
Standard deviation, 98
Stars
   shooting. *See* Meteors (shooting stars)
   variable, 88, 366
Stratosphere, 271–275, 319
Streamers, 183. *See also* Aurorae
Suess effect, 182, 195, 218. *See also* Wiggles
Sum, of magnitudes, 7, 12 (fig.)
Summer, 42, 232, 247, 250, 259, 272 passim, 280 (table), 324, 334, 341 passim, 356
Sun. *See also* Rotation, solar
   blue, 42
   globe, 71
   night, aurora, 130
   quiet, 1, 181, 239–240
Sun and Climate 17, 27 (ref.), 149n, 234, 292
Sung (A.D. 933–1279), 65, 123–129
Sunspots. *See also* Cycles, and wave trains
   Africa, 39
   ancient, 30–38, 102–105
   areas, 75, 87
   blood, 31, 36
   Chinese, 51–69, 109–112
   curves, 3–7
   drawings, 34, 179 (fig.), 187n

   duration of, 52, 70, 90–91
   earliest, 30–38
   1500+, 14 (table)
   4th century, 111 (fig.)
   Islamic, 39, 43–50
   Korean, 52, 57–61, 69
   latitude, 70, 90–91, 171
   medieval, 40–50, 51–59, 63 (figs.)
   9th century, 31–32, 41, 317
   numbers, 10 (table), 17 (fig.), 75, 375 (col. *i*)
   observations, xi, 70–71, 74–81
   since 650 B.C., 354 (tables), 368 (table)
Supernova, 98, 127
Superoscillation, tropical pressure, 234 (list), 257
Sweden, 313 (fig.), 335 (table), 343–345, 359
Switzerland, 75, 175, 192, 343, 360. *See also* Bern, Switzerland; Observatory, Zürich (Swiss)
Synchrony, 105, 114, 285, 334, 364 (fig.). *See also* Teleconnection
Syria, 39, 79, 173, 317

T'ang (A.D. 618–906), 33, 67 (fig.)
Tasman Sea, 18 (map), 94, 323 (map). *See also* Australasia
Teleconnection, 281, 335 (fig.), 342 (fig.), 363 (table), 364 (figs.)
Telescope, 33, 70, 170, 177, 183 (fig.)
Temperature
   cycles, 21 (fig.), 24 (fig.)
   England 1659+, 280 (table and fig.), 282n
   floating anomalies, 1100+, 245 (table)
   fluctuations, 24 (fig.), 245–246, 284 (fig.)
   Late Glacial 341–342 (figs.)
   sea, 252, 286
   subpolar, 24 (fig.), 259 (table)
   and sunspots, 24 (fig.), 37, 248, 254
   summer, 600+, 42, 230, 232, 250
   today, 292
   tropical, 248 (fig.), 254 (table), 320
Tests, 334, 341, 361, 362n
Thirty-year changes, 8 (fig.), 23, 242, 267, 289
   barometric *(P)*, 230, 234, 242, 250, 258 (table)
   radiocarbon, 196
   sunspots *(S)*, 93
   temperature *(T)*, 230, 242, 250
   varves, 334
Time of Rise *(T)*, and Fall *(U)*, 2-3, 12 (fig.), 93 (fig.), 146, 154–161 (table), 318 (table)
T-max. *See* Thirty-year changes
Tokyo, 78
Torch, aurora or comet, 123
Trade cycles, 227, 241
Transit, of planets in front of sun, 45, 47n, 50, 74
Tree-line, 334, 345, 356
Tree-rings, 22, 25, 115n, 229, 250–267, 357, 363 (fig.)
   colors, 344
   drought-sensitive, 260, 264, 269, 334
   England, 260, 261 (fig.), 323, 364n
   errors, 38, 362, 364n
   Germany, 346 (table), 359n
   Ireland, 361, 362
   missing, 269 (table), 277 (table), 318, 325
   Scandinavia, 350

since A.D. 900, 346 (fig.), 360
U.S., 11, 22 (fig.), 245 (table), 252–266, 337
    (fig.), 346–347 (figs.), 359n
Triennial cycles (3-year), 230, 279, 281–282 (fig.),
    323 (map), 339 (fig.), 347, 364 (fig.), 399
    1800/1900, 281 (fig.)
    tropics, 290
Tropics, 254, 316. *See also* Equator; Quasibiennial
    oscillations (QBO); Southern oscillation
    (S.O.)
Troposphere, upper, 146, 320. *See also* Upper high
Turning points, of sunspots, 1500+, 1, 14 (table),
    254 (table)
Turkey, 300, 344, 364 (figs.)

Umbra, 177
Upper high, 93, 146, 231, 275 (fig.), 287, 310 (fig.),
    320 (fig.). *See also* Indian Ocean
Upper ride (Lapland), 232
Upper troposphere, 146, 320. *See also* Upper high
Upper trough, 94, 231, 287, 299. *See also* Davis
    Strait
U.S.
    Alaska, 251, 256
    climate, 244, 250–267
    Gulf states, 292
    Nevada, 347
    New England, 146, 333–357
    semi-arid, 251, 268–273
    southwest, 230, 251, 270
    Virginia, 33
    west coast, 232, 245–247 (table)
USSR, 18 (fig.), 77–78, 192, 231, 289, 363 (figs.)
    Saki, Crimea, 303, 352, 360

Variability, of sun, 239–240
Varves, 363 (fig.). *See also* Lake Van (Armenia);
    USSR, Saki, Crimea
    African, 320
    Canadian, 281, 335 (table), 359–360
    charts, 339, 342, 363 (figs.)
    cycles, 308 (fig.), 348
    distal, 358
    Finland, 313, 335 (table), 341, 352, 359
    fluvial, 303, 334
    Germany, 360
    Holocene, 336, 360
    Late Glacial, 281, 305, 322
    magnetism, 76, 299–300, 345

map, 363 (map)
marine, 352
medieval, 313 (fig.)
Precambrian, 25
proximal, 356
Russian. *See* USSR, Saki, Crimea
Sweden, 303, 345, 359–360
U.S., 335 (table), 360
Venus, transits, 43, 47n, 361
Vintages (proxy data for summer temperature),
    132, 277–278 (figs.)
Visions, 98, 103, 111 (fig.), 116, 317
Volcanoes, 344 (table), 361
    A.D. 536, 122
    A.D. 1883, 254, 257, 271 (fig.), 272
    44/43 B.C., 31, 41

Weather, 226, 228, 242, 295, 300
    600–1600, 245–246, 249n, 284 (fig.)
    1600–1960, 244–245 (figs.)
    and sunspots, 17, 229, 248, 279–280 (table), 291
        (fig.), 293–295, 326 (fig.)
    and very weak cycles, 22, 230, 280 (table)
    U.S., 1880+, 253 (table), 264 (table)
Wheat prices, 227–228
Wiggles, 23, 217–223 (figs.), 298, 337. *See also*
    Radiocarbon, deviation
Wind
    since 1650, 243–246 (figs.)
    1836/1850, in U.S., 263 (table)
    1850/1960, 264, 286
    solar, 97, 101, 186, 233
Wine harvests, 132, 277–278 (fig.), 245–246 (table),
    299
Winters, 249n, 250, 346–347, 351
Wobble, of upper high, 231, 270
Wolf minimum, (1280–1350), 365 (table)
Wolf numbers, 1, 75

X-ray, 273, 289, 358

Yüan (A.D. 1260–1368), 67 (fig.)
Yugoslavia (Dalmatia), 360
Yunnan. *See* Observatory, Yunnan

Zechstein, 299, 329
Zodiacal light, 124, 127, 183
Zone leap, solar latitudes, 91 (fig.), 192
Zürich, 1-2, 11, 75, 175

About the Editor

D. JUSTIN SCHOVE chose mathematics and physics for his first degree, history for his teacher's diploma, and geography for his higher degrees. Sunspots, tree-rings, and weather interested him while still at school and he served as a meteorological officer in the Royal Air Force. His *Spectrum of Time* project led to his receiving a large collection of historical auroral records from different parts of the world and he used them to pinpoint the dates and magnitudes of sunspot cycles in the past. A speaker at international conferences in various disciplines, he is a member of the International Committee on Calibration of the Radiocarbon Dating Time-Scale, the Institute of Historical Research, and head of an INQUA section on Teleconnection of Varves. Schove is also a fellow of the Royal Geographical Society and the Royal Astronomical Society. He has published over 100 papers, mainly on sunspot cycles, tree-rings, weather-history, and chronology. Other books in progress are *The Spectrum of Time* and *Eclipses and Comets, A.D. 1-1000* (with Dr. A. Fletcher, Boydell and Brewer, Suffolk, England), and he is editing (with Professor R. W. Fairbridge) *Ice-cores, Varves and Tree-rings* (Balkema).

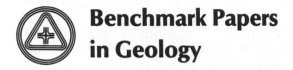

Benchmark Papers
in Geology

Series Editor: Rhodes W. Fairbridge
Columbia University

Volume

1 ENVIRONMENTAL GEOMORPHOLOGY AND LANDSCAPE
    CONSERVATION, Volume I: Prior to 1900 / *Donald R. Coates*
2 RIVER MORPHOLOGY / *Stanley A. Schumm*
3 SPITS AND BARS / *Maurice L. Schwartz*
4 TEKTITES / *Virgil E. Barnes and Mildred A. Barnes*
5 GEOCHRONOLOGY: Radiometric Dating of Rocks and Minerals /
    C. T. Harper
6 SLOPE MORPHOLOGY / *Stanley A. Schumm and M. Paul Mosely*
7 MARINE EVAPORITES: Origin, Diagenesis, and Geochemistry /
    Douglas W. Kirkland and Robert Evans
8 ENVIRONMENTAL GEOMORPHOLOGY AND LANDSCAPE
    CONSERVATION, Volume III: Non-Urban Areas / *Donald R. Coates*
9 BARRIER ISLANDS / *Maurice L. Schwartz*
10 GLACIAL ISOSTASY / *John T. Andrews*
11 GEOCHEMISTRY OF GERMANIUM / *Jon N. Weber*
12 ENVIRONMENTAL GEOMORPHOLOGY AND LANDSCAPE
    CONSERVATION, Volume II: Urban Areas / *Donald R. Coates*
13 PHILOSOPHY OF GEOHISTORY: 1785–1970 / *Claude C. Albritton, Jr.*
14 GEOCHEMISTRY AND THE ORIGIN OF LIFE / *Keith A. Kvenvolden*
15 SEDIMENTARY ROCKS: Concepts and History / *Albert V. Carozzi*
16 GEOCHEMISTRY OF WATER / *Yasushi Kitano*
17 METAMORPHISM AND PLATE TECTONIC REGIMES / *W. G. Ernst*
18 GEOCHEMISTRY OF IRON / *Henry Lepp*
19 SUBDUCTION ZONE METAMORPHISM / *W. G. Ernst*
20 PLAYAS AND DRIED LAKES: Occurrence and Development /
    James T. Neal
21 GLACIAL DEPOSITS / *Richard P. Goldthwait*
22 PLANATION SURFACES: Peneplains, Pediplains, and Etchplains /
    George F. Adams
23 GEOCHEMISTRY OF BORON / *C. T. Walker*
24 SUBMARINE CANYONS AND DEEP-SEA FANS: Modern and Ancient /
    J. H. McD. Whitaker
25 ENVIRONMENTAL GEOLOGY / *Frederick Betz, Jr.*
26 LOESS: Lithology and Genesis / *Ian J. Smalley*

27 PERIGLACIAL PROCESSES / *Cuchlaine A. M. King*
28 LANDFORMS AND GEOMORPHOLOGY: Concepts and History /
     Cuchlaine A. M. King
29 METALLOGENY AND GLOBAL TECTONICS / *Wilfred Walker*
30 HOLOCENE TIDAL SEDIMENTATION / *George deVries Klein*
31 PALEOBIOGEOGRAPHY / *Charles A. Ross*
32 MECHANICS OF THRUST FAULTS AND DÉCOLLEMENT / *Barry Voight*
33 WEST INDIES ISLAND ARCS / *Peter H. Mattson*
34 CRYSTAL FORM AND STRUCTURE / *Cecil J. Schneer*
35 OCEANOGRAPHY: Concepts and History / *Margaret B. Deacon*
36 METEORITE CRATERS / *G. J. H. McCall*
37 STATISTICAL ANALYSIS IN GEOLOGY / *John M. Cubitt and
     Stephen Henley*
38 AIR PHOTOGRAPHY AND COASTAL PROBLEMS / *Mohamed T.
     El-Ashry*
39 BEACH PROCESSES AND COASTAL HYDRODYNAMICS / *John S.
     Fisher and Robert Dolan*
40 DIAGENESIS OF DEEP-SEA BIOGENIC SEDIMENTS / *Gerrit J.
     van der Lingen*
41 DRAINAGE BASIN MORPHOLOGY / *Stanley A. Schumm*
42 COASTAL SEDIMENTATION / *Donald J. P. Swift and Harold D. Palmer*
43 ANCIENT CONTINENTAL DEPOSITS / *Franklyn B. Van Houten*
44 MINERAL DEPOSITS, CONTINENTAL DRIFT AND PLATE TECTONICS /
     J. B. Wright
45 SEA WATER: Cycles of the Major Elements / *James I. Drever*
46 PALYNOLOGY, PART I: Spores and Pollen / *Marjorie D. Muir and
     William A. S. Sarjeant*
47 PALYNOLOGY, PART II: Dinoflagellates, Acritarchs, and Other
     Microfossils / *Marjorie D. Muir and William A. S. Sarjeant*
48 GEOLOGY OF THE PLANET MARS / *Vivien Gornitz*
49 GEOCHEMISTRY OF BISMUTH / *Ernest E. Angino and David T. Long*
50 ASTROBLEMES—CRYPTOEXPLOSION STRUCTURES / *G. J. H. McCall*
51 NORTH AMERICAN GEOLOGY: Early Writings / *Robert Hazen*
52 GEOCHEMISTRY OF ORGANIC MOLECULES / *Keith A. Kvenvolden*
53 TETHYS: The Ancestral Mediterranean / *Peter Sonnenfeld*
54 MAGNETIC STRATIGRAPHY OF SEDIMENTS / *James P. Kennett*
55 CATASTROPHIC FLOODING: The Origin of the Channeled Scabland /
     Victor R. Baker
56 SEAFLOOR SPREADING CENTERS: Hydrothermal Systems / *Peter A.
     Rona and Robert P. Lowell*
57 MEGACYCLES: Long-Term Episodicity in Earth and Planetary History /
     G. E. Williams
58 OVERWASH PROCESSES / *Stephen P. Leatherman*
59 KARST GEOMORPHOLOGY / *M. M. Sweeting*
60 RIFT VALLEYS: Afro-Arabian / *A. M. Quennell*
61 MODERN CONCEPTS OF OCEANOGRAPHY / *G. E. R. Deacon and
     Margaret B. Deacon*

62 OROGENY / *John G. Dennis*
63 EROSION AND SEDIMENT YIELD / *J. B. Laronne and M. P. Mosley*
64 GEOSYNCLINES: Concept and Place Within Plate Tectonics /
    F. L. Schwab
65 DOLOMITIZATION / *Donald H. Zenger and S. J. Mazzullo*
66 OPHIOLITIC AND RELATED MELANGES / *G. J. H. McCall*
67 ECONOMIC EVALUATION OF MINERAL PROPERTY / *Sam L.
    VanLandingham*
68 SUNSPOT CYCLES / *D. Justin Schove*
69 MINING GEOLOGY / *Willard C. Lacy*
70 MINERAL EXPLORATION / *Willard C. Lacy*
71 HUMAN IMPACT ON THE PHYSICAL ENVIRONMENT /
    Frederick Betz
72 PHYSICAL HYDROGEOLOGY / *R. Allan Freeze and William Back*
73 CHEMICAL HYDROGEOLOGY / *William Back and R. Allan Freeze*
74 MODERN CARBONATE ENVIRONMENTS / *Ajit Bhattacharyya and
    G. M. Friedman*
75 FABRIC OF DUCTILE STRAIN / *Mel Stauffer*
76 TERRESTRIAL TRACE-FOSSILS / *William A. S. Sarjeant*
77 GEOLOGY OF COAL / *Charles Ross and June R. P. Ross*
78 NANNOFOSSIL BIOSTRATIGRAPHY / *Bilal U. Haq*
79 CALCAREOUS NANNOPLANKTON / *Bilal U. Haq*
80 RIVER NETWORKS / *Richard S. Jarvis and Michael J. Woldenberg*